PATHWAY ANALYSIS FOR DRUG DISCOVERY

Wiley Series on Technologies for the Pharmaceutical Industry

Sean Ekins, Series Editor

Computational Toxicology: Risk Assessment for Pharmaceutical and Environmental Chemicals
Edited by Sean Ekins

Pharmaceutical Applications of Raman Spectroscopy
Edited by Slobodan Šašić

Pathway Analysis for Drug Discovery: Computational Infrastructure and Applications
Edited by Anton Yuryev

PATHWAY ANALYSIS FOR DRUG DISCOVERY

Computational Infrastructure and Applications

Edited by

ANTON YURYEV
Ariadne Genomics, Inc.
Rockville, Maryland

WILEY

A JOHN WILEY & SONS, INC., PUBLICATION

Published by John Wiley & Sons, Inc., Hoboken, New Jersey.
Published simultaneously in Canada.

Library of Congress Cataloging-in-Publication Data:
Pathway analysis for drug discovery : computational infrastructure and applications / [edited by] Anton Yuryev.
p. ; cm. — (Wiley series on technologies for the pharmaceutical industry)
Includes bibliographical references and index.
ISBN 978-0-470-10705-8 (cloth)
1. Drug development–Data processing. 2. DNA microarrays—Data processing.
3. Computational biology. I. Yuryev, Anton. II. Series.
[DNLM: 1. Drug Design. 2. Computational Biology. 3. Microarray Analysis—methods. QV 744 P297 2008]
RM301.25.P374 2008
615′.190285—dc22

2008015107

Printed in the United States of America.

10 9 8 7 6 5 4 3 2 1

CONTENTS

PREFACE

This book is a compilation of articles by pioneers of pathway analysis. While providing some solutions and the state-of-the-art overview, the book formulates many questions that yet to be addressed by the scientific community. The phrase "drug discovery" in the title was intended to emphasize the pragmatic approach for the book. Pathway analysis attempts on the enormous task of formalizing the molecular biological knowledge to make it suitable for predictive computation. Pathway analysis is currently in its infancy and requires the framework for thinking and development of useful applications. This framework can only be based on practical solutions that have a direct impact on human life and well-being of society. Improving human health and optimizing biological organisms for human needs are two main practical applications of molecular biology that rapidly move it away from being an academic discipline toward the application science in commercial industry.

Drug discovery industry is likely to benefit most from pathway analysis. There are two major computational challenges in the drug development. First is calculating the structure of drug molecules that have specific and predictable protein targets and therefore predictable biological effects. Second is calculating the biological effects themselves. Both tasks require extensive computational resources but use very different fundamental principals. The structural drug design is based on physics of molecular structure and interaction while calculating biological effects is based on the analysis of information flow or pathways. Even though the information flows inside the living cell through the physical interaction network, the physics of the molecular interactions has limited effect on biological pathways. Instead, the combinatorial effect from players in the protein community network determines the biological outcome of the drug treatment, disease progression, and healthy signaling throughout the human body. The computational complexity in structural drug design is due to the large number of atoms participating in the molecular interaction, while the complexity of pathway analysis is due to the large number of processes that occur in a single human cell, multiplied by the large number of tissues in human organism.

Currently used approach of "computing" the drug effects using live organisms such as animal models and patients in clinical trials is expensive, error prone, and can be viewed as unethical. Therefore, everyone appears to believe that *in silico* predictions should be the safest and most economical way of improving the drug discovery pipeline. Because the excuse of having insufficient computational resources rapidly vanishes into the history, the book attempts to summarize critical elements that are necessary for successful *in silico* pathway analysis for drug development.

Anton Yuryev

CONTRIBUTORS

Gordana Apic, Cambridge Cell Networks Ltd., St John's Innovation Centre, Cambridge CB4 0WS, United Kingdom; gordana.apic@camcellnet.com

Michael J. Birrer, MD, PhD, National Cancer Institute, Center for Cancer Research, 37 Convent Drive, Bldg 37, Room 1130, Bethesda, MD 20892; birrerm@bprb.nci.nih.gov

Fedor Bokov, Ariadne Genomics Inc, 9430 Key West avenue, Suite 113, Rockville, MD 20850; masf@ariadnegenomics.com

Andrej Bugrim, GeneGo, Inc., 500 Renaissance Drive, #106, St Joseph, MI 49085; andrej@genego.com

Bojana Cosovic, Cambridge Cell Networks Ltd, St John's Innovation Centre, Cowley Road, CB4 0WS Cambridge, United Kingdom; Bojana.Cosovic@camcellnet.com

Nikolai Daraselia, Ariadne Genomics Inc, 9430 Key West avenue, Suite 113, Rockville, MD 20850; nikolai@ariadnegenomics.com

Sreenivas Devidas, PhD, Vice President Business Development, GVK BioSciences, 5457 Twin Knolls Road, Suite 101, Columbia, MD 21045; sreeni.devidas@gvkbio.com

Zoltan Dezso, GeneGo, Inc., 500 Renaissance Drive, #106, St Joseph, MI 49085; zoltan@genego.com

Sean Ekins, PhD, DSc, Collaborations In Chemistry, 601 Runnymede Ave, Jenkintown, PA 19046, USA; ekinssean@yahoo.com

John Farley, MD, Associate Professor, Department of Obstetrics and Gynecology, Uniformed Services University of the Health Sciences, 4301 Jones Bridge Road, Bethesda, MD 20814; jfarley@usuhs.mil

Craig N. Giroux, Institute of Environmental Health Sciences, Wayne State University, 2727 Second Avenue, Detroit, MI 48201; cgiroux@genetics.wayne.edu

Iaroslav Ispolatov, PhD, Departamento de Fisica, Universidad de Santiago de Chile, Casilla 302, Correo 2, Santiago, Chile. Ariadne Genomics Inc., 9430 Key West Avenue, Suite 113, Rockville, MD 20850; slava@ariadnegenomics.com

Andrey Kalinin, Ariadne Genomics Inc, 9430 Key West avenue, Suite 113, Rockville, MD 20850; kalinin@ariadnegenomics.com

Mark P. Kühnel, EMBL-Heidelberg, Cell Biology and Biocomputing Meyerhofstrasse 1, 69117 Heidelberg, Germany; mark.kuehnel@camcellnet.com

Sergei Maslov, Department of Physics, Brookhaven National Laboratory, Upton, NY 11973; maslov@bnl.gov

Goran Medic, Cambridge Cell Networks Ltd, St John's Innovation Centre, Cowley Road, CB4 0WS Cambridge, United Kingdom; Goran.Medic@camcellnet.com

Alexander Nikitin, Ariadne Genomics Inc., 9430 Key West Avenue, Suite 113, Rockville, MD 20850; shura@ariadnegenomics.com

Tatiana Nikolskaya, GeneGo, Inc., 500 Renaissance Drive, #106, St Joseph, MI 49085; tatiana@genego.com

Yuri Nikolsky, GeneGo, Inc., 500 Renaissance Drive, #106, St Joseph, MI 49085; yuri@genego.com

Laurent L. Ozbun, PhD, Center for Cancer Research, National Cancer Institute, 37 Convent Drive, Bldg 37, Room 1130, Bethesda, MD 20892; ozbunl@mail.nih.gov

Robert B. Russell, EMBL-Heidelberg Biocomputing Meyerhofstrasse 169117 Heidelberg, Germany. Email: russell@embl.de

Sergey Simakov, Moscow Institute of Physics and Technology, Department of Applied Mathematics, Instituskii Lane, 9, Dolgoprudny, Russia; simakovss@ya.ru

Andrey Y. Sivachenko, PhD, Senior Staff Scientist, Ariadne Genomics, Inc. 9430 Key West Ave. #113, Rockville, MD 20850; andrey.sivachenko@gmail.com

Todor Vujasinovic, Helios Biosciences, 8, Avenue du Général Sarrail, 94010 Créteil, France; todor.vujasinovic@heliosbiosciences.com

Anton Yuryev, PhD, Ariadne Genomics Inc., 9430 Key West Avenue, Suite 113, Rockville, MD 20850; ayuryev@ariadnegenomics.com

André Siniša Žampera, Helios Biosciences, 8, Avenue du Général Sarrail, 94010 Créteil, France; sinisa@heliosbiosciences.com

1

INTRODUCTION TO PATHWAY ANALYSIS

ANTON YURYEV

Table of Contents

1.1 INTRODUCTION

Pathway analysis is a rapidly developing discipline that combines software tools, database models, and computational algorithms—all of which help molecular biologists to convert molecular interaction data into a set of computational models. The models are developed for better prediction of cell behavior in response to a drug, nutrients, or other outside stimuli. The

Pathway Analysis for Drug Discovery, Edited by Anton Yuryev

development of pathway analysis was triggered by the expansion of high-throughput methods and the completion of human genome sequencing project. Because of these technological advances, the emphasis of molecular biology has shifted from reductionism to system integration. Suddenly, nearly all of the components of a living cell became known and the new goal of "putting them all together" into a working computational model of the living cell is awaiting the scientific community. This model must be built by a consensus effort of all molecular biologists and will be constantly refined for a significant period of time. The first and most important goal of pathway analysis is to provide tools and infrastructure that facilitate building a consensus cell model by the collective effort of the scientific community. These tools must enable adequate data exchange, automatic data integration, communication with central public depositories of pathways, and molecular interaction information supporting consensus knowledge base building. In this review, I discuss current approaches for constructing a molecular interaction database, explain the available pathway analysis methods from the drug discovery point of view, and place pathway analysis into the historical perspective of advances in molecular biology.

1.2 METHODS TO CONSTRUCT THE PATHWAY ANALYSIS KNOWLEDGE BASE

Several layers of consensus information are necessary for pathway analysis: (1) a generally agreed-upon list of molecules; (2) a consensus global molecular interaction network; and (3) a collection of consensus pathways for known biological processes. Even though significant portion of the work to create the molecular "inventory" of the human organism has been accomplished by sequencing the human genome, it is still far from being completed. The alternative splicing and protein modification isoforms are yet to be fully cataloged. This next major challenge for this knowledge level is being addressed now by development of exon and phosphorylation microarray technologies. The consensus interaction network is being created by a combination of efforts in high-throughput experiments, prediction of interactions, and classical molecular biological and genetic techniques aimed at elucidating the function of individual proteins. Because the results of numerous small-scale experiments usually are available only in the form of scientific publications and because no central depository for molecular interactions exists in the scientific community, special text-mining techniques have been developed to extract this information into machine-readable format [1].

Every available technique to record interactions for a global network database has some degree of a false-positive rate. High-throughput methods for detection of protein–protein interactions, such as a two-hybrid screen or an identification of protein complexes by co-immunoprecipitation followed by mass spectrometry, are currently being reassessed in a panic due to an apparently worrisome 50–70% false-positive rate [2]. Even though the sources of errors are well understood for these methods, the only way to reduce them at present is to scale down these experiments, effectively reducing them back to one of the laboratory techniques and preventing their high-throughput application. Consequently, attempts to improve the reliability of these methods continue amidst criticism [3]. As a reconciliation note for all high-throughput methods, I emphasize that proteins were designed by nature to interact with each other so as to provide the structural backbone for a living organism. They literally stick to each other to be alive. Therefore, it is not surprising that every protein can interact with many different partners, and many interactions that appear as false positives are in fact true physical interactions. Yet, many of these interactions are not biologically meaningful. Some of them never occur inside the living cell because two proteins never meet each other in space or time. Even if an interaction does occur *in vivo*, it simply may not perform a biological function—it is not followed by the cascade of molecular events that is called a "cell process." In my opinion, high-throughput methods probably produce mostly correct interaction data, but additional evaluation is necessary to sort out biologically functional and meaningful interactions. The identification of biologically meaningful interactions is a separate task from measuring them. If this is the case, our frustrated energy should be diverted away from these methods and refocused on remedying our general inability to understand what each interaction means for cell physiology. In Chapter 3, about automatic pathway inference, I show that measuring and recording regulatory interactions is one way to calculate biological meaning for physical interactions.

The prediction of physical interactions is typically accomplished using sequence homology [4,5], and regulatory interactions can be calculated from time-series gene expression data using Bayesian inference [6,7]. By holding the powerful promise to "reverse engineer" biological objects, Bayesian inference has been even proclaimed as the principal method of systems biology. However, it currently suffers from the noisy and dispersed experimental data available for analysis, a lack of understanding how to construct good training sets [8], and the emerging realization of the plasticity of biological regulatory networks [9].

Peer-reviewed scientific literature is still the most important source of reliable molecular interaction data. The sheer number of scientific publications and the fact that they are written in machine non-readable format necessitated the development of methods to extract this information into machine-readable format that can be used by computational algorithms. These attempts focused on the manual recording of interactions into a database by an army of curators and, at the same time, on the development of natural language processing algorithms to read scientific papers automatically. Manual curation turned out to be a slow and expensive process that is not error-free: humans do make mistakes when both writing and reading papers. The rate of manual recording appears unable to keep up with the rate of new articles being published.

Automatic extraction of the interaction from peer-reviewed scientific literature faces its own challenges. The main advantage of automatic text-mining algorithms is the speed that allows the processing of hundreds of thousands of articles in minutes. For example, MedScan technology from Ariadne Genomics can process the entire PubMed database that contains more than 14 million abstracts in 2 days on a regular personal computer. This speed allows comprehensive coverage of the entire body of scientific literature, as well as a fairly good assessment of interaction confidence. Interaction confidence can be estimated using the frequency with which the interaction was recovered from the literature. The high misinterpretation rate that occurs during fact extraction is the biggest problem of text-mining technologies. Natural language processing algorithms that use a full sentence parsing approach have the lowest error rate, but they also have the lowest recovery rate among all methods for automatic extraction of interactions [1]. Because errors are evenly distributed among all sentences, false-positive interactions appear as interactions with a low number of supporting references. This technique can be used as a filter to eliminate erroneous interactions at the end of an extraction. Unfortunately, it effectively increases the error rate among relations with a low reference count. Recently discovered true interactions by definition have a low number of references. Thus, automatic extraction methods make the effective confidence of novel interactions even lower due to contamination by false-positive facts. For example, some linguistic patterns used by the MedScan technology extract interactions with one supporting reference with only 70% accuracy. Hence, it is difficult to use the extracted data immediately for analyzing experimental data and building pathways. Additional efforts to curate automatically extracted data are required prior to an analysis such as this one [10,11].

Improvement of all methods for recording of molecular interactions will have to continue for some time until a clear winner can emerge. The most likely outcome, however, is that all interactions for human proteins will be found by the combined effort of all these methods before a winner is determined. After the best method is apparent, interactions for new organisms will be predicted mostly from the consensus interaction network for the human organism and other model organisms. Despite a seemingly overwhelming challenge to measure all physical interactions between proteins in the human genome, this goal will be achieved within the life span of most readers of this book. Indeed, the total number of unique interactions between N proteins is equal to $N(N-1)/2$. There are 30,000–35,000 genes in the human genome, making the total number of all possible pairwise interactions around 500 million. This number is the upper estimate for human interactome size, which includes both true- and false-positive interactions regardless of the methods used to measure and record them to the global database. The interactions for alternatively spliced proteins can be found relatively easily by calculation using protein sequence information, known interactions of the longest isoforms, and interactions from the homologs to determine the protein interacting domains. Thus, even though alternatively splicing greatly increases the number of possible protein–protein interactions, measuring and recording the interactions between splicing isoforms should not require much time and investment. To measure all 500 million possible interactions, each of 500,000 molecular biologists will have to measure about 100 interactions to achieve this goal in 1 year. Assuming there are about 50,000 research projects worldwide actually measuring protein–protein interactions, all physical interactions will be measured in 10 years. The speed of measuring the new physical interaction should gradually increase, and the actual number of interactions is smaller than 500 million. Therefore, 10 years is a very safe upper estimate for time required for recording of all physical interactions.

1.3 ORGANIZATIONAL CHALLENGES FOR CONSTRUCTING THE KNOWLEDGE BASE FOR PATHWAY ANALYSIS

The major barrier separating humankind from measuring all physical protein–protein interactions in its own species is the lack of organization and communication among individual scientists. I have reason to assume that this book will not change this situation, so research will continue as usual by measuring the same interactions multiple times in different laboratories that are trying to prove or disprove each other's theories. Typically, the authors of these studies publish only interactions that seem to support their respective favorite scientific theory or model in hopes of securing funding in the future. High-throughput methods will also continue to contribute to a significant amount of interaction data while attempting to improve their accuracy. Taking the opportunity given to me by this book, I want to join the call to release all interaction data into the public domain [2]. The relatively small organizational challenges that accompany this call include having a central authority to maintain an interaction depository, the tools for data submission and building a consensus global network database, and a method for calculating the confidence of a specific interaction from supporting evidence submitted by multiple sources.

Several public institutions have taken an early lead in the attempt to become a central authority for pathway and molecular interaction databases. Kyoto University provides the Kyoto Encyclopedia of Genes and Genomes (KEGG) database curated by its own staff. Its main disadvantage is that a system cannot be used as a depositary by external users outside KEGG. The Signal Transduction Knowledge Environment (STKE) database is maintained by the American Association for the Advancement of Science (AAAS) and contains a collection of pathways curated by scientists considered to be the top experts in the field. This database contains a small collection of highly reliable and canonical pathways and accurately reflects the current state of the art of the pathway analysis field: very few pathways are actually known and experimentally verified at present. The slow rate of curation and the absence of any formal method to create pathways, including the absence of universal identifiers for pathway components, are the main disadvantages of the STKE database. Unfortunately, the usual leaders in storage of biological information, the National Center for Biotechnology Information (NCBI) in the United States and the European Bioinformatics Institute (EBI) in the European Union, seem to be overwhelmed by the amount of sequencing data they need to maintain. Currently, they lag behind in creating a central resource for pathway information. NCBI, for example, has limited itself by integrating protein physical interaction information from public databases: Biomolecular Interaction Network Database (BIND), Human Protein Reference Database (HPRD), BioGRID, and EcoCyc. Moreover, the constant introduction of new protein identifiers by these organizations unnecessarily complicates the issue even further. Among other public sources of pathway information, I must mention the Reactome database maintained by Cold Spring Harbor Laboratory in collaboration with EBI and Gene Ontology, and the Database of Interacting Proteins (DIP) at the University of California at Los Angeles. Currently, the largest pathway and molecular interaction databases are only available commercially from privately held companies such as Ingenuity Systems, GeneGo, and Ariadne Genomics.

1.4 FROM MOLECULAR INTERACTION DATABASE TO PATHWAY COLLECTION

The collection of physical interactions is not, however, the pathway database. It merely provides the underlying network or pool of interactions necessary for pathway and network building. This pool is not likely to be 100% accurate because of all the reasons I previously mentioned. Software tools and methods developed for pathway building must take into account the reliability of each interaction when working with such a database. At this point, it is worth emphasizing the difference between a network and a pathway, because both can be built by the same software tools. A network represents a static image of all possible physical and/or regulatory interactions between biological entities, while a pathway represents how the information propagates through the network. Because information propagation is a directional process, a pathway must have entry nodes where the information flow starts and terminal points where the information flow ends. The pathway components represent the sequence of molecular events in space or time while the biological process is occurring. A network, in contrast to a pathway, can contain any relation or entity, including those that do not participate in the information flow. For example, a physical interaction network can include structural interactions, while regulatory networks can include indirect relations that actually are mediated by a set of physical interactions. Network analysis can provide important insights into biological functions. It can, for example, identify major regulators and targets in a biological process that appear as hubs—nodes with many connections for this process in the network. Hubs also can provide an idea about the information flows within the network, starting from major hub regulators and propagating toward major hub targets. Network analysis can also identify protein complexes involved in a process. Yet, a network cannot be used for dynamic modeling because it lacks one essential ingredient of any pathway: an initial signal or input.

Because a pathway is a way to represent specific events that take place after exposing a cell to an extracellular signal or environmental condition, the main task of pathway analysis is to establish methods for converting network information into a pathway. Any biological pathway is essentially an abstraction or an approximation describing the major channels of information flow through the physical interaction network. The goal of pathway inference is generating a diagram simple enough to be used in kinetic simulations yet adequately describing what happens inside a cell after the stimulation. That known canonical pathways represent the preferred path through the network was proposed some time ago [12]. If this premise is true, then a pathway is simply a path through hubs in the global regulatory network. The evidence taken from scientific literature certainly supports this view: essentially every component in any currently known canonical pathway is a hub in the global and regulatory network. Hubs should be also the first experimentally detectable components of a pathway because they are the best targets for pathway inhibition, which is the favored method to study pathways *in vivo*. For this reason, the currently known connectivity of a hub must be artificially elevated relative to other "non-hub" proteins in both physical and regulatory networks derived from experimental literature: historically, researchers first identified hubs as principal pathway components and then began identifying other proteins interacting with them. Thus, in reality, the relative connectivity of hubs may not be as high as it appears to be in the currently known network. Nevertheless, currently known hubs will still probably remain hubs even after the entire network is known.

Figure 1.1 represents the principal task of pathway analysis: converting a network into a pathway suitable for dynamic modeling. The input for pathway analysis is a network and a stimulus used to invoke the information flow. The output of pathway analysis is a pathway suitable for dynamic modeling with adequate predictive power. Where does the input network come from? From another class of high-throughput experiments aimed at measuring the state of an entire biological system, such as gene expression microarray experiments. In spite of being noisy like all other biological high-throughput methods, they provide a snapshot of a system upon exposing cells to a stimulus. Ideally, the state of a cell must be measured on several different levels such as cell transcriptome, proteome, and metabolome [13]. The current state of the art, however, produces only gene expression data with acceptable quantity and accuracy. Yet recent developments in proteomic methods measuring protein

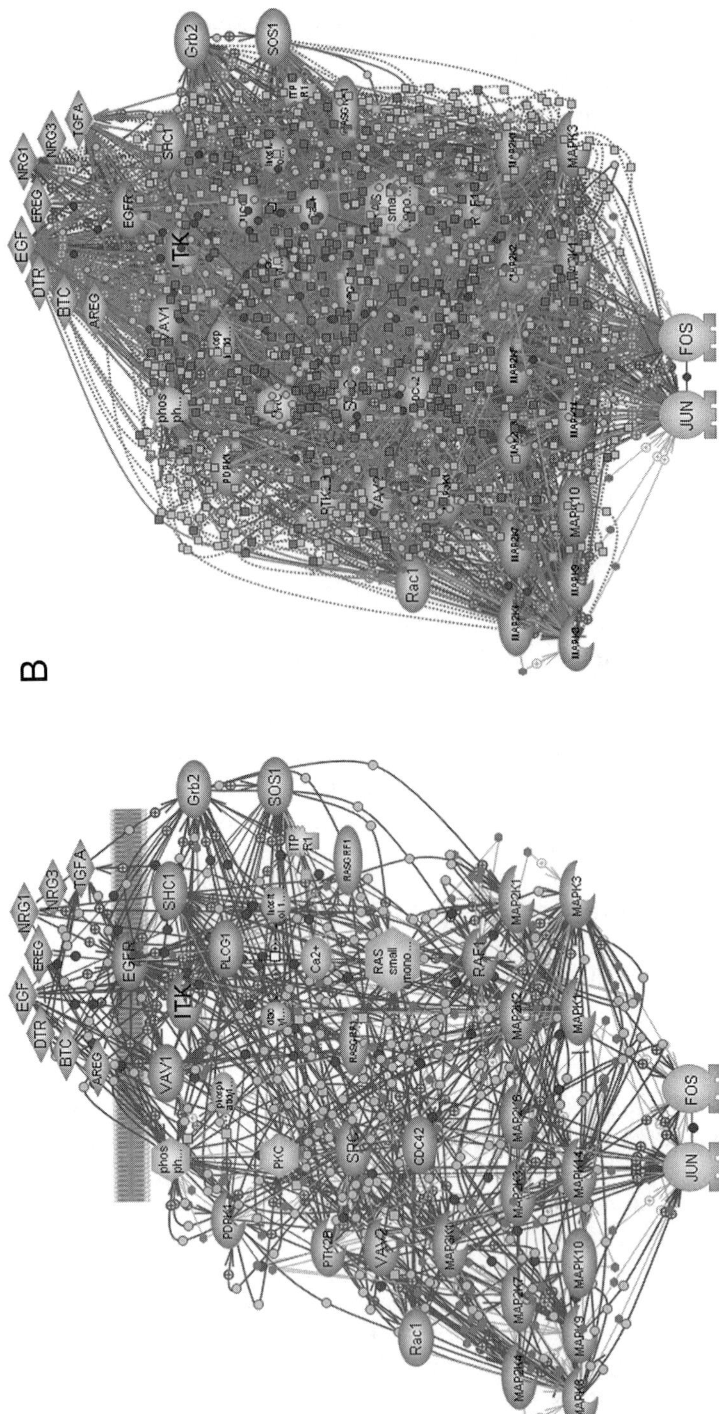

Figure 1.1 Canonical EGFR pathway shown in three ways: (A) physical interaction network; (B) physical interaction network combined with regulatory network; (C) as a pathway ready for kinetic modeling; and (D) kinetic model of MAP kinase cascade, which is a portion of EGFR pathway. One way to visualize the goal of pathway analysis is to imagine that it changes the contrast of the network image to make the main information flow more visible by hiding relations that are nonessential for depicting the principal information flow. See color insert.

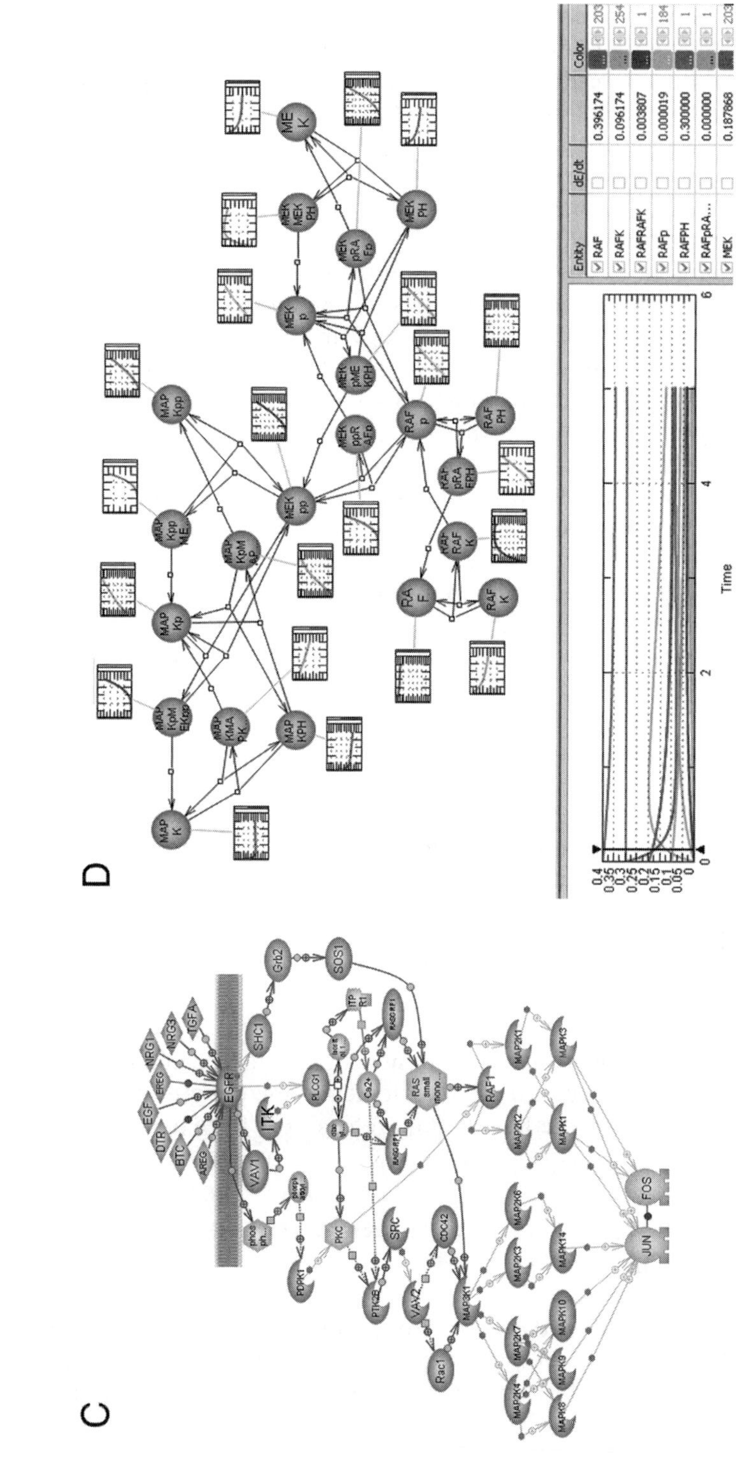

Figure 1.1 *(Continued)*

concentration and modification already necessitate integration of information from different experimental types into software for pathway analysis. The important workflow described above must be supported by all pathway analysis software: it must provide access to the molecular interaction database, permit analysis of high-throughput data to identify molecular networks appearing in response to an experiment, and subsequently allow calculation of pathways suitable for molecular modeling.

An additional approach for building pathways for the human organism is by using orthologous pathway information from model organisms and paralogous information about known pathways in human tissues. One of the most important achievements of network biology in the last decade is providing further support to the duplication-divergence theory of molecular evolution (see reference [14] and references therein). The best way to evolve is to duplicate an existing mechanism and then modify one or both copies to develop new functions while keeping older functions in one of the copies, if necessary. Evolution constantly duplicates individual genes and occasionally makes a copy of entire genomes in order to mutate genes later and to develop new interactions and functions [15]. As further evidence in support of the duplication-divergence model of evolution, current efforts in studying model organisms provide crucial insights into general rules for the modular and pathway organization of a cell. They have revealed and will reveal more of the conserved, "must have" mechanisms in molecular signaling and cell physiology [16]. The following list enumerates currently known conserved principles of pathway organization:

- Pathway sub-compartmentalization using clustering in physical interaction networks [17] and scaffolding [18,19]
- Fast decay of crosstalk mediated by binding interactions [20]
- Feedback loops providing positive self-activation of a pathway [18,21]
- Feed-forward loops providing noise tolerance for a pathway [22,23]
- Cross-pathway inhibition [18,22,24]

In addition, model organisms reveal the conserved molecular interaction and regulatory blocks necessary for biologically meaningful propagation of information [25], such as the MAPK-kinase cascade, for example [18].

1.5 PATHWAY ANALYSIS SOFTWARE AND THE SCIENTIFIC COMMUNITY

While providing the tools and methods for pathway building and data analysis, pathway analysis software provides additional important functions for scientific enterprise: enabling fast communication, data exchange, and education among members of the scientific community. It has been recognized that graphical and other visual information is more effective than text for learning molecular biological concepts [26]. For example, the image of the double-helix DNA structure is the most common form used by people to learn, think, and teach about DNA. This image migrates from one textbook to another. Any text describing the double-helix is merely a caption for the image. Similarly, diagram visualization in pathway analysis software allows scientists to exchange information about biological networks and work with them more efficiently. I want to demonstrate how important visualization is for pathway analysis by suggesting the following virtual experiment. First, take any pathway diagram that you find in this book and describe it in writing. Be warned that this exercise may be very boring. Second, find two of your colleagues who have never seen this pathway before. Show the diagram to the first colleague and show the text to the second one. Give both of them the same amount of time to inspect and memorize the pathway information. Then, ask them both to reproduce the pathway. You will discover that the colleague who saw the pathway will reproduce it more accurately than the person who read about the same pathway. Now, try to draw the same pathway yourself and you will realize that it is much faster to describe a pathway as text than to make a good drawing of it. Describing a pathway is easier because you most likely have a text processor program on your computer but do not have an application for pathway drawing. Imagine, however, that this pathway-drawing program exists and also enables you to send a pathway diagram to your colleagues so that they can reproduce an exact copy of your pathway, add their information to it, or compare it to their own experimental results or to other pathways. It should now be readily apparent that a pathway analysis tool can increase your ability to communicate with the scientific community and speed up your collaboration projects many times over.

Biologists usually visualize three major classes of processes as diagrams: biochemical pathways, molecular signaling cascades, and various cellular mechanisms such as a cell cycle and apoptosis. As all scientific papers are written these days using computers, pathway diagrams are drawn with computer programs as well. The degree of sophistication of these pathway-drawing programs ranges from simple vector graphics drawing tools like Microsoft PowerPoint to the database programs like Pathway Studio from Ariadne Genomics that link every pathway to the underlying global molecular interaction network and to the functional annotation of biological molecules. At the same time, these programs allow the comparison of thousands of data points from high-throughput experiments with the pathway collection in the database. The increased sophistication of pathway drawing tools has put the term "pathway analysis" on the same level with other scientific methods and disciplines that study the propagation of information inside the living cell, such as: systems biology, molecular network analysis, and dynamic modeling or kinetic simulation. In the remaining part of this introduction, I will position pathway analysis relative to these approaches in an attempt to show how pathway analysis both differs from and complements them.

1.6 PATHWAY ANALYSIS AND SYSTEMS BIOLOGY

In short, systems biology is a discipline and pathway analysis is one of its methods. Historically, the term "systems biology" was used as an umbrella to describe various attempts to understand and to model the behavior of an entire cell or organism. Since our current biological knowledge is still incomplete, systems biology focuses on the development of computational methods for analysis of high-throughput data and on designing databases and data models to store and refine the information necessary for achieving the ultimate goal of modeling cell physiology. Pathway analysis is not different in this respect from any other methods of systems biology. It allows compiling, maintaining, classification, and utilization of pathway information. The reason why pathway analysis must be isolated from other methods is evident from the following estimates. There are more than 520 signaling ligands in the human genome and 232 tissues in the human organism. Even though many tissues do not have receptors for every ligand, one hormone often can bind different types of receptors and sometimes activate different pathways [27]. Therefore, we can estimate about 100,000 signaling diagrams that are necessary in order to have a comprehensive collection of signaling pathways for the human organism. Variations of about 50 canonical biochemical pathways in 232 human tissues add another 10,000 diagrams. The number of various intracellular and physiological intercellular processes can be estimated to be about 1,000. This estimate increases the number of necessary diagrams to roughly about 200,000. Finally, there are about 2,500 complex diseases that are usually depicted as diagrams of defective pathways and cellular processes. We also must remember that pharmaceutical research typically uses animal models that require a separate pathway collection for each model organism: mouse, rat, dog, etc. All of the previous numbers estimate a daunting collection of nearly 500,000 pathways needed to build a comprehensive database for drug discovery. To create and maintain this vast pathway collection, we need a rather sophisticated software infrastructure. This software must provide interactive access to pathway diagrams, enable fast and seamless data exchange between users and databases, and allow the comparison of pathway collection with high-throughput and other experimental data. One of the goals for pathway analysis is to devise strategies and algorithms to compose such a collection and to develop an appropriate software infrastructure for its maintenance, for its utilization in analysis, and for drug discovery. The effort to create this pathway collection for drug discovery is comparable in scale to the Human Genome Project (HGP). Continuing this analogy, I would say that the current technological level of pathway analysis is about the same as the level of sequence analysis soon after Frederick Sanger proposed his method for DNA sequencing.

1.7 PATHWAY ANALYSIS AND NETWORK ANALYSIS

Briefly, network analysis studies the global properties of biological networks, while pathway analysis studies the propagation of information through a network. I have already described the pathway diagram as an intermediate step between isolating the network modules and building the dynamic model of a biological process, as shown in Figure 1.1. The method of network analysis gained momentum in 1999 after a publication by Hartwell et al. [28] that presented an intuitively clear paradigm about the modular organization inside a living cell. Since then, biological networks have undergone intense investigation. The recent efforts to analyze biological networks have yielded several important findings relevant to pathway analysis. I will list them here, since this book does not cover network analysis in much detail. First, it has been demonstrated that both physical and regulatory biological networks indeed have a modular structure that correlates well with known functional modules, as defined by humans in protein annotation [17]. Second, it was found that biological protein networks tend to have power law degree distribution, meaning that they have hubs—highly connected proteins with many interaction partners. This discovery appears to be true for both physical and regulatory networks (Figure 1.2). The main reason for having a hub or scale-free network topology is to provide robustness to the network so it does not break into independent components when a link or node is occasionally removed [29]. A node or link removal can occur because of the following: genetic mutations in evolution; somatic mutations occurring in a disease and throughout the life of the organism; and environmental conditions such as diet, trauma, and stress. The scale-free topology of a network allows biological systems to randomly try new ways of evolutionary adaptation without the danger of disintegration and to survive rather significant damage during a disease.

1.8 PATHWAY ANALYSIS OF DISEASE

Network biology logic suggests that, in order to become stable and robust, a disease network must acquire a scale-free topology with hubs. A disease is developed if its sub-network or module that carries out the malignant function becomes robust and perpetual. These perpetual networks may take years to develop in an organism. Genetic predisposition, diet, lifestyle, infection, accidental trauma, and stress all may contribute to the development of a disease network. Because of these multiple contributions, the networks causing the same disease most likely differ among individual patients, thus necessitating personalized drug intervention. To cause the same disease in different

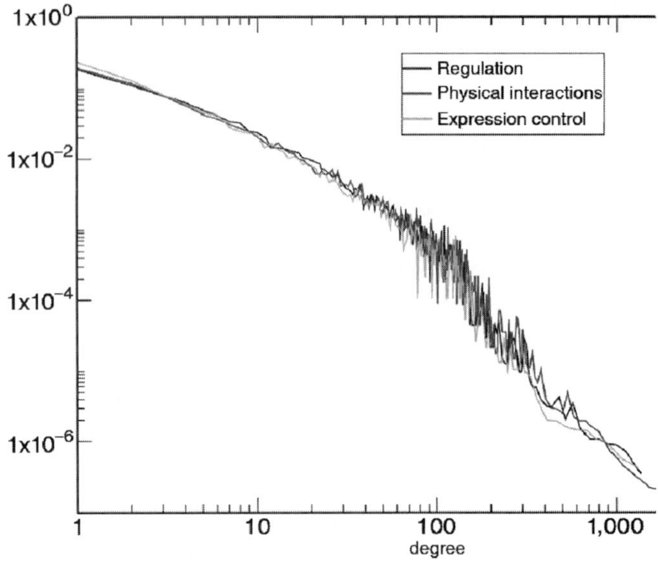

Figure 1.2 Degree distribution among known physical and regulatory interactions in ResNet database from Ariadne Genomics. The networks were built by automatically extracting relations from scientific literature using MedScan, a natural language processing technology [16].

individuals, individualized disease networks must, however, have in common the misdirected information flow that should occur through the sharing of the pathway components. Since hubs are the best targets to disrupt a scale-free network, it is yet to be seen whether individualized disease networks have similar hub composition or they overlap only in the "peripheral" nodes that are responsible for malignancy. Once formed, the disease network should be resilient to drugs precisely because of the robustness that has enabled its existence in the first place [30]. Therefore, from the network biology perspective, a drug design strategy must be a strategy to disrupt the robustness of a disease network. The network disruption itself cannot be adequate and must be supplemented by other changes that will make the disruption irreversible. Irreversibility can be achieved by either using other drugs or changes in lifestyle and diet. It is highly unlikely, however, that drugs will restore the original "normal" *network*, due to the general complexity and evolutionary nature of the development of the malignant network. Nonetheless, drugs must restore the "healthy" information flow, which is a desirable clinical outcome. Therefore, from the pathway analysis perspective, drugs must redirect a malignant information flow to restore the normal, healthy *pathway*. In many cases, this restoration can mean returning to the original pathway disrupted by a disease. In some cases, for example, in diabetes or obesity, it may mean "improving" the original pathway and making it even "healthier."

I must mention that no disease network has been completely established, and the previous paragraph is only a speculation. This book will offer examples of the current efforts toward building such networks. Here I want to mention the networks recently identified for inherited ataxia [31] and the angiogenic switch in human pancreatic cancer [32]. Most complex diseases, such as cancer or diabetes, are multistep processes of malignant transformation. Each step is probably characterized by a different robust malignant network. Therefore, it is not entirely correct to speak about a cancer disease network, for example. It is quite appropriate to speak about and study a cancer pathway, however. This pathway represents the path of cancer development that has a start, a direction, and a final stage. Every step on this pathway represents a biological process whose defect causes the progression of the disease (Figure 1.3). Each step has its own malignant network. This is another example of the difference between a pathway diagram and a network diagram: the depiction of disease history as a chain of events in time.

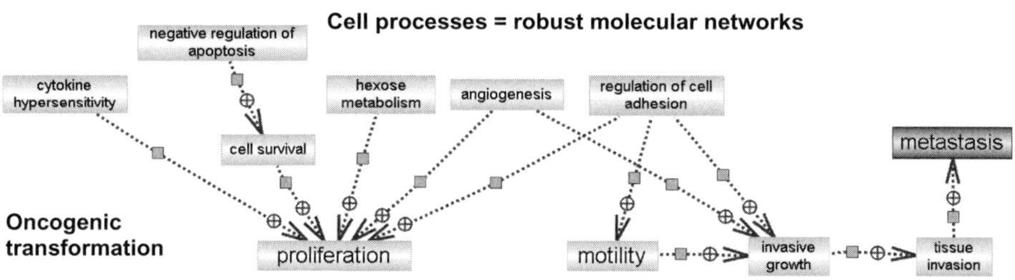

Figure 1.3 Typical cancer development pathway. This diagram depicts the sequence of cell transformation events occurring from the onset of the disease until the death of the patient. Each transformation is triggered by defects in one or several cellular processes shown at the top level of the diagram. Depending of the type of cancer, some steps may be irrelevant or skipped. For example, angiogenesis is not necessary for oncogenic transformation of blood cells. The exact order of events contributing to malignant cell proliferation is not known. In principle, this sequence can vary in different patients or cancer types, but tumor-induced angiogenesis occurs at the later stages of cancer after micro-tumors have reached a certain size. Angiogenesis in the context of cancer development means the induction of blood vessels growth by tumors. Therefore, the angiogenesis network in cancer cells mediates the production of cytokines responsible for angiogenesis in endothelial cells and the angiogenic switch network in endothelial cells mediates proliferation and differentiation of endothelial cells.

A defect in every process contributing to cancer development occurs through the establishment of the robust molecular network performing the malignant function in the cell [73]. These networks are beginning to be identified [31] for each step of cancer progression. In principle, these networks should be different in every human tissue, and each tissue may have several networks for each cancer step depending on individual genetic and environmental factors that lead to the development of cancer in the first place.

1.9 PATHWAY ANALYSIS AND DYNAMIC MODELING IN DRUG DEVELOPMENT

In essence, dynamic modeling or pathway kinetic simulation requires building a pathway diagram prior to developing a kinetic model. The mathematical model developed from experimental data is considered to be the final triumph of the effort to understand biological processes. The main criterion for successful mathematical modeling is the correct simulation of experimentally observed behavior. For this reason, good models can be built only when a system or a process is studied well enough to have known reproducible behavior in response to a stimulus. Very few examples of such processes in molecular biology exist, imposing a major limitation on the development of kinetic models. Experimentally observed cycling, oscillations, and threshold behavior appear as the most attractive targets for kinetic modeling. Hence, the first models developed in molecular biology were for the cell cycle [33], the circadian cycle [34], and the oscillation in NF-kappaB pathway [35]. Recently, the models simulating development of apoptosis [36–41], signaling in the EGFR receptor [42], and the JAK-STAT pathway [43] were developed. The criteria for a successful apoptosis model were selected based on known drug effects [44,45], threshold behavior or bistability between cell apoptotic responses and survival responses to cytotoxic stress [39], and tissue-specific differences in bistable (irreversible) and monostable (reversible) apoptotic responses [40]. For the EGFR signaling model, the criterion for success was the known activation profile of EGFR downstream targets. For the JAK-STAT pathway, the criterion for success was the cycling of STAT protein between cytoplasm and nuclei.

A mathematical theory for modeling of signaling pathways has been put forward by Heinrich et al. [46]. The theory was designed to model rate, duration, and amplitude of a signal in linear kinase-phosphatase cascades, coupled to feedback interactions and crosstalk with other signaling pathways. Undoubtedly, these modeling attempts contribute to our general understanding of biological processes dynamics. For example, they showed that phosphatases affected the rate and duration of signaling, whereas kinases controlled signal amplitude in the EGFR pathway [46]; that RAF-1 signaling was the most important regulator of EGFR phosphorylation [47]; that EGF-induced responses were stable over a wide range of ligand concentration; and that the initial velocity of receptor activation determines signaling efficiency through the EGFR pathway [42].

The behavior of a biological system, by definition, must be probabilistic in order to cope with novel environmental factors previously unmet in evolution. The only way a cell can find the optimal response to a novel stimulus is by presenting all possible responses while looking for the first optimal one that may become selected. For example, that cell transcriptional response to previously unknown challenges is fundamentally random was recently shown for yeast [9]. Drug treatment, by all means, represents the unmet challenge for a cell and therefore the cell response to a drug must be a stochastic process, which varies among the cell population of one or several human tissues. Transcriptional and epigenetic reprogramming of an entire system in response to a drug may provide strong limitations on using dynamic models in drug development. Drug-induced cell reprogramming can be caused by both the inhibition of an intended drug target and the side effects of a drug. If a drug causes global transcriptional reprogramming in a cell, its efficacy cannot be predicted by solely using the kinetic models. Instead, the new cell state first must be determined through using experimental measurements in every individual patient. The reprogrammed cell will have changed relative concentrations of proteins involved in the process that is to be modeled. Therefore, mechanistic dynamic modeling can be useful in evaluating how close the new cell state and the desired clinical outcome are by calculating the dynamics of affected processes in a new state. In the case when a drug does not cause the global reprogramming, or reprogramming is mild, it should be possible to predict the cell response by a dynamic model. Due to the variability of the initial conditions among cells in the body, the biologically meaningful outcome of any modeling should be a solution space of all possible cell responses to a drug treatment with a specific probability score assigned to every response curve in the solution space. Even though computational approaches to address this problem are being developed [48–50], they currently suffer from the general lack of knowledge about intracellular events, which is necessary for creating the model. These approaches also have knowledge gaps about cell behavior that is necessary for model validation.

A global cell model is the ultimate goal of pathway analysis. Its general complexity with millions of freely adjustable parameters probably will require more experimental constraints than modern molecular biology can ever provide. Thus, the uncertainty due to the lack of knowledge about the system must be added to the fundamental variability of the cell response, making the predictions by available models even less reliable. Nevertheless, current efforts will ultimately yield a computational model for the entire cell. The useful application of the global cell models, however, will be made in the even more distant future than the creation of complete database of molecular interactions for the human cell. Therefore, the best success stories of dynamic modeling are likely to happen beyond the life span of most readers of this book.

One goal of current dynamic modeling efforts is to isolate the minimal set of components and relations that can be used to correctly predict the behavior of a system and then to predict a system response, such as drug action, to the perturbations. While creating a model that uses pathway analysis methods, one can identify essential interactions from the pool of all cellular interactions mediating the modeled process. Once principal interactions are identified from the pool of all known interactions, it may become evident that the pathway is incomplete and actually misses some interactions. Pathway analysis can point an investigator to missing pathway components and help in the design of an appropriate experiment for identifying missing components. An assumption that the correct evaluation and prediction of drug efficacy requires the kinetic simulation of pathways containing drug targets automatically necessitates simulating hundreds of thousands of pathways for the same reasons described in previous sections. Such a large number of models will require automatic assembly using pathway analysis tools. As you will see in upcoming chapters of this book, pathway analysis may provide some alternatives to brute force kinetic modeling in evaluating drug efficacy and toxicity, as well as in selecting drug targets.

1.10 STEADY-STATE ANALYSIS OF METABOLIC NETWORKS

Flux balance analysis (FBA) [51] and extreme pathway analysis [52] are two main methods of global analysis of metabolic networks developed by Bernard Palsson and his colleagues in the last 10 years. They take advantage of a basic physical principle of mass balance with the reasonable assumption that intracellular metabolic reactions are in a steady state and therefore have constant fluxes. Certainly, metabolism changes multiple times throughout the day in the human organism [53]. These changes are rapid and the periods between changes can be viewed as a steady state. Steady-state analysis predicts a set of allowed metabolic phenotypes or phenotypic planes within the space of all possible fluxes while avoiding any kinetic modeling. The allowed flux space is calculated using standard linear algebraic methods and has the form of a convex polyhedral cone. The cone edges are called "extreme pathways" [54] and the space between them is a phenotypic plane [55]. The initial FBA solution space turned out to be very large, and the problem of calculating extreme pathways was found to be NP-hard. For genome-scale networks, this solution space has proven to be infeasible [56]. For those reasons, additional constraints on thermodynamics [57], expression regulation [58], compartmentalization [59], maximum capacity, and reaction irreversibility [60] had to be taken into account to narrow down the solution space.

In my view, metabolic reconstruction represents the most advanced method available for pathway analysis. It uses the power of the complete genome sequence, modern mathematical and computational approaches to model metabolism using basic physical principles. Yet, it possesses several limitations for drug discovery that do not allow it to become the main focus of this book. First, metabolic control in multicellular organisms is under tight hormonal control. Second, the scale of changes detectable by this method may be too detailed and irrelevant for drug discovery. It is more important to predict global metabolism homeostasis of an entire organism than the metabolism of the local cell population. The difficulties of finding a physiological interpretation of the results from extreme pathway analysis were acknowledged by Bernard Palsson himself [61].

FBA has proven to be a very good method to model the metabolic state and to predict growth conditions and metabolite yields of the microorganisms

[62,63]. But the applications of FBA to drug discovery are yet to be found. FBA proponents suggested using this technique to model the metabolic state of mammalian mitochondria [64], red blood cells [56], and even the classical JAK-STAT signaling pathway [65]. One obvious application of FBA in drug discovery is predicting the response of microbial flora in the human body to drugs. The human body contains about 1.5 kilograms of symbiotically living microorganisms. Bacteria such as E. coli [58], H. influenza [66], and H. pylori [67] are among them and were modeled by Palsson and his colleagues. A better understanding of bacterial metabolism will likely help to develop better drugs and antibiotics against diseases linked to bacteria, such as diarrhea and gastritis.

Metabolic control analysis (MCA) is complementary to the FBA method for modeling biochemical reactions in a steady state. Its goal is to calculate a control coefficient for each enzymatic step in a pathway, reflecting the extent to which the component is rate-limiting [68]. MCA has been used to identify transketolase as a rate-limiting enzyme responsible for thiamine deficiency in the Ehrlich's ascites tumor model [69]. Interestingly, the authors of this paper failed to explain why a very high concentration of thiamine exhibits an inhibitory effect on the tumor, which is the opposite of its effect predicted by MCA and observed at modest concentrations. The metabolic control analysis methodology has been recently modified for transient signaling through the classical MAP kinase cascade [47,70].

1.11 SO WHAT IS PATHWAY ANALYSIS?

To answer this question, I want to remind you that pathway analysis is a method or a tool. As with any tool, it has a finite life span: it has appeared out of necessity to solve several biological problems currently facing the scientific community. Once these problems are solved, most pathway analysis methods will become obsolete, as it happened with many sequence analysis methods after human genome sequencing was completed. I have outlined problems that have to be solved by pathway analysis in previous sections and will summarize them in the next paragraph, while putting them into historical and social perspective.

In order to build a comprehensive computational model of the living cell, we need to know all of the principal components of the system plus the interactions that enable information flow through the system. The sequencing of the human genome has moved our understanding of the molecular mechanisms that govern life processes from the phase when many components and their interactions were unknown to the phase when most components are known but the interactions between them largely are not. More and more molecular interactions are being discovered as a result of the development of new high-throughput methods to measure individual interactions and because of the individual efforts of biologists worldwide. In 2004, biologists estimated that the discovering of all possible biological interactions would take about 20 years [71]. This estimate is consistent with my previous estimates of having these goals completed within the life span of most readers of this book. While working toward this goal for another 15–17 years, we also must learn how to use this information to better diagnose diseases and to improve our drugs. Let me direct you again to Figure 1.1 for an illustration of the ultimate goal of pathway analysis: converting a huge, unstructured molecular interaction network into well-defined modules describing the major information flows that can be used for predictive dynamic modeling. These modules will describe individual biological processes and signaling cascades with enough accuracy so that they can be combined into the computational model of an entire cell. The global cell model will ultimately consist of these modules linked by inter-model interactions and will predict cell response to a drug. Because generally speaking, drug development and the state of public health are the most important practical validations for any achievements in biological science, the goal of pathway analysis must be to yield improvements in drug development and disease prevention.

The community of billions of interacting proteins yields the visible living structure that we call the cell. Community of cells in turn forms human organism. Both organism and an individual cell can also be described as an information system that constantly follows and responds to environmental signals. The information inside a cell is propagated through physical interaction network. Therefore, a disease can be described as broken pathways or as a disruption of a physical structure maintained by the network of interacting proteins. Effort in pathway analysis shall eventually yield the representation of human organism as an information system via the collection of pathways. This collection will allow quick assessment of the personal health state and efficiency of a drug action.

The collection of hundreds of thousand of pathways can exist and be used only in electronic form, which makes pathway analysis a discipline of computer science. However, the roots of pathway analysis are in experimental molecular biology and genetics. The classical deductive path in molecular biology starts with the definition of a phenotype or biological process. In principle, it is the most important step that defines the overall goal of research. The rest of the effort is "simply" elucidation of the molecular mechanisms driving the process or explaining the phenotype. The word "simply" is in quotation marks because, in reality, the understanding of the molecular mechanism for every biological process once was a daunting task that involved thousands of researchers, millions of dollars spent on salaries and reagents, and usually produced several Nobel laureates. The magnitude of the effort was multiplied by the general lack of understanding of basic principles governing the function and structure of biological molecules.

The biological knowledge base has dramatically changed after Human Genome Project (HGP) was finished. First, all missing pathway components have suddenly become known. Second, the structure and function of proteins have suddenly become predictable using sequence homology and new computational algorithms. Finally, protein–protein interactions have also become abundant and predictable with a certain amount of accuracy, thanks to the development of high-throughput methods. Currently, more than 70% of all known physical protein–protein interactions for human proteins were measured by high-throughput experiments. By my estimate, there are about 200,000 experimentally determined protein–protein interactions and more than 5,000,000 predicted interactions for human proteins. The next challenge is to create a multitude of pathways and dynamic models. The lack of tools and talent to develop pathway infrastructure was indicated in 2000 by the same Bernard Palsson [72]. The situation has changed 7 years later in part due to Palsson's efforts, but also due to the efforts of the scientific community, several commercial bioinformatics companies, and scientists in big pharmaceutical companies. With these changes in mind, this book provides an overview of current tools, methods, and software for pathway analysis.

REFERENCES

1. Daraselia N, Yuryev A, Egorov S, Novichkova S, Nikitin A, Mazo I. Extracting human protein interactions from MEDLINE using a full-sentence parser. Bioinformatics 2004;20(5):604–611. Epub 2004 Jan 22.
2. Hart G, Ramani A, Marcotte E. How complete are current yeast and human protein-interaction networks? Genome Biol 2006;7:120.
3. Ewing RM, et al. Large-scale mapping of human protein–protein interactions by mass spectrometry. Mol Syst Biol 2007;3:89.
4. Kotelnikova E, Kalinin A, Yuryev A, Maslov S. Prediction of protein–protein interactions on the basis of evolutionary conservation of protein functions. Evol Bioinform 2007;3:323.
5. Yu H, Luscombe NM, Lu HX, Zhu X, Xia Y, Han JD, Bertin N, Chung S, Vidal M, Gerstein M. Annotation transfer between genomes: protein-protein interologs and protein-DNA regulogs. Genome Res 2004;14(6):1107–1118.
6. Heron EA, Finkenstadt B, Rand DA. Bayesian inference for dynamic transcriptional regulation; the Hes1 system as a case study. Bioinformatics 2007; 23:2596–2603.
7. Csete ME, Doyle JC. Reverse engineering of biological complexity. Science 2002;295(5560):1664–1669.
8. Michoel T, Maere S, Bonnet E, Joshi A, Saeys Y, van den Bulcke T, Van Leemput K, Van Remortel P, Kuiper M, Marchal K, Van De Peer Y. Validating module network learning algorithms using simulated data. BMC Bioinformatics 2007;May 3;8(Suppl 2):S5.
9. Stern S, Dror T, Stolovicki E, Brenner N, Braun E. Genome-wide transcriptional plasticity underlies cellular adaptation to novel challenge. Mol Syst Biol 2007;3:106. Epub 2007 Apr 24.

10. Yuryev A, Mulyukov Z, Kotelnikova E, Maslov S, Egorov S, Nikitin A, Daraselia N, Mazo I. Automatic pathway building in biological association networks. BMC Bioinformatics 2006;Mar 24;7:171.
11. Rodriguez-Esteban R, Iossifov I, Rzhetsky A. Imitating manual curation of text-mined facts in biomedicine. PLoS Comput Biol 2006;2(9):e118.
12. Marcotte EM. The path not taken. Nat Biotechnol 2001;19:826.
13. Ideker T, Thorsson V, Ranish JA, Christmas R, Buhler J, Eng JK, Bumgarner R, Goodlett DR, Aebersold R, Hood L. Integrated genomic and proteomic analyses of a systematically perturbed metabolic network. Science 2001;292(5518): 929–934.
14. Ispolatov I, Krapivsky PL, Yuryev A. Duplication-divergence model of protein interaction network. Phys Rev E Stat Nonlin Soft Matter Phys 2005;71(6 Pt 1):061911.
15. Sankoff D. Gene and genome duplication. Curr Opin Genet Dev 2001;11(6): 681–684.
16. Schwartz MA, Madhani HD. Principles of MAP kinase signaling specificity in *Saccharomyces cerevisiae*. Annu Rev Genet 2004;38(1):725–748.
17. Daraselia N, Yuryev A, Egorov S, Mazo I, Ispolatov I. Automatic extraction of gene ontology annotation and its correlation with clusters in protein networks. BMC Bioinformatics 2007;8:243.
18. Morrison D, Davis R. Regulation of MAP kinase signaling modules by scaffold proteins in mammals. Annu Rev Cell Dev Biol 2003;19(1):91–118.
19. Levchenko A, Bruck J, Sternberg P. Scaffold proteins may biphasically affect the levels of mitogen-activated protein kinase signaling and reduce its threshold properties. Proc Natl Acad Sci U S A 2000;97(11):5818–5823.
20. Maslov S, Ispolatov I. Propagation of large concentration changes in reversible protein-binding networks. Proc Natl Acad Sci U S A 2007;104(34):13655–13660.
21. Yeung K, Janosch P, McFerran B, Rose D, Mischak H, Sedivy J, Kolch W. Mechanism of suppression of the Raf/MEK/Extracellular signal-regulated kinase pathway by the RAF kinase inhibitor protein. Mol Cell Biol 2000;20:3079–3085.
22. Acar M, Becskei A, van Oudenaarden A. Enhancement of cellular memory by reducing stochastic transitions. Nature 2005;435(7039):228–232.
23. Brandman O, Ferrell J, Li R, Meyer T. Interlinked fast and slow positive feedback loops drive reliable cell decisions. Science 2005;310(5747):496–498.
24. Dang V, Bohn C, Bolotin-Fukuhara M, Daignan-Fornier B. The CCAAT box-binding factor stimulates ammonium assimilation in Saccharomyces cerevisiae, defining a new cross-pathway regulation between nitrogen and carbon metabolisms. J Bacteriol 1996;178(7):1842–1849.
25. Sharan R, Suthram S, Kelley RM, Kuhn T, McCuine S, Uetz P, Sittler T, Karp RM, Ideker T. Conserved patterns of protein interaction in multiple species. Proc Natl Acad Sci U S A 2005;102(6):1974–1979.
26. Mayer RE, Hegarty M, Mayer S, Campbell J. When static media promote active learning: annotated illustrations versus narrated animations in multimedia instruction. J Exp Psychol Appl 2005;11(4):256–265.
27. Deupi X, Kobilka B. Activation of G protein-coupled receptors. Adv Protein Chem 2007;74:137–166.

28. Hartwell LH, Hopfield JJ, Leibler S, Murray AW. From molecular to modular cell biology. Nature 1999;402:C47–52.
29. Barabási A-L, Oltvai Z. Network biology: understanding the cell's functional organization. Nat Rev 2004;5:101.
30. Kitano H. A robustness-based approach to systems-oriented drug design. Nat Rev Drug Discov 2007;6(3):202–210. Epub 2007 Feb 23.
31. Lim J, Hao T, Shaw C, Patel A, Szabó G, Rual J-F, Fisk C, Li N, Smolyar A, Hill DE, Barabási A-L, Vidal M, Zoghbi HY. A protein-protein interaction network for human inherited ataxias and disorders of Purkinje cell degeneration. Cell 2006;125:801–814.
32. Abdollahi A, Schwager C, Kleeff J, Esposito I, Domhan S, Peschke P, Hauser K, Hahnfeldt P, Hlatky L, Debus J, Peters JM, Friess H, Folkman J, Huber PE. Transcriptional network governing the angiogenic switch in human pancreatic cancer. Proc Natl Acad Sci U S A 2007;104(31):12890–12895.
33. Novak B, Tyson JJ. Modeling the control of DNA replication in fission yeast. Proc Natl Acad Sci U S A 1997;94(17):9147–9152.
34. Leloup JC, Gonze D, Goldbeter A. Limit cycle models for circadian rhythms based on transcriptional regulation in Drosophila and Neurospora. J Biol Rhythms 1999;14(6):433–448.
35. Covert MW, Leung TH, Gaston JE, Baltimore D. Achieving stability of lipopolysaccharide-induced NF-kappaB activation. Science 2005;309(5742):1854–1857.
36. Krammer PH, Kaminski M, Kiessling M, Gulow K. No life without death. Adv Cancer Res 2007;97C:111–138.
37. Fussenegger M, Bailey JE, Varner J. A mathematical model of caspase function in apoptosis. Nat Biotechnol 2000;18(7):768–774.
38. Bentele M, Lavrik I, Ulrich M, Stosser S, Heermann DW, Kalthoff H, et al. Mathematical modeling reveals threshold mechanism in CD95-induced apoptosis. J Cell Biol 2004;166(6):839–851.
39. Bagci EZ, Vodovotz Y, Billiar TR, Ermentrout GB, Bahar I. Bistability in apoptosis: roles of bax, bcl-2, and mitochondrial permeability transition pores. Biophys J 2006;90(5):1546–1559.
40. Legewie S, Bluthgen N, Herzel H. Mathematical modeling identifies inhibitors of apoptosis as mediators of positive feedback and bistability. PLoS Comput Biol 2006;2(9):e120.
41. Hua F, Cornejo MG, Cardone MH, Stokes CL, Lauffenburger DA. Effects of Bcl-2 levels on Fas signaling-induced caspase-3 activation: molecular genetic tests of computational model predictions. J Immunol 2005;175(2):985–995.
42. Schoeberl B, Eichler-Jonsson C, Gilles ED, Muller G. Computational modeling of the dynamics of the MAP kinase cascade activated by surface and internalized EGF receptors. Nat Biotechnol 2002;20(4):370–375.
43. Swameye I, Muller TG, Timmer J, Sandra O, Klingmüller U. Identification of nucleocytoplasmic cycling as a remote sensor in cellular signaling by databased modeling. Proc Natl Acad Sci U S A 2003;100(3):1028–1033.
44. Li B, Dou QP. Bax degradation by the ubiquitin/proteasome-dependent pathway: involvement in tumor survival and progression. Proc Natl Acad Sci U S A 2000;97(8):3850–3855.

45. Katiyar SK, Roy AM, Baliga MS. Silymarin induces apoptosis primarily through a p53-dependent pathway involving Bcl-2/Bax, cytochrome c release, and caspase activation. Mol Cancer Ther 2005;4(2):207–216.
46. Heinrich R, Neel BG, Rapoport TA. Mathematical models of protein kinase signal transduction. Mol Cell 2002;9(5):957–970.
47. Hornberg JJ, Binder B, Bruggeman FJ, Schoeberl B, Heinrich R, Westerhoff HV. Control of MAPK signaling: from complexity to what really matters. Oncogene 2005;24(36):5533–5542.
48. Christopher R, Dhiman A, Fox J, Gendelman R, Haberitcher T, Kagle D, Spizz G, Khalil IG, Hill C. Data-driven computer simulation of human cancer cell. Ann N Y Acad Sci 2004;1020:132–153.
49. Aksenov SV, Church B, Dhiman A, Georgieva A, Sarangapani R, Helmlinger G, Khalil IG. An integrated approach for inference and mechanistic modeling for advancing drug development. FEBS Lett 2005;579(8):1878–1883.
50. Meng TC, Somani S, Dhar P. Modeling and simulation of biological systems with stochasticity. In Silico Biol 2004;4(3):293–309.
51. Schilling CH, Edwards JS, Palsson BO. Toward metabolic phenomics: analysis of genomic data using flux balances. Biotechnol Prog 1999;15(3):288–295.
52. Schilling CH, Letscher D, Palsson BO. Theory for the systemic definition of metabolic pathways and their use in interpreting metabolic function from a pathway-oriented perspective. J Theor Biol 2000;203(3):229–248.
53. Albrecht U, Eichele G. The mammalian circadian clock. Curr Opin Genet Dev 2003;13(3):271–277.
54. Schilling CH, Edwards JS, Letscher D, Palsson BO. Combining pathway analysis with flux balance analysis for the comprehensive study of metabolic systems. Biotechnol Bioeng 2000–2001;71(4):286–306.
55. Edwards JS, Ramakrishna R, Palsson BO. Characterizing phenotypic plasticity: a phase plane analysis. Engineering in Medicine and Biology. BMES/EMBS Conference, 1999 Proceedings of the First Joint 2:1217.
56. Barrett CL, Price ND, Palsson BO. Network-level analysis of metabolic regulation in the human red blood cell using random sampling and singular value decomposition. BMC Bioinformatics 2006;7:132.
57. Henry CS, Broadbelt LJ, Hatzimanikatis V. Thermodynamics-based metabolic flux analysis. Biophys J 2007;92(5):1792–1805. Epub 2006 Dec 15.
58. Covert MW, Palsson BO. Constraints-based models: regulation of gene expression reduces the steady-state solution space. J Theor Biol 2003;221(3):309–325.
59. Feist AM, Henry CS, Reed JL, Krummenacker M, Joyce AR, Karp PD, Broadbelt LJ, Hatzimanikatis V, Palsson BO. A genome-scale metabolic reconstruction for Escherichia coli K-12 MG1655 that accounts for 1260 ORFs and thermodynamic information. Mol Syst Biol 2007;3:121. Epub 2007 Jun 26.
60. Price ND, Papin JA, Schilling CH, Palsson BO. Genome-scale microbial in silico models: the constraints-based approach. Trends Biotechnol 2003;21: 162–169.
61. Wiback SJ, Palsson BO. Extreme pathway analysis of human red blood cell metabolism. Biophys J 2002;83:808–818.

62. Edwards JS, Palsson BO. The *Escherichia coli* MG1655 in silico metabolic genotype: its definition, characteristics, and capabilities. Proc Natl Acad Sci U S A 2000;97:5528–5533.

63. Alvarez-Vasquez F, Sims K, Cowart L, Okamoto Y, Voit E, et al. Simulation and validation of modelled sphingolipid metabolism in Saccharomyces cerevisiae. Nature 2005;433:425–430.

64. Ramakrishna R, Edwards JS, McCulloch A, Palsson BO. Flux-balance analysis of mitochondrial energy metabolism: consequences of systemic stoichiometric constraints. Am J Physiol Regul Integr Comp Physiol 2001;280(3):R695–704.

65. Papin JA, Palsson BO. The JAK-STAT signaling network in the human B-cell: an extreme signaling pathway analysis. Biophys J 2004;87(1):37–46.

66. Papin JA, Price ND, Edwards JS, Palsson BO. The genome-scale metabolic extreme pathway structure in *Haemophilus influenzae* shows significant network redundancy. J Theor Biol 2002;215(1):67–82.

67. Price ND, Papin JA, Palsson BO. Determination of redundancy and systems properties of the metabolic network of *Helicobacter pylori* using genome-scale extreme pathway analysis. Genome Res 2002;12(5):760–769.

68. Hornberg JJ, Bruggeman FJ, Bakker BM, Westerhoff HV. Metabolic control analysis to identify optimal drug targets. Prog Drug Res 2007;64:171,173–189.

69. Comin-Anduix B, Boren J, Martinez S, Moro C, Centelles JJ, Trebukhina R, Petushok N, Lee WN, Boros LG, Cascante M. The effect of thiamine supplementation on tumour proliferation. A metabolic control analysis study. Eur J Biochem 2001;268(15):4177–4182.

70. Hornberg JJ, Bruggeman FJ, Binder B, Geest CR, De Vaate AJ, Lankelma J, et al. Principles behind the multifarious control of signal transduction. ERK phosphorylation and kinase/phosphatase control. FEBS J 2005;272(1):244–258.

71. Aloy P, Russell RB. Ten thousand interactions for the molecular biologist. Nat Biotechnol 2004;22(10):1317–1321.

72. Palsson B. The challenges of in silico biology. Nat Biotechnol 2000;18:1147–1150.

73. Kitano H. The theory of biological robustness and its implication in cancer. Ernst Schering Res Found Workshop 2007;(61):69–88.

2

SOFTWARE INFRASTRUCTURE AND DATA MODEL FOR PATHWAY ANALYSIS

FEDOR BOKOV AND ANTON YURYEV

Table of Contents

2.1 INTRODUCTION

The comprehensive software solution for pathway analysis consists of three major parts:

Pathway Analysis for Drug Discovery, Edited by Anton Yuryev

1. Database for storage of molecular interaction network, collection of pathways, and molecular annotation and classifications (ontologies).
2. Algorithms allowing the navigation of the network in the databases, the statistical analysis of the high-throughput data, and the pathway inference and modeling.
3. Software client interface for pathway and network visualization and layout.

The global interaction network is usually stored as a collection of individual mostly binary interactions; pathways are stored as subnetworks.

The implementation details of these three components vary in different solutions but usually follow the standard three-tier client–server architecture of the web-based or enterprise class software consisting of (1) database engine; (2) application server middle layer running algorithms, interfacing with the database and the third party products; and (3) graphical user interface for pathway visualization, data editing, and network navigation. Despite the occasional efforts to develop dedicated open-source software, languages [1,2], and database engines [3] for bioinformatics all currently available solutions for pathway analysis use the existing commercial software technologies. They implement schema for storage and retrieval of molecular network and annotation in standard commercial relational database engines such as Oracle (Oracle Inc), SQL server (Microsoft), or DB2 (IBM). The middle layer can be implemented using Java application server developed in Tomcat container or by using variety of J2EE certified commercial application server platforms such as Weblogic (BEA) or WebSphere (IBM). .NET platform from Microsoft, which is alternative to J2EE has not been used yet in any pathway analysis software but its utilization is imminent. The graphical user interface for network visualization can be implemented in Java (IPA software from Ingenuity, open source Cytoscape software), in C++ (Pathway Studio software from Ariadne Genomics). Some applications use available technologies for web browsers: web client interface for Pathway Studio Enterprise from Ariadne Genomics uses ActiveX technology from Microsoft, MetaCore from GeneGO uses Flash animation technology from Adobe. The open source Cytoscape program (www.cytoscape.org) is the only software that does not use the database as a backend but loads the entire network form a file [4]. This approach allows faster network navigation compared with the network stored in the database but essentially reduces the network data to read-only access.

2.2 BACKEND

2.2.1 Data Model for Network Storage

The standard data model for network storage in the relational database represents the network as a set of *Entity1-relation-Entity2* triplets, where *relation* object stores the description of the relationship between objects *Entity1* and *Entity2*. Thus, relation encodes for network edge and *Entity* encodes for network vertices or nodes. Various annotation fields for *Entities* and *relations* are stored in other tables linked to the network objects via global database identifier. This data model is similar to triplets in the Resource Description Framework (RDF), which serves as the foundation for Semantic Web. In both theory and practice, the network of biological interactions is heterogeneous and contains different types of relations between different types of biological entities. The classification of both entities types and relations types slightly varies from one data model to another; however, all of them contain (1) proteins and small molecules (drugs and metabolites) as separate entity types; and (2) physical interactions, protein modification, and regulatory interactions as separate relations types. The example of the data model used in Pathway Studio software from Ariadne Genomics is shown in Table 2.1.

2.2.2 Container Entities and Relations

As much as any human knowledge, the biological knowledge is hierarchical. The examples of such hierarchy can be found in every major biological database such as Gene Ontology [5] and KEGG [6]. This hierarchy can be and should be used in pathway analysis to reduce the complexity of large networks produced by the analysis of the high-throughput data. The most popular use of knowledge hierarchy is the representation of a pathway containing sub-pathways as nodes in the diagram. To avoid overloading the pathway diagram, most people prefer to represent a sub-pathway as a node connected to the proteins in the super-pathway. Cytokine signaling pathway through NF-kappaB cascade can be shown as a regulatory path from cytokine receptor to at least eight protein components of NF-kappaB cascade, or alternatively as one relation between receptor and one pathway entity, representing the entire NF-kappaB cascade (Figure 2.1). Another example of knowledge hierarchy is a link between entity representing functional protein class and its downstream targets. The fact that the 14-3-3 proteins bind RAF1 kinase can be presented as a collection of 11 relations between 11 14-3-3 isoforms or as just one relation between 14-3-3 functional class entity and RAF1 protein (Figure 2.2). In both examples, the functional class entity or NF-kappaB cascade entity must store the information about the internal content: their protein members and how they regulate each other in case of NF-kappaB sub-pathway. Such multicomponent entities are called *container entities* in the context of data model or *complex nodes* in the context of graphical pathway visualization. The relation connecting container entity must store the information on how its individual components connect to other entities in the super-pathway. Therefore, the relation should be also a container. Having both container and non-container entities unnecessarily overcomplicates the data model and necessitates implementation of separate routines to manage two types of entities. Therefore, it is more efficient to have all entities and relations as containers having

TABLE 2.1 Data Model and Statistics for the ResNet 5.0 Database from Ariadne Genomics

Name	Description	Statistics
Protein	Represents everything that is encoded by a single gene: protein with all its splicing isoforms, mRNA and DNA region	106,749
Small Molecule	Represents any small molecular weight chemical (metabolites and drugs) and nonbiological polymers	47,852
Functional Class	Represents protein functional class containing more than one protein member in any biological organism	4,563
Complex	Represents protein complex	461
Cell Process	Represents biological process	1,094
Treatment	Represents environmental stimuli or condition	17
Disease	Represents disease and organism malfunctions	2,301
Pathway	Represents biological pathway: collection of entities and relations	749
Group	Represents any group of entities of different types	8,638
Direct physical relations		
Binding	Represents direct physical interaction	55,683
ProtModification	Represents reaction of protein modification. Has mechanism property specifying the exact type of modification.	16,479
PromoterBinding	Represents binding of a protein to a gene promoter	5,635
DirectRegulation	Represents regulation by means of physical interaction	725
Chemical Reaction	Represents chemical reaction	13
Regulatory relations		
Expression	Represents regulation of gene expression or protein levels. Has mechanism property specifying mechanism of regulation: expression, degradation.	147,160
MolTransport	Represents regulation of molecular transport. Has mechanism property specifying mechanism of regulation: export, import.	62,046
MolSynthesis	Represents regulation of small molecule levels	117,867
Correlation	Represents correlation between concentration of two entities	0
Regulation	Represents unspecified regulatory event	850,692

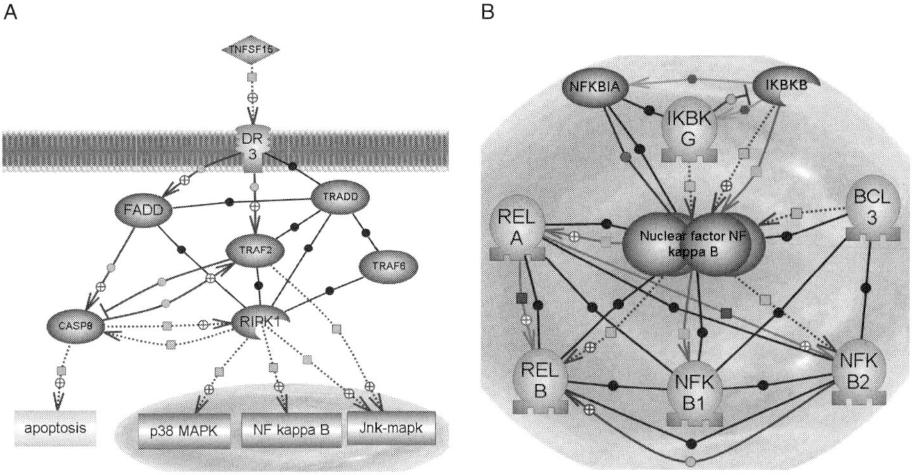

Figure 2.1 (A) Cytokine signaling pathway for VEGI from TNF ligand superfamily shows activation of three sub-pathways: p38-MAPK, NF-kappaB, and JNK-MAPK signaling cascades. (B) NF-kappaB sub-pathway from the previous Figure can be viewed in a separate window.

the common set of routines to open or close a complex node on the graph. Container entities can represent any type of biological object including cellular organelles, biological polymers and cytoskeleton, and even different cell types, tissues, and organs. Biological processes are essentially pathways and therefore containers as well. At present, there is no software that uses the container concept to big extent. Therefore, potential performance problems due to large containers such as cellular organelle are not yet understood. Theoretically the number of possible complex nodes in the database should grow exponentially when all possible combinations of the components can be included into a complex node. However, the actual number of biologically meaningful complex entities seems to be within reasonable range that can be handled by modern hardware.

Another way to use complex nodes is for changing the level of detailization or granularity in the network diagram. The large biological networks are routinely produced during the analysis of the high-throughput data containing measurements for thousands of proteins or metabolites, or when pathways are built by automatically extracting facts from the large number of articles. One way to understand the information flow and functional organization of the large networks is to change the detailization level from relations between numerous individual proteins, to relations between functional classes or sub-pathways containing the proteins in the network. For example, most hormone signaling pathways can be represented at a very high detailization level as a simple diagram with three protein localization class nodes: Extracellular proteins→Membrane proteins→Intracellular proteins, or alternatively as diagram

A

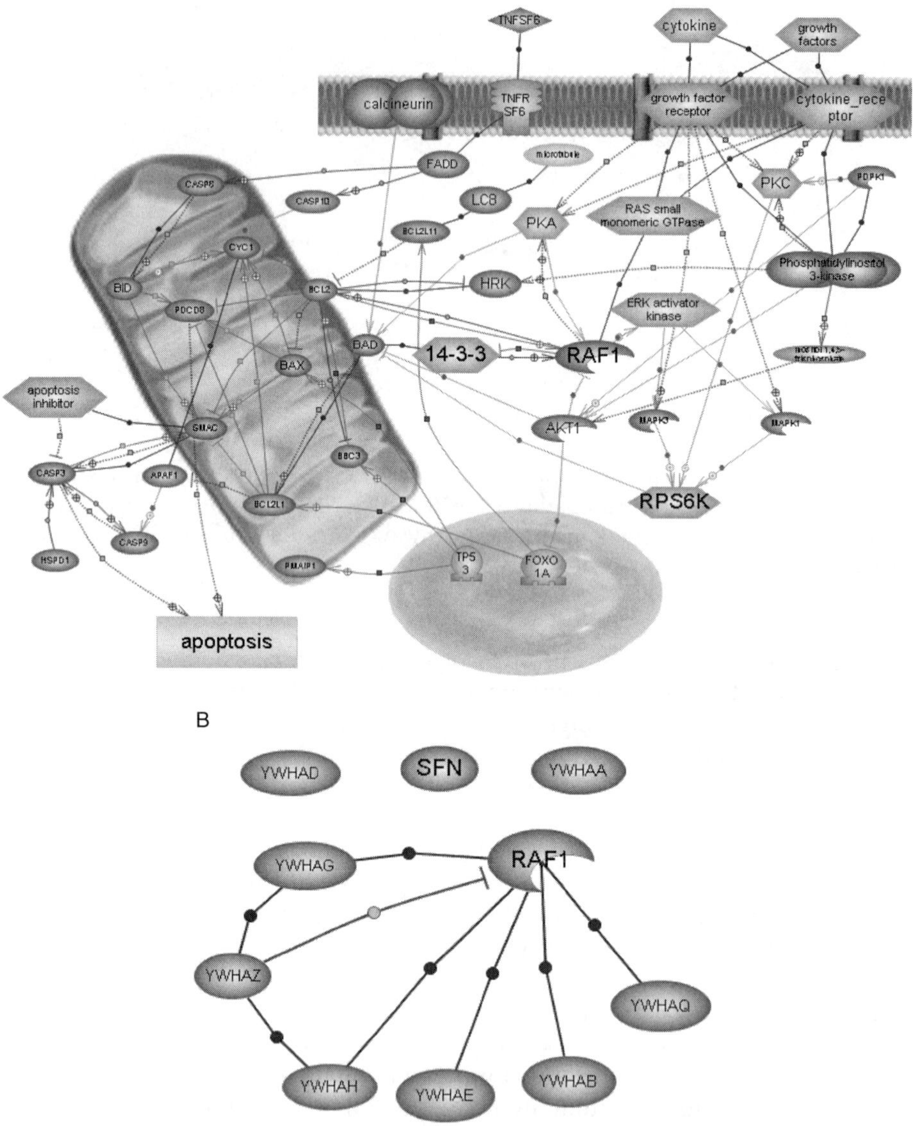

Figure 2.2 (A) Graph depicting mitochondrial control of apoptosis showing several protein functional classes as orange octagons. (B) Protein members of 14-3-3 protein functional class interacting with RAF kinase. The same information is shown as binary interaction between 14-3-3 complex nodes and RAF kinase on panel A.

containing functional class Ligand→Receptors→Scaffolds→Kinases→Transcription factors. On the other hand, the detailization of pathways containing proteins can be increased by displaying the information about interaction of individual splicing isoforms, and even one further down by displaying the protein interaction domains and/or protein modification forms for each isoform.

The implementation of rules automatically increasing the network detailization by specifying more "microscopic" details about the interaction in the graphical user interface is a straightforward task. The layout algorithm simply replaces container entity with its members and container relation with member relations. The opposite algorithm decreasing the detailization level by collapsing the network into higher-level entities is more complicated. If a database contains a lot of containers entities, the large network can be collapsed using several different combinations of containers (Figure 2.3). Therefore, the collapsing strategy should be specified prior to network collapse. The collapsing strategies can be stored and recycled for different networks, or designed on the fly from the list of all containers possible for a given network. More sophisticated algorithms that automatically calculate optimal collapsing strategies, based on the available network topology using user-defined optimization criteria, can be envisioned. For example, such algorithms can collapse a network by optimizing for the number of container entities or relation in the collapsed network. Some collapsing strategies can be allowed to use only "known" container entities that already exist in the database. Other strategies can be allowed creating new containers while trying to match them as close as possible with existing container relations.

2.2.3 Equivalence of Complex Entities and Hierarchical Classifications

The Gene Ontology (GO) classification is an example of acyclic hierarchical graph where one child node can belong to multiple parent nodes. The GO groups are connected with links representing two types of term relations: "is-a" and "part-of" [2]. It is a custom to represent GO graph as tree of groups [7–9]. It is also possible to represent GO classification as a set of three complex nodes corresponding to three GO branches: biological classification, molecular function, and cellular component. Each complex node in this representation contains child groups that are also complex nodes and entire ontology is essentially a "Russian doll" of complex nodes. Such representation of hierarchical ontologies as complex nodes helps to avoid developing a dedicated software code to store hierarchical classifications as a separate entity. The special layout algorithm arranging complex nodes representing ontology in the familiar tree-like view can be also useful for laying out other complex nodes and therefore its code can be recycled.

There are already two examples of algorithms that use the hierarchical knowledge to reduce the complexity of the observed gene expression patterns.

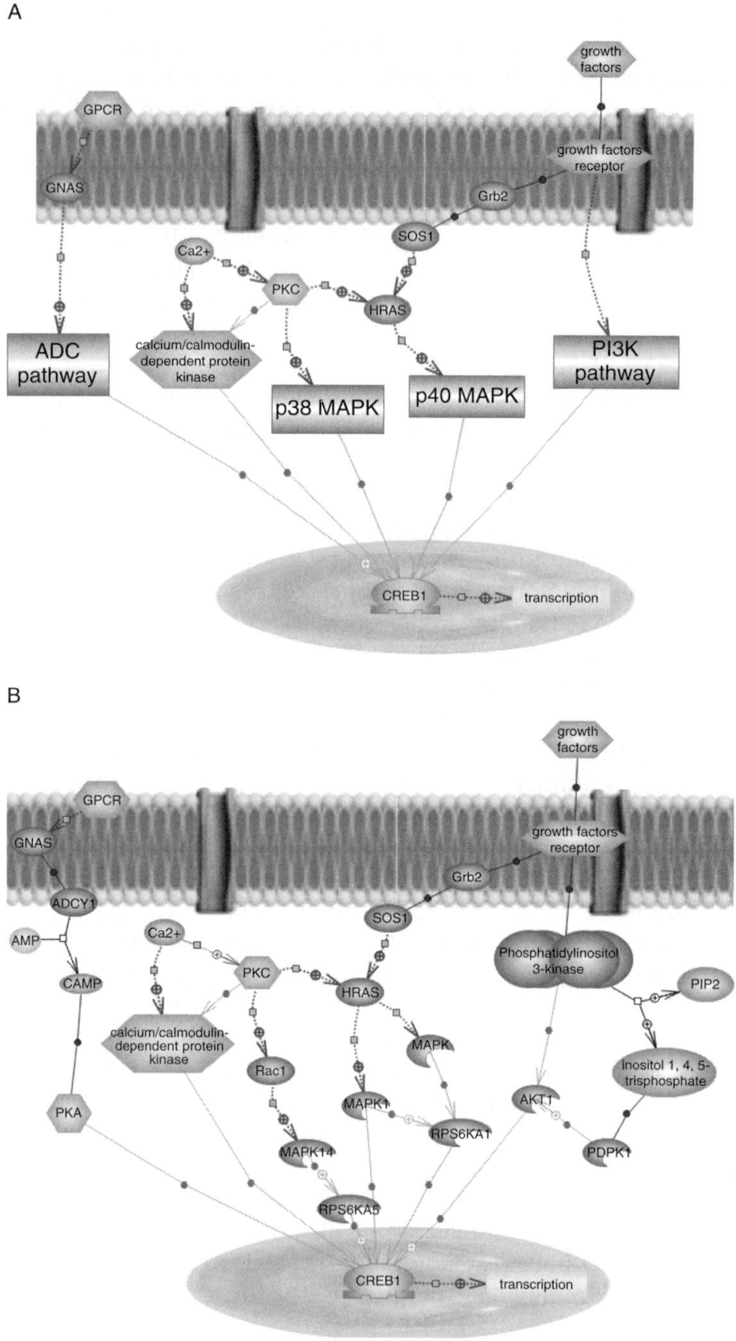

Figure 2.3 Different collapsing strategies of CREB activation pathway. (A) Using sub-pathways. (B) Using functional classes. See color insert.

Gene set enrichment analysis (GSEA) [10] uses Gene Ontology or any other predefined gene sets to collect differentially expressed genes into functional groups or biological processes. Network enrichment analysis (NEA) finds the differentially expressed subnetworks in the large network database [11]. The subnetworks are built using the predefined network queries from known regulatory and physical relations in the network database. Algorithm can use different queries and the network types to construct biologically meaningful subnetworks.

2.2.4 Data Model for Nonbinary Relations

Most biological interactions are binary, i.e. they occur between two entities. There are few examples, when the relations connect more than two entities. The measurements of the protein–protein interaction using mass-spectrometry analysis of co-immunoprecipitates are typically recorded as a spoke model where all detected proteins are connected by one relation [12]. This reflects the fact that the knowledge about the exact binary interactions in complexes measured by this technique is incomplete.

Another example of nonbinary relation is control-of-control relation symbolizing that an entity regulates the relations between another two entities. For instance, if a protein A blocks interaction between other two proteins B and C, it can be shown as A blocking the B–C relations [13]. Control-of-control relation can also symbolize the formation of tertiary complex where the interaction of the third protein depends on the presence of heterodimer formed by the first two proteins. In reality, however, such control-of-control relations are always mediated by a set of binary interactions. Protein A blocks B–C interaction because it regulates either B or C individually therefore preventing formation of B–C complex. Tertiary complexes are formed by means of several weaker binary interactions between individual components, and therefore the protein complexes can be presented as a set of densely linked protein communities rather than as a sequence of binding events. Alternatively, complex can be represented as container node describing the steps how the complex is formed from its protein components. To simplify the analysis and navigation of the biological network, most relations can be and should be represented as a set of binary interactions.

2.2.5 Database Schema and Molecular Annotation

The optimized database schema is very important for performance of pathway analysis software because most algorithms require frequent access to the database for network navigation. Ideally, the network triplets should be stored in one table to enable fast retrieval of the entire network required for some algorithms such as network enrichment analysis for example. However, other solutions that store network nodes and controls separately can also provide the sufficient performance for most network navigation tasks. The example of such solution is shown in Figure 2.4 illustrating schema for Pathway Studio database from Ariadne Genomics Inc. The complete network retrieval in this case can be optimized on the level of the database interface layer of the application server by implementing the dedicated SQL query. The lazy loading strategy implemented in the data providing interface is another technique used for optimizing the database access. The rapid database access can be

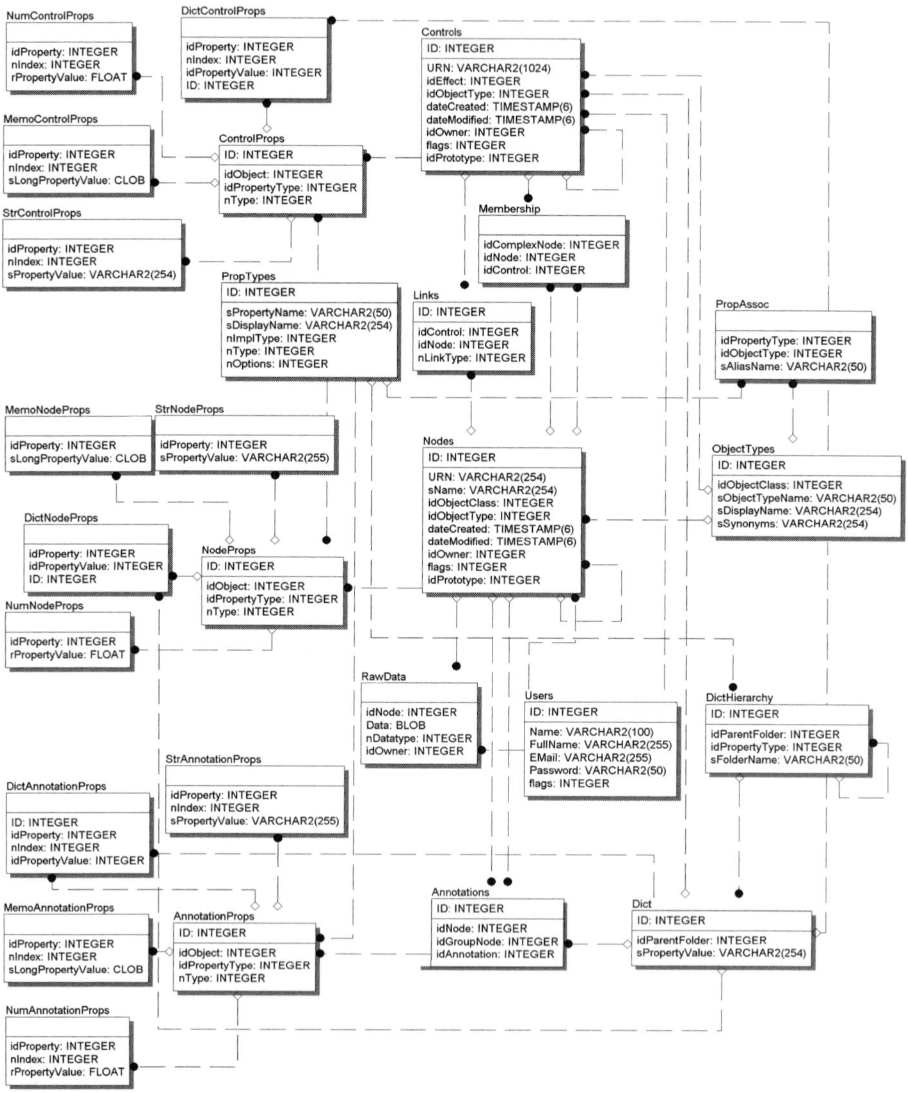

Figure 2.4 Pathway Studio database schema stores biological network in the relational database.

further accelerated by optimizing the hardware configuration. Disk arrays in stripe configurations RAID0 or RAID1+0 can significantly increase the performance of the application. EMC storage arrays such as Symmetrix must also boost the performance of the network database.

Besides the table that stores relations between entities, all other tables in the relational database schema store the annotation for objects and relations. The node annotation usually includes name and various identifiers for biological objects from other databases to maintain the data integrity when importing a network or annotation from other databases. The database can also store some annotation necessary for algorithms. For example, the cell localization layout requires the annotation of entities with subcellular localization, pathway inference algorithms require annotation with protein function, and the algorithms for navigation and query of the network require information about type of relation, its direction, effect and mechanism as well as type of the node to be used in the network construction. It becomes clear that the biological molecular interactions must be annotated with the confidence score, indicating the reliability of supporting evidence as well as the strength of the physical interaction [14]. This information can be used by the algorithm to adjust the confidence of the visualized network and to recalculate the confidence based on additional evidence.

2.2.6 Export/Import Utilities

It is highly beneficial for any software designed for pathway analysis to have data export/import utility into XML format. XML format is the best way to exchange data between databases and to transform one data model into another. These two tasks are likely to become maintenance routine for molecular interaction knowledge bases due to the rapid accumulation of the data for biological networks, constant increase in the number of studied organisms, and numerous public and commercial databases. The perception of the scientific database as a static, rarely changed data source containing highly accurate reference information persists among many biologists. There is no doubt that such "textbook" database is highly important for both research and educational purposes. However, the research databases allowing frequent changes of the content while maintaining the data integrity are likely to gain a widespread use in the scientific community, providing the relative ease of software use and adequate user education. These databases and complementary software are necessary to enable the high-throughput research analyzing hundreds and thousands of different biological molecules at a time. They must allow necessary flexibility to build and test their private hypothesis and validate them with proprietary experimental data without contaminating the "textbook" database with unverified information. The necessity to have research databases demands the functionality to exchange data between databases, which in turn bring us to a very important point of data integrity and universal entity identifiers in biology.

The database content is stored in relational databases in multiple tables which are linked to each other via database identifiers. This standard practice requires identifiers in one database to be identical to identifiers in another database to guarantee data integrity during information exchange between databases. Thus, the universal or global identifiers must be exported into the database exchange XML format. Ariadne Genomics has adopted the Universal Resource Name (URN) as the major identifier for its XML file format. The concept of URN was introduced by W3 consortium to start building the foundation framework for semantic web [15]. While URNs play a role of global entity identifiers for Ariadne's technology, each database contains internal integer indexes used by software for performing SQL queries. This is done to optimize database navigation performance since numerical indexes enable faster data access in the tables. The database importer uses URNs supplied by XML file to convert them into Object Identifiers that have a scope of the imported database only. The relationship between URN and Object ID is stored in the Nodes table (Figure 2.4) and used during export to assign URN to entities in the XML file. Rules describing how to form Ariadne's URNs are described in "Ariadne URNs" document available from Ariadne's web site www.ariadnegenomics.com. The URNs are usually formed using the identifier from one of the major biological databases such as Entrez Gene or Pubchem with added prefix, symbolizing the source of the identifier (i.e. database name). For example, the standard URN for proteins is formed as "urn:ag-llid:123" where 123 is Entrez Gene ID and "llid" stands for LocusLink ID, which is the old name of Entrez Gene database. If data are imported into Ariadne's database using the standard database importer for RNEF XML, the data from the objects with the identical URNs will be merged in the database, i.e. annotation and all relations described in XML will be assigned to the object with the same URN in the database. If the URN does not exist in the database, the software will create a new object in the database. Figure 2.5 shows the example of the ResNet exchange XML format developed by Ariadne Genomics for the data exchange between databases.

2.3 ALGORITHMIC MIDDLE LAYER

2.3.1 Software Architecture: Application Server

Key components of the Java application server for Pathway Studio are shown in Figure 2.6. *Data provider* provides access to the data objects stored in the database, such as entities, relations, experiments, saved analysis results, temporary objects, etc. It utilizes Object-Relational mapping (ORM) framework for Java Persistence API from Hibernate (http://www.hibernate.org) to map object model to relational database (tables). Data provider exposes several high level services, such as Object Database, Folder Database, Search Service, Temporary Storage, etc., which in turn use Data Access Objects (DAOs) to

```
<resnet>
 <nodes>
    <node local_id="N1" urn="urn:agi-llid:50">
    <attr name="NodeType" value="Protein"/>
    <attr name="Name" value="ACO2"/>
   </node>
   <node local_id="N2" urn="2395">
    <attr name="NodeType" value="Protein"/>
    <attr name="Name" value="FXN"/>
   </node>
 </nodes>
 <controls>
   <control local_id="L1">
    <link type="in-out" ref="N1" />
    <link type="in" ref="N2" />
    <attr name="ControlType" value="DirectRegulation" />
    <attr name="mref" value="16713569" />
    </control>
 </controls>
</resnet>
```


Figure 2.5 Example of RNEF XML format developed by Ariadne Genomics to exchange network data between different databases. ⟨resnet⟩ section is used to describe interaction between two entities. It contains entity and relation annotation and stores information about the direction of the interaction. The complete description of the XML format is available from Ariadne Genomics' web site.

retrieve/store the data in the database. The data provider's purpose is to transform data stored in the relational database into Java objects used in the middle layer for running algorithms and communication with end-user interfaces. *Synchronous Operations service* implements generic framework to run short user interface and data manipulation operations, such as Delete object, Rename object, etc. Each button in UI corresponds to separate operation in Synchronous Operations framework. *Asynchronous Tools service* and *Tool Servers* implement generic framework to run long-running tools, such as network navigation and statistical analysis algorithms. This framework allows running algorithms in a distributed environment on multiple servers. Application server includes one embedded Tool Server. Any number of additional external Tool Servers can be added to relieve the load from the application server.

2.3.2 Application Programming Interface (API)

API and CPU load distribution are two major reasons for having three-tier architecture software solutions. Multi-tier architecture allows application to run each tier on a separate computer. Addition of asynchronous distributive tool service to the application server layer also allows running individual algorithms or applications in batch mode on additional CPUs. This allows

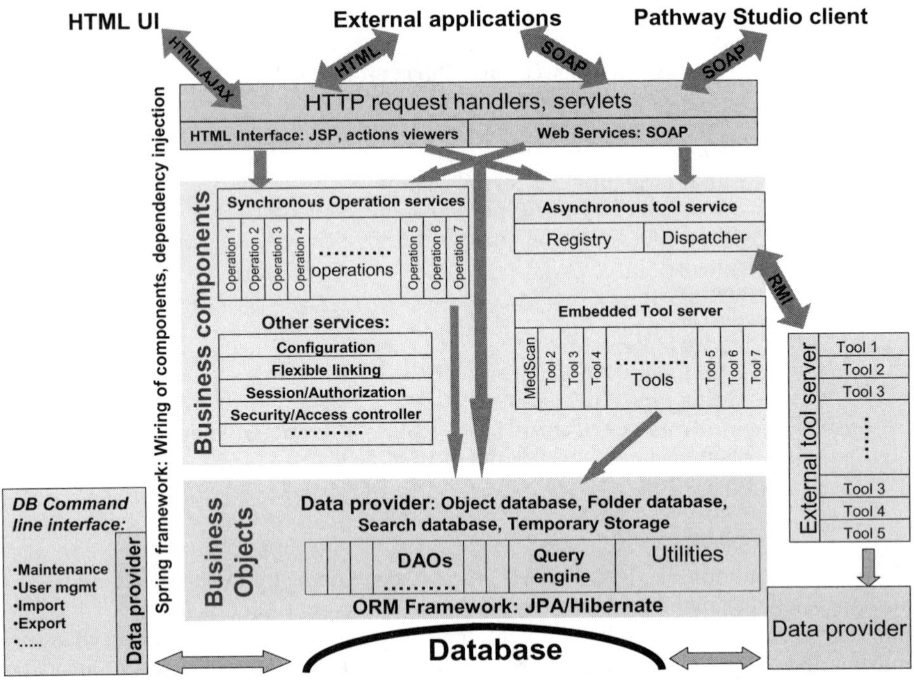

Figure 2.6 Pathway studio application server schema. SOAP, Simple Object Access Protocol; JSP, Java Server pages; AJAX, Asynchronous JavaScript and XML; DAO, Data Access object; ORM, Object-Relational mapping; JPA, Java Persistence API; RMI, Remote Method Invocation.

multiple users accessing the same database and running different algorithms and services at one time without competing for hardware resources.

API makes software open to customization and integration. The principal functionality of the pathway analysis software is storage, retrieval, and visualization of molecular network data. It must provide interface for various algorithms that use these data for statistical analysis and pathway building to access the data and then ability to display algorithm results. The new algorithms are destined to appear in the future while pathway analysis field matures. It is absolutely unnecessary to have dedicated pathway analysis software for every new algorithm. Therefore, the comprehensive solution must provide API to ensure scalability and flexibility of the system and its usefulness in the future. The algorithms for pathway analysis have two major types of input: collection of pathways and ontologies in the database and the entire network. The output of these algorithms is the list of existing pathways or the list of new pathways generated by the algorithm with assigned statistical score. Other output types include new annotation for relations and entities in the database and graph layout for pathways and subnetworks. The annotation can include statistical scores assigned to entities and relations by the algorithm. Sometimes such scores can be only relevant within a scope of a pathway processed by the algorithm. In this case, this annotation must have a scope of a subnetwork and displayed as local property not available outside the scored pathway.

Besides allowing integration of new and proprietary algorithms for pathway analysis, API allows seamless integration of the software with the third party software into existing software infrastructure. The input into pathway analysis software usually comes from the statistical packages for microarray data analysis in the form of gene lists or gene or protein numerical data such as gene expression values. The output of the algorithm should be exported into other statistical packages, graphics, and imaging software for pathway drawing and annotation, and into software for pathway simulation and dynamic modeling.

2.4 CLIENT INTERFACE FOR NETWORK AND PATHWAY ANALYSIS

Client interface for pathway analysis software performs three major tasks:

1. Graphical visualization networks and pathways
2. Visualization and navigation of the hierarchical classifications such as pathway collections or protein functional classification
3. Presentation of algorithm results

"Pathway" and "network" are two names that are used most often to describe communities of proteins and metabolites performing common function. Perhaps it is worth to spell out the distinction between them here because it is important for description of automatic layout algorithms. **Pathway** representation usually requires demonstrating the flow of information, free energy, or metabolites as a series of consecutive steps. Pathway is a diagram depicting the sequence of events either in space or time. Therefore, it is important to show the start or input on a pathway diagram as well as the output or terminal nodes, or in rare cases the cycling of information. **Networks** in biology are usually characterized by hubs—nodes that have the most connection in the networks and by clusters—a subset of nodes that have higher link density between themselves as compared with the rest of network. As described in Chapter 3, there are algorithms that can predict pathways from the networks; however, software for pathway analysis must visualize both types of information. Different algorithms for automatic layouts can be used for either network or pathways.

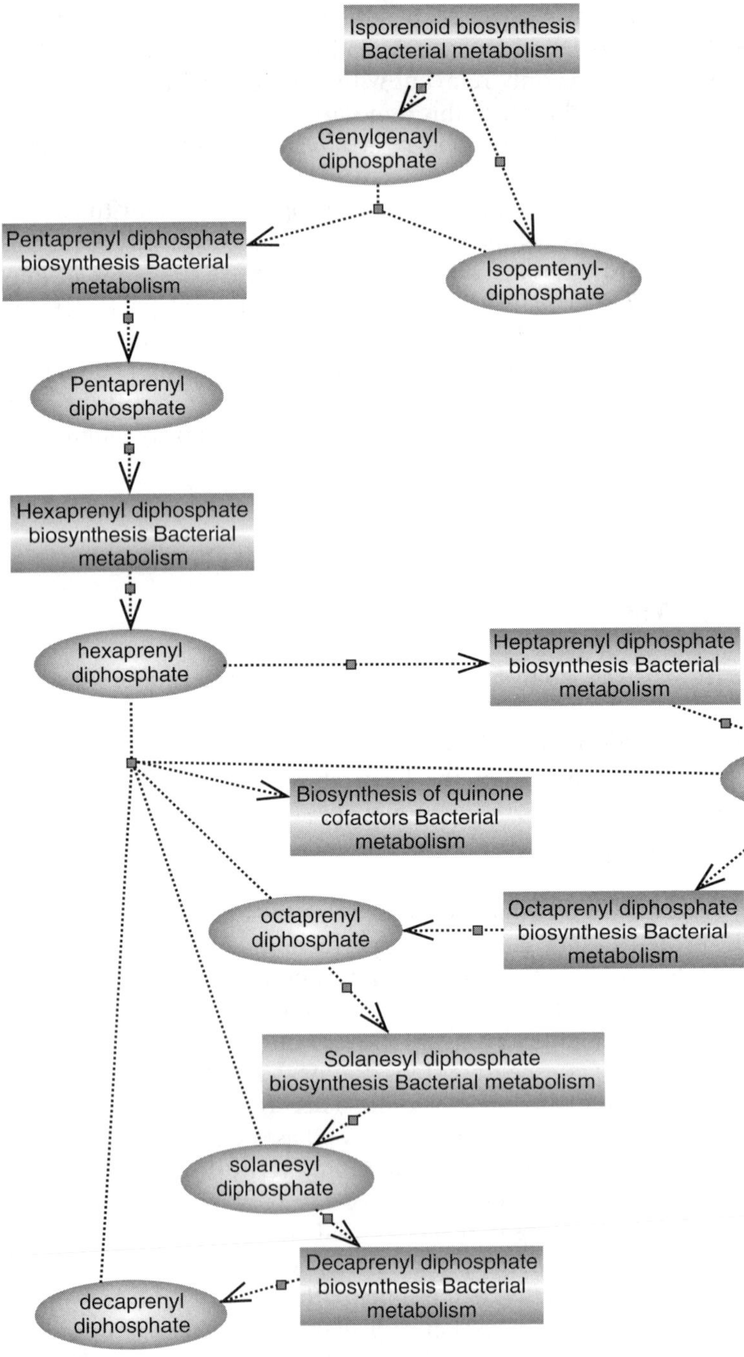

Figure 2.7 Example of hierarchical layout. All-trans isoprenoid chain biosynthesis pathway in Bacterial metabolism.

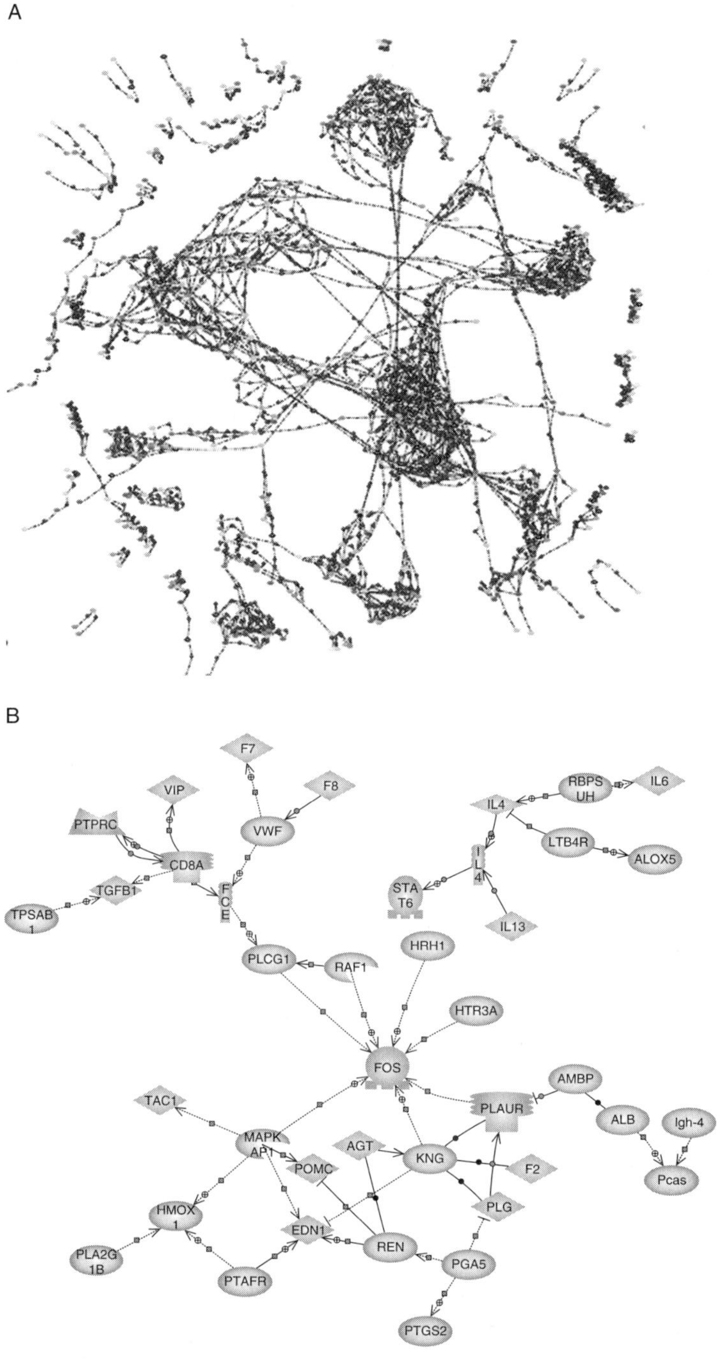

Figure 2.8 (A) Example of network clustering using direct-force layout. Gene expression correlation network clustered to identify proteins with correlated expression profile. The link between two proteins in such network represents the fact of correlation between expression profiles of two genes. (B) Direct-force layout identifies hubs in anaphylaxis network.

2.4.1 Graphical Layout Algorithms for Pathways Visualization

Hierarchical layout [14] works best for directed graph and is very popular to draw biological pathways. It assigns a layer or score for each node based on the number of the nodes it has upstream and downstream in the directed graph. Then, the nodes that have the smallest number of upstream nodes are put on top of the graph and nodes that have the largest number of upstream nodes are put at the bottom of the graph. It does not do a good job for laying out graph with loops and cycles but is superb for laying out directed acyclic graphs. The algorithm is most appropriate for laying out classical regulatory signaling kinase cascades starting from a receptor and metabolic cascade. Once applied, it will show you the nodes where the information flow begins by placing them on top of the graph, it will place the most downstream nodes at the bottom of the graph. You can see the example of hierarchical layout on Figure 2.7.

Another popular method to draw biological pathways is by arranging nodes by *cellular localization*. The popularity of the method is based on the notion that information in the cell must flow from the membrane toward the nucleus. Environmental signals are collected by membrane receptors that transmit their information to cytoplasm and nuclear protein to invoke the cell response to stimuli. The implementation of the algorithm for automatic layout by cell localization requires information about protein cell localization stored in the database. The algorithm can either simply arrange nodes with the same cell localization on different levels in the graph, or use organelle images with assigned cell localization to associate node position with the position of the organelle. In the latter case, algorithm matches the cell localization annotation between organelle and proteins. Examples of layout by cell localization in Pathway Studio software from Ariadne Genomics is shown in Figures 2.1A and 2.2A.

Direct-force algorithm is the principal algorithm to lay out the biological networks. By providing the appropriate network structure, the algorithm can easily identify hubs in the network placing them in the center of a graph, or arrange densely linked communities of proteins in the network clusters (Figure 2.8).

REFERENCES

1. Mangalam H. The Bio* toolkits—a brief overview. Brief Bioinform 2002;3(3):296–302.
2. Stajich JE, et al. The Bioperl toolkit: Perl modules for the life sciences. Genome Res 2002;12(10):1611–1618.
3. Michalickova K, et al. SeqHound: biological sequence and structure database as a platform for bioinformatics research. BMC Bioinformatics 2002;3:32.

4. Shannon P, et al. Cytoscape: a software environment for integrated models of biomolecular interaction networks. Genome Res 2003;13:2498–2504.
5. Ashburner M, et al: Gene ontology: tool for the unification of biology. Nat Genet 2000;25:25–29.
6. Kanehisa M, Goto S. KEGG: kyoto encyclopedia of genes and genomes. Nucleic Acids Res 2000;28(1):27–30.
7. Zeeberg BR, Feng W, Wang G, Wang MD, Fojo AT, Sunshine M, Narasimhan S, Kane DW, Reinhold WC, Lababidi S, Bussey KJ, Riss J, Barrett JC, Weinstein JN. GoMiner: a resource for biological interpretation of genomic and proteomic data. Genome Biol 2003;4(4):R28.
8. Liu H, Hu ZZ, Wu CH. DynGO: a tool for visualizing and mining of Gene Ontology and its associations. BMC Bioinformatics 2005;6:201.
9. The Gene Ontology (GO) project in 2006. Gene Ontology Consortium. Nucleic Acids Res 2006;34(Database issue):D322–326.
10. Subramanian A, Tamayo P, Mootha VK, Mukherjee S, Ebert BL, Gillette MA, Paulovich A, Pomeroy SL, Golub TR, Lander ES, Mesirov JP. Gene set enrichment analysis: a knowledge-based approach for interpreting genome-wide expression profiles. Proc Natl Acad Sci U S A 2005;102(43):15545–15550.
11. Sivachenko AY, Yuryev A, Daraselia N, Mazo I. Molecular networks in microarray analysis. J Bioinform Comput Biol 2007;5(2b):429–456.
12. Bader GD, Hogue CW. Analyzing yeast protein-protein interaction data obtained from different sources. Nat Biotechnol 2002;20(10):991–997.
13. Kohn KW. Molecular interaction map of the mammalian cell cycle control and DNA repair systems. Mol Biol Cell 1999;10(8):2703–2734.
14. von Mering C, Huynen M, Jaeggi D, Schmidt S, Bork P, Snel B. STRING: a database of predicted functional associations between proteins. Nucleic Acids Res 2003;31(1):258–261.
15. URIs, URLs, and URNs: Clarifications and Recommendations 1.0. Report from the joint W3C/IETF URI Planning Interest Group. W3C Note. 21 September 2001. Available at http://www.w3.org/TR/uri-clarification/

3

AUTOMATIC PATHWAY INFERENCE IN HETEROGENEOUS BIOLOGICAL ASSOCIATION NETWORKS

ANTON YURYEV, ANDREY KALININ, AND NIKOLAI DARASELIA

Table of Contents

3.1 INTRODUCTION

Biological association network (*BAN*) is a term describing a database of molecular interactions and associations with other biological concepts such as

Pathway Analysis for Drug Discovery, Edited by Anton Yuryev

cellular and physiological processes, diseases, and phenotypes. Both associations and interactions are represented as links or relations of different types between molecules and concepts. The association of proteins with biological processes can be also stored in the form of molecular annotation. Currently, biological association network databases are available from several commercial companies: GeneGo, Ingenuity Systems, and Ariadne Genomics. The first two companies rely on manual curation to create database content, while Ariadne Genomics uses MedScan natural language processing technology for automatic extraction of associations from scientific literature. There are two major classes of relationships in BANs: regulatory relations and physical interactions. Regulatory relations between biological molecules may be both direct and indirect. Molecules are always connected to concepts by regulatory relations that symbolize the involvement of a molecule in a concept or effect of a process on the molecule. Physical interactions typically have subtypes such as physical binding, protein modification, promoter binding, and chemical reaction. Major subtypes of regulatory interactions include expression regulation, regulation of metabolite synthesis, regulation of molecular transport and translocation, genetic interactions, and regulatory events with unknown mechanisms. All relations, with the exception of physical and genetic interactions, have directions showing the upstream regulator and the downstream target.

The recording of an interaction or association into the BAN database attempts to formalize findings from typical biological scientific publications in which authors describe the various ways that their favorite molecules interact with each other and play roles in biological processes. Therefore, regardless of the recording method—using natural language processing technology or by manual curation—a biological association network is essentially the formalized representation of scientific literature. BANs can be supplemented with the interaction data from other manually curated databases such as KEGG, Entrez Gene, Reactome, and BioGRID. Predicted and high-throughput physical interactions can be added to the BAN to enhance its content.

While providing the raw digest of molecular interactions described in the literature, the biological association network can be used for evaluation and cross-validation of interactions and provide evidence for further automatic and manual curation of the interaction data [1]. A BAN also can be used for automatic and manual building of pathways from interactions stored in the database. This chapter describes several approaches for automatic pathway building in BANs. The quality of the resulting pathways depends on the quality of data in a BAN and on the quality of molecular annotation used in some of the approaches.

3.2 PATHWAY BUILDING IN MOLECULAR BIOLOGY

As described in this section, every algorithm for pathway building in a BAN consists of two major steps: (1) finding nodes connected in the regulatory and association networks and, (2) connecting nodes found in the first step with physical interaction relationships. This process is essentially an automation of the literature review performed by scientists who compile results from various papers to build a pathway. It is based on the intuitively clear fact that every regulatory interaction inside the cell must be mediated by a set of physical interactions. Experimental approaches for pathway identification in molecular biology include searching for molecules that respond to the excitation or inhibition of a molecule in focus and then explaining how the response propagates through physical interactions from the stimulus toward the downstream targets. The stimulus can be a hormone treatment, a specific protein inhibitor such as a drug, a dominant negative mutant, siRNA, or a targeted gene knockout. Current methods for detecting proteins that respond to a stimulus usually include various microarray technologies such as gene expression arrays, phosphorylation arrays, proteomics experiments, and various smaller-scale molecular biological techniques. Manual pathway building involves the compilation and comparison of results from various publications to create a consensus diagram consistent with a majority of the papers.

This approach is used to create pathways in the STKE database (http://stke.sciencemag.org), in which every pathway is built manually by an editing authority—a scientist who is considered an expert in the field. Every scientific report usually provides one or a few individual interactions and sometimes the measurement of one interaction by various methodologies both *in vivo* and *in vitro*. Therefore, the manual building of a single pathway may sometimes involve reviewing hundreds of papers. The STKE database has a collection of about 100 consensus experimentally proven pathways for various model organisms. The task of the STKE authority is to review and curate the pathways on a regular basis and update it with new information published in peer-review journals. Undoubtedly, this is the most accurate and conservative approach to generate pathway collection, yet it is the slowest and the most expensive way to do it. Taking into account the cost of numerous experiments performed to identify and prove each interaction, the cost of every STKE pathway can be easily estimated at several million dollars. Therefore, the classical approach for pathway building cannot achieve the goal of building the collection of 500,000 human pathways necessary for drug discovery. New approaches for automatic pathway reconstruction or inference are required to meet this goal.

Because STKE pathways are manually curated, they also tend to be relatively small, containing no more than 50–100 molecules. It becomes clear, however, that signaling and functional blocks may contain a few hundred or even thousands of molecules. For example, the original modeling of NF-kappaB oscillations in a Toll-like receptor (TLR) pathway was performed using 17 parameters [2]. The most recent model of this pathway includes 652 species and 444 reactions (340 proteins, 79 simple molecules and other entities), yet its authors still acknowledge that their network is far from completion [3]. The current model for the EGFR signaling pathway contains 211 reactions and 322 species [4]. The increased size of biological functional blocks provides additional motivation for developing algorithms for automatic pathway inference.

The STKE approach, however, is the absolutely necessary first step for building a comprehensive pathway collection for the human organism. It provides insights into general principles of molecular organization in pathways and produces a collection of conserved molecular signaling blocks that are reused by the evolution of various organisms. The basic principles of biological pathway organization can be used as additional rules and constraints for algorithms, while conserved regulatory blocks can simplify and accelerate pathway building and allow the sorting out of irrelevant or nonessential physical interactions. In the following sections, we describe several approaches for automatic pathway inference in biological association networks using the principle of explaining the regulatory and association relations by physical interactions.

3.3 ALGORITHMS

All algorithms described in this section rely on network navigation tools available in pathway analysis software. These tools allow different strategies for expansion of a heterogeneous BAN network by combining different graph queries with different types of relations. Due to the noise and incompleteness of available biological association networks, these algorithms currently require some manual intervention. The global biological association network usually contains a dense area with many relations for well-studied proteins. This area tends to have a high noise level due to mistakes made during the recording of facts and due to errors in publications. The area that contains proteins that are not very well studied, on the other hand, suffers from incomplete information. Therefore, manual curation is required to either remove false or unreliable facts or to add missing links to a pathway.

While building pathways in real BANs, the last step of the algorithms—connecting nodes with either regulatory or physical interactions—almost always produces a major interaction network plus a set of unconnected nodes. The nodes cannot be connected to the major network either because they are not well studied and do not have enough relationships in the database or because they may have nothing to do with the biological process or the regulome in the first place. Both errors in recording regulatory interactions and those made during experiments can add false connections between a protein and an original concept used for pathway building, such as a ligand-receptor pair, cell processes, or a disease. If a protein or a metabolite is falsely linked to a biological concept, it has fewer chances to become part of the physical interaction network formed by proteins that are truly involved in the concept. Therefore, the second algorithm step serves as a filter to remove false or unreliable relations from a pathway.

3.3.1 Regulome Pathways

The regulome pathway explains how the information from a molecule propagates toward its downstream targets in a regulatory network through the physical interaction network. The input for the regulome pathway is one or two functionally related proteins, for example, a ligand and its receptor. The downstream targets are found simply as proteins regulated by the original molecule(s). Experimentally, this is done by inhibiting or activating the selected molecule and measuring which proteins change activity in response to the stimuli. In a BAN, this experimental evidence is recorded as regulatory interactions. Figure 3.1 illustrates how the direction of information flow can be determined in an ideal BAN, where all regulatory and physical interactions are known and accurate. In reality, however, not all links necessary for building of the regulome pathway are known. Therefore, in order to calculate the most likely direction of information flow, the nodes in a pathway should be scored based on their in-degree to out-degree ratio. The presence of feedback regulatory loops in a BAN further complicates the calculation. Therefore, it is highly desirable to detect and delete feedback loops prior to building a regulome pathway.

The detection and removal of feedback loops is equivalent to the minimum feedback arc set problem in graph theory where a directed acyclic graph must be built by removing a minimal number of links from the directed graph [5]. It was shown that the solution can be found only with certain accuracy using the constant-factor approximation algorithm; that is, the minimum feedback arc set problem is APX-hard [6]. We found that this task can be interpreted as an optimization problem and tackled by probabilistic methods such as simulated annealing. The details of this procedure are described in Chapter 4.

Once feedback loops are removed, the in-degree to out-degree ratio can be used to arrange molecules by the direction of information flow. The original protein(s) used for finding the downstream targets must have an in-degree equal to zero and will have the highest score that puts it at the top of an information flowchart. The end nodes terminating the information flow in a regulome pathway must have an out-degree equal to zero (Figure 3.3). Alternatively, the end nodes can be set manually, based on the biology of a pathway. For example, transcription factors can be specified as end nodes in regulomes for a ligand-receptor pair.

After the sequence of regulatory events is identified, the next step in building a regulome pathway is to connect nodes by physical interactions. In an ideal BAN, interactions between nodes at different regulation levels are the interactions that participate in the information flow, while interactions between nodes at the same regulatory level indicate a physical complex or scaffolding interactions in a pathway. In reality, however, noisy and incomplete BANs sometimes make it impossible to differentiate between "structural" physical interactions that hold the pathway components together

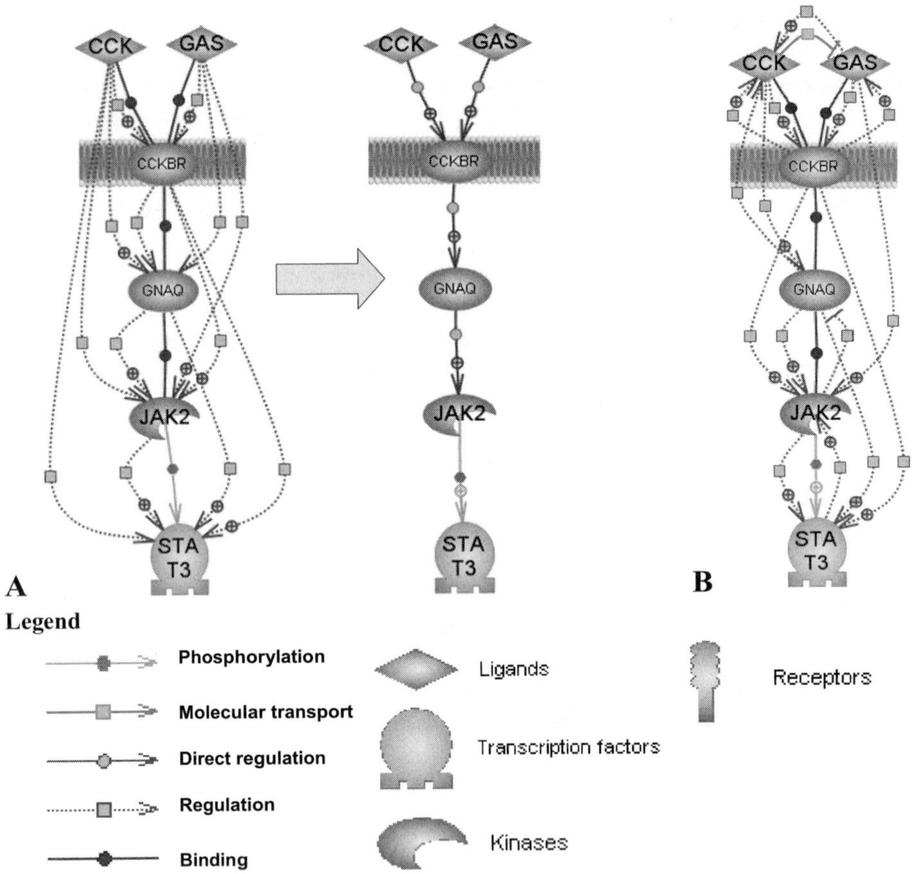

Figure 3.1 The basic principles of pathway building in a biological association network are similar to the manual compilation of experimental evidence for a consensus pathway. The order of components for information flow can be determined from the combination of regulatory interaction. Upstream molecules must have only outgoing relations to all downstream pathway components. After the order of components is determined from a regulatory network, the components must be connected by physical interactions that mediate signal propagation. (A) The relations are shown in an ideal BAN that contains a complete and accurate set of experimental interactions. (B) The relations between proteins in the same pathway are present in the ResNet 5 database. ResNet 5 contains biological association networks automatically extracted from scientific publications using MedScan, a natural language processing technology. Notice multiple feedback loops and incomplete regulatory network that complicate pathway calculation in a real-life BAN. See color insert.

and the "regulatory" physical interactions actually mediating the information flow.

An example of a real-life regulome pathway is shown in Figure 3.3. Notice that the regulome pathway correctly identifies major pathway components but

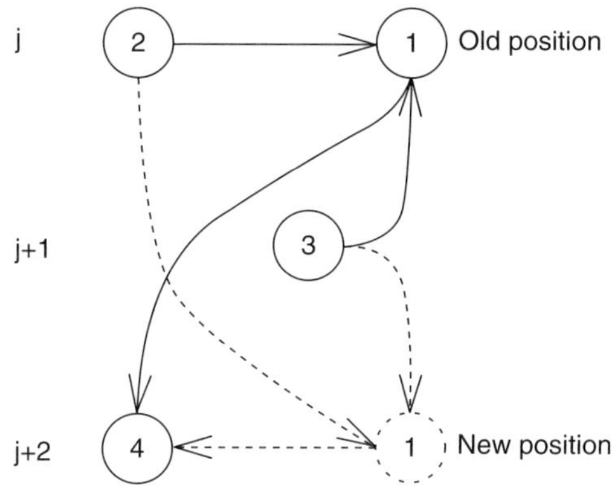

Figure 3.2 Node 1 with two incoming and one outgoing links is selected to move from its current position on level j to a new position on level j + 2. The associated energy difference is $\Delta E = -1 - 1 + 1 = -1$, where two −1 contributions come from making (2, 1) and (3, 1) links hierarchical, and the single +1 contribution comes from turning the link (1, 4) from hierarchical to nonhierarchical.

essentially fails to find the direction of information flow due to the multiple feedback loops that were not removed prior to pathway building. The only way to visualize the information flow for this pathway is by laying proteins in the pathway by cell localization to show the orientation of pathway relative to the direction of the major information flow in the cell—from plasma membrane to nucleus.

3.3.2 Signaling Line Pathways

The signaling line algorithm can be considered either a modification or an alternative to the regulome method. The goal of the algorithm is finding individual "signaling channels" from cell receptors to transcription factors in a BAN. Each "signaling channel" contains exactly one receptor and one transcription factor, a chain of intracellular effectors transmitting the signal, and a set of ligands activating the receptor. The algorithm uses five groups from protein functional annotation: extracellular ligands, receptors, transcription factors, nuclear receptors, and effectors (all others). The algorithm works in three steps:

(1) For each individual receptor, a set of target transcription factors is identified in the regulatory network.
(2) For each receptor-transcription factor pair, an optimal "signaling channel" is calculated in the physical interaction network.
(3) Appropriate ligands interacting with the receptor are added to complete the pathway.

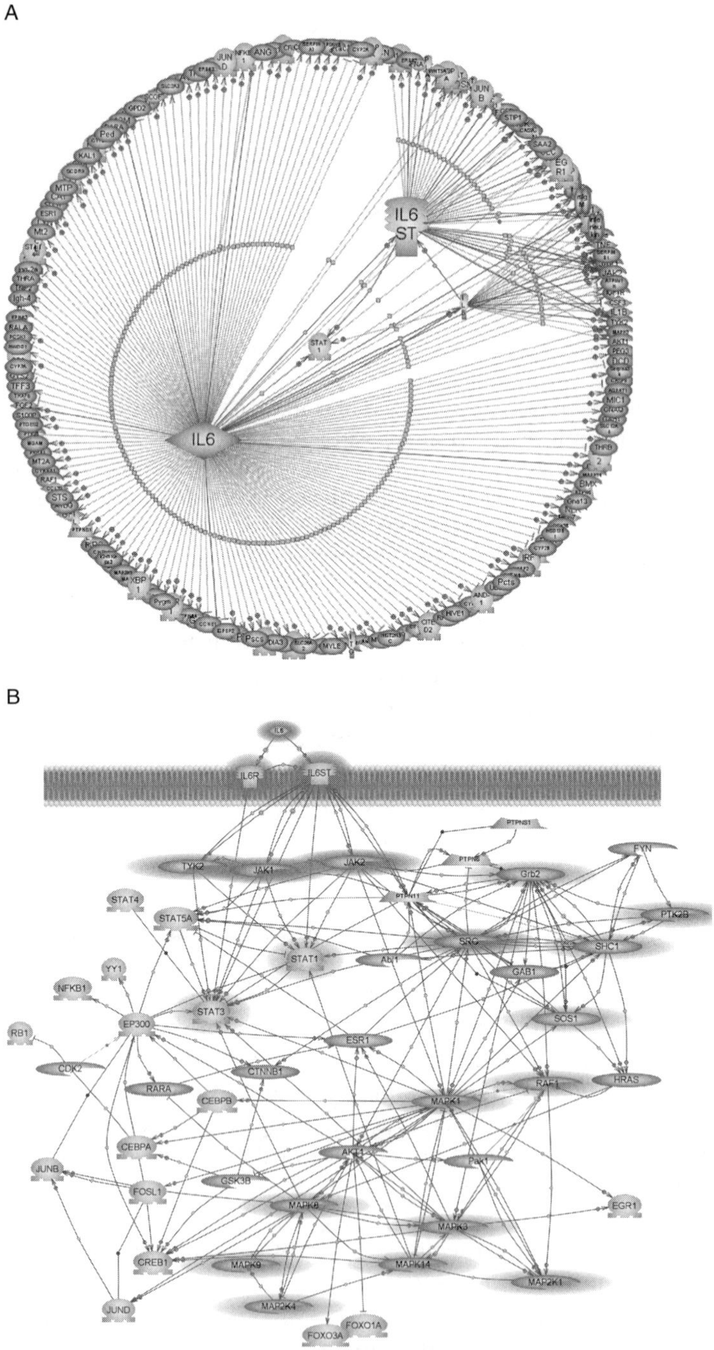

Figure 3.3 An example of a regulome pathway. (A) All proteins found as downstream targets of IL6 or its receptor IL6R in a BAN. (B) IL6 targets connected by physical interactions found in a BAN. The classical JAK-STAT pathway is shown in blue, and the MAP kinase pathway is shown in green. Only transcription targets were allowed to be targets-only proteins during the construction. The network trimming procedure was performed as described in the literature [1].

The first step is performed by recording all transcription factors regulated by the receptor. The second step utilizes graph theory algorithms to calculate a signaling pathway. Only "effector" proteins are allowed to participate in signal transduction from the receptor to the transcription factor, and only physical interaction relationships are allowed to be used for connecting effector proteins in the pathway. Prior to pathway calculation, each relation must have a weight indicating the reliability of the interaction and relative likelihood of information flow through the link. For example, the weight can be based on a number of references for each relationship as a measure of its reliability and can take into account a sequence similarity to a paralogous protein that has similar relation as a measure for information flow likelihood. The latter score is proportional to the number of close homologues of A and B, which are also connected by the same type of relationship.

During the second step, the shortest path based on link weights between the receptor and the transcription factor is calculated using Dijkstra's algorithm [7]. This approach demonstrates the generation of valid signaling paths, but they tend to be shorter than real paths, because the shortest path algorithm tends to omit some components of signal transduction channels, especially if they are parts of multi-protein complexes. To overcome this problem, the shortest path can be augmented by additional proteins and relationships if they are part of multi-protein complexes. Complexes can be defined as densely connected protein–protein interaction subnetworks around a pair of adjacent proteins directly linked in the original shortest path. After the augmentation step, the minimal spanning tree (MST) calculation algorithm is applied to the resulting graph, and the single tree branch that leads from the receptor to transcription factor is selected as a signaling channel pathway. The augmentation-MST calculation step was found to lengthen the originally calculated shortest paths while including relevant signal transduction components missed by Dijkstra's algorithm.

The third step is accomplished by adding all the extracellular ligands connected to the receptor by protein–protein interaction relations to every channel pathway. At the end of the process, all pathways must be manually curated to make sure they include the complete conserved signaling blocks such as MAP kinase cascade, JAK/STAT cascade, IKK-NF-kappaB pathway, ADC/cAMP/PKA cascade, PLC/DAG/PKC, etc. We have identified 34 conserved signaling blocks altogether (Table 3.1). The combination of all signaling line pathways from one receptor should, in principle, yield the regulome pathway for the same receptor in an ideal BAN. An example of a signaling line pathway is shown in Figure 3.4.

3.3.3 Biological Processes Pathways

In BANs containing biological processes as entities, the involvement of individual proteins and metabolites in a cell process is shown as a regulatory relation between a protein and an entity symbolizing the process. In

TABLE 3.1 A List of Experimentally Proven Signaling Modules Used for Building Signaling Line Pathways

Module Components	Description
ADC>cAMP>PKA	Protein kinase A activation by cyclo-AMP
PKA>RAP1A>B-Raf>MEK1/2>ERK1/2	ERK activation by PKA
GUCY>cGMP>PKG	Protein kinase G activation by cyclo-GMP
PLC>DAG>PKC	Protein kinase C activation by diacylglycerol
PLC>IP3>ITPR>Ca++>PKC	Protein kinase C activation via Ca++ release
Ca^{2+}>RasGRFs>Ras	Ras activation via Ca++ release
Shc1>GRB2>SOS1>Ras	Ras activation by Shc1-SOS1 complex
Ras>Raf>MEK1/2>ERK1/2	ERK activation by Ras
Ras>CDC42/Rak>PAK>MAP3Ks>MKK4/7>JNKs	JUN kinase activation by Ras
Ras>CDC42/Rak>PAK>MAP3Ks>MKK3/6>p38 MAPKs	p38 MAP kinase activation by Ras
VAV>CDC42/Rak>PAK>MAP3Ks>MKK4/7>JNKs	JUN kinase activation by Vav
VAV>CDC42/Rak>PAK>MAP3Ks>MKK3/6>p38 MAPKs	p38 MAP kinase activation by Vav
MEKK1/MLK3>MEK1/2>ERK1/2	ERK activation by MAP kinase kinase
MEKK2>MEK5>ERK5	ERK5 activation
NIK>IKK>IkB>NFkB	NF-kB activation pathway
PI3K>PIP3>PDPK>Akt	Akt activation by Phosphoinositide-3-kinase
Act>GSK3β>β-Catenin	β-Catenin activation by Act
Dvl>AXIN/FRAT>GSK3β>β-Catenin	β-Catenin activation by Dvl
PI3K>PIP3>PDPK>PKC	Protein kinase C activation by Phosphoinositide-3-kinase
PKC>Ras	Ras activation by Protein kinase C
Akt>IKK>IkB>NFkB	NF-kB activation by Akt
Akt>NOS>NO	Nitric oxide synthase activation by Akt
MyD88>IRAKs>TRAFs>MAP3Ks>MAP2Ks>JNKs/p38 MAPKs	MAP kinase activation by TRAFs
TRAFs>IKK>NF-kB	NF-kB activation by TRAFs
MAP3Ks>IKK>NF-kB	NF-kB activation by MAP kinases
MAP4Ks>MAP3Ks	MEKK kinase activation by MEKKK kinases
JAK>STAT	Jak-Stat pathway
JAK>SHP2>GRB2>SOS1>Ras	Ras activation by JAK kinases
JAK>PI3K	Phosphoinositide-3-kinase activation by JAK kinases
BMPR/ACVR>SMAD1/5/8	SMAD activation by BMP receptor
TGFBR/ACVR>SMAD2/3	SMAD activation by TGFβ receptor
Gq-proteins>PLC	Phospholipase C activation by Gq-proteins
Gs-proteins >ADC	Adenylate cyclase activation by Gs-proteins

TABLE 3.1 A List of Experimentally Proven Signaling Modules Used for Building Signaling Line Pathways (*continued*)

Module Components	Description
Gi-proteins >PI3K	Phosphoinositide-3-kinase activation by Gi-proteins
SRC>PLC	Phospholipase C activation by SRC kinase
SRC>ADC	Adenylate cyclase activation by SRC kinase
SRC>PI3K	
SRC>Shc1	Shc activation by SRC kinase
SRC>Vav	Vav activation by SRC kinase

The direction of the arrow indicates the direction of information flow within the module. Chemicals abbreviations: IP3 = inositol 1,4,5-trisphosphate; PIP3 = phosphatidylinositol-3,4,5-trisphosphate; cAMP = cyclo-AMP; cGMP—cyclo-GMP; NO = nitric oxide; DAG = diacylglycerol.

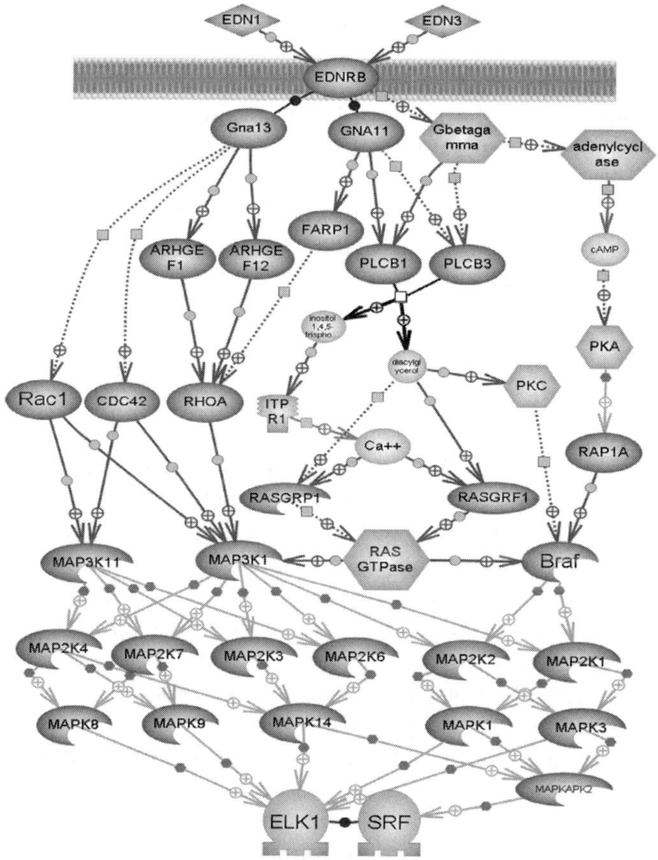

Figure 3.4 An example of a signaling line pathway showing an information flow path from endothelin receptor B (ENDB) toward serum response factor (SRF).

experimental molecular biology, the "cell process" is the reproducible cell behavior that can be detected using a dedicated functional cell-based assay [8,9]. Often, cell-based assays measure an activity of a biomarker indicative of the process. The biomarker can be an enzymatic activity [10], a protein expression [11], or a change in cell morphology or phenotype [12,13]. The experimental results of these assays are reported in the literature and consequently recorded into a BAN as a regulatory relation between a protein and a cell process.

An alternative to graphical representation in BAN, proteins can be annotated with a cell process name. In this case, the input entity list for a pathway building algorithm can be generated directly from protein functional annotations, such as Gene Ontology biological processes [14]. A protein functional annotation can be compiled by manual curation or by automatic text mining of scientific literature [15]. Both methods have the same problems as recording methods for BANs: manual curation is slow and therefore incomplete, automatic annotation has a higher level of false positives. Therefore, similar precautions must be taken when building pathways from the functional annotation, as in building a pathway from a BAN.

Once the set of proteins involved in a cell process is identified, the next step is to determine the direction of information flow in a process. This can be done the same way as for regulome pathways by connecting all proteins by regulatory interactions, deleting feedback loops, and scoring the nodes using the in-degree/out-degree ratio. The direction of information flow can be also set manually using biological knowledge. The physical interactions mediating information flow are determined in the last step. An example of a pathway built using Gene Ontology information is shown in Figure 3.5.

3.3.4 Disease Networks

Disease networks can be built the same way as pathways for biological processes by using a disease entity in a BAN or using protein disease annotation to generate the initial input for pathway construction. The experimental input for disease networks are proteins whose activity and amount changes in patients as compared to healthy individuals. Usually, the experimental approach to identify such proteins is measuring differentially expressed genes [16]. More proteins can be added by identifying downstream targets physically linked to the differentially expressed genes. A similar approach to the identification of direction of information flow in regulome pathways might prove useful in determining the malignant information flow in a disease network. Since very few disease networks have been identified so far, it is not clear how useful this approach can be.

At least two experimentally proven disease networks have been described so far in the literature. It is worth describing in detail the approaches for building them in this section. The network for the angiogenic switch was reported and proven experimentally by Abdollahi et al. [16]. The authors' goal was to

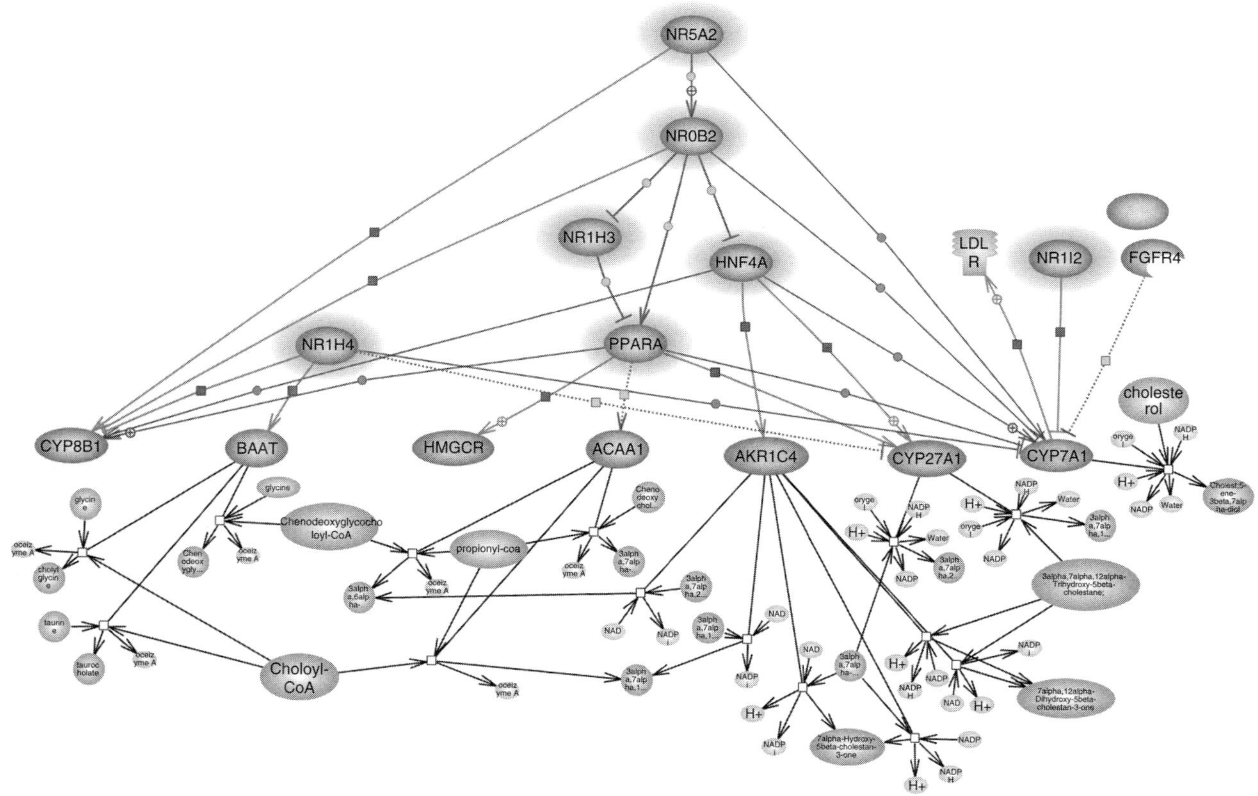

Figure 3.5 An example of a biological process pathway built from protein annotation with the Gene Ontology term "bile acid biosynthesis" (GO:0006699). Note that the GO annotation contains nuclear receptors (highlighted in blue) that regulate the gene expression of enzymes, catalyzing the biochemical pathway for bile acid biosynthesis. The pathway was built essentially as described by Caldwell [13]. Chemical reactions were added manually after the construction of the regulatory and physical interaction network. See color insert.

understand the mechanisms that trigger the growth of blood vessels inside tumors. The angiogenesis happens at late stages of cancer and eventually promotes metastases. VEGF and bFGF are hormones that promote angiogenesis, and endostatin is a hormone that inhibits it. To identify proteins involved in establishing the angiogenic switch, the authors selected genes that were up-regulated in response to VEGF and bFGF, and were down-regulated in response to endostatin in epithelial cells. The well-defined input protein list was one important ingredient for success. Then, the proteins were connected using interactions from a BAN, which allowed only interactions that were confirmed by *in vivo* experiments. To validate their approach, the authors performed the targeted gene knockout in mice in one of the hubs in the angiogenic switch network—PPARδ protein. They showed that the angiogenesis was inhibited around micro-tumors embedded into PPARδ–/– mice.

Another example of a disease network was reported for ataxia genes [17]. The authors performed a targeted two-hybrid screen to identify interaction partners for 54 proteins genetically linked to inherited ataxia. They also expanded the network by two to three steps relative to ataxia proteins with interactions reported in the literature or predicted by protein sequence homology. They have validated the network by confirming some interactions by *in vivo* co-immunoprecipitation, by successfully predicting a novel ataxia gene, and by finding genes modifying ataxia in an animal model among proteins of their network. Interestingly, they found that many original ataxia proteins were not hubs in the reported network. An example of a disease network built using information available from a BAN is shown in Figure 3.6.

3.4 ADDITIONAL SOURCES OF FUNCTIONALLY RELATED PROTEINS

The first step in automatic pathway building is connecting entities in the regulatory and association networks. If a BAN contains entities symbolizing biological concepts such as cell processes or diseases, the information about molecules participating in these concepts is readily available in the form of links between molecules and the concepts. There are two additional methods to identify functionally related molecules: (1) using functional annotation and (2) from dedicated experiments. We have already mentioned these approaches in the previous section and use the following section to compare them.

3.4.1 Functional Protein Annotation

Functional annotation essentially represents the same information as links between proteins and cell processes in a biological association network. Their difference is simply in data representation due to the difference in methods for gathering and recording information about protein function. Gene Ontology is the most accurate currently available functional annotation that is *de facto* the gold standard in molecular biology [14]. Because it is manually

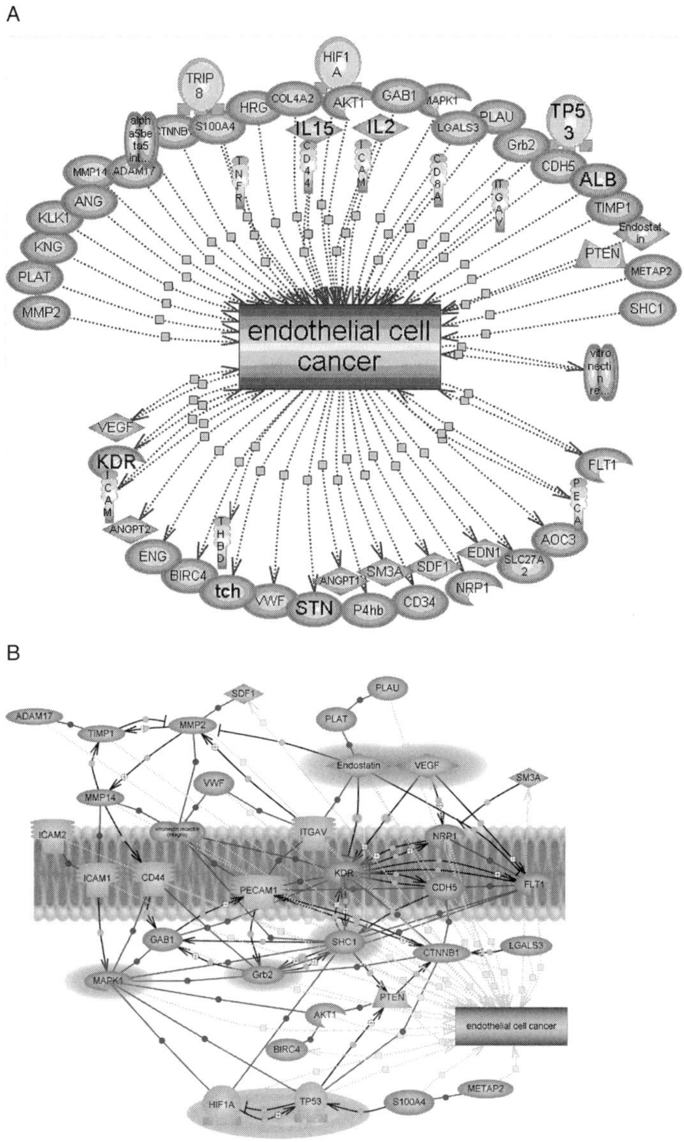

Figure 3.6 An example of a disease network built for endothelial cancer from relationships available in the ResNet 5 biological association network. (A). The first step of pathway building—identification of proteins linked to endothelial cancer in a BAN. (B). The second step of pathway building—connecting proteins into a physical interaction network. Note that about half of the proteins originally linked to the disease in the BAN could not be connected to the major network and were removed. The hubs in the endothelial cancer network (vitronectin receptor, MMP2, and GRB2) point to major proteins involved in the development of this type of cancer.

curated, Gene Ontology suffers from a slow rate of information gathering and curator subjectivity, and is far from being complete [15]. Manually curated BANs also suffer from incompleteness and subjectivity, while BANs automatically extracted using text mining algorithms usually have comprehensive coverage of functional protein association but also contain noisy data.

Because, in reality, both molecular annotation and BAN currently complement each other, the proteins involved in a biological process should be searched for in both BANs and Gene Ontology. Both types of functional annotation are likely to merge in the future; however, since a biological process is an artificial concept representing the current view of cell function by humans, the list of biological processes will constantly evolve and must be regularly updated.

3.4.2 Experimentally Derived Functional Protein Association

If data about protein association are not available from the literature, a BAN, or a molecular annotation, an experimental approach must be used. Experiments can be aimed at determining the function of an individual protein as in aforementioned cell-based assays, or they can be conducted in a high-throughput manner in order to identify a large group of proteins involved in a process. There are several important considerations when a protein set derived by the latter approach for pathway building is used. First, high-throughput experiments tend to have a high level of noise. Second, criteria for selecting genes that change their behavior in a cell process are not always obvious. For example, there are still debates about how to select differentially expressed genes in a microarray experiment [18]. There are at least three criteria for selection: the ratio of expression values indicating the strength of the differential expression, the p-value indicating the confidence of the differential expression calculated from multiple experimental sample replicas for each gene, and a false discovery rate indicating the level of noise in a group of differentially expressed genes. To add to the confusion, the p-value and the false discovery rate can be calculated by several statistical algorithms. It is worth mentioning here that building a network from high-throughput data is also a way to filter out noise: true differentially expressed genes are expected to be functionally linked in a BAN, while genes erroneously identified as differentially expressed should not have any connection to the network of differentially expressed genes. In our experience, several different cutoffs should be tried in order to identify a network cluster formed by the differentially expressed genes. The optimal cutoff identifies the well-defined cluster in a network that is the community of densely linked proteins.

The networks built from experimental data provide important insights about major regulators and targets among responsive genes that appear as network hubs. Nevertheless, for building a pathway that demonstrates information flow, one must perform a time-course experiment. If no time-course data are available, pathway analysis tools can identify the molecular network formed by responsive genes in the experiment, but they will not reveal the direction of information flow without additional data. Additional information can be obtained from a time-course experiment or can be already present in the database.

3.4.3 Finding Functionally Related Proteins by Text Mining

One more way to gather a list of functionally related proteins for input to pathway building is by extracting the list of entities from scientific literature gathered on a specific topic. If an available BAN does not contain your disease or cell process of interest, it can be useful to collect proteins mentioned in all publications about the process of interest and then use them as input for pathway building. Proteins can be collected manually or using text mining technology that is capable of recognizing protein names.

3.5 BIOLOGICAL RULES AND CONSTRAINTS FOR PATHWAY BUILDING

Recent studies of signaling pathways in model organisms have identified several rules for *in vivo* pathway regulation (see [21] and the following references). They are listed with an explanation of how each rule can be used to improve algorithm construction.

(1) The information inside a cell propagates by means of physical interactions and catalytic reactions. This self-explanatory rule is derived from basic principles of molecular biology. It dictates that the final pathway must consist of only physical interactions and catalytic reactions, even though regulatory interactions can be used during pathway construction. This rule provides the theoretical foundation for the second major step in pathway building—connecting nodes found in regulatory network by physical interactions.

(2) Pathway sub-compartmentalization using clustering in a physical interaction network [15] and scaffolding [19,20]. The components of a pathway must form a cluster in both physical and regulatory interaction networks, creating a community of densely interconnected nodes. If proteins used for algorithm input do not form a network cluster, the biological function of such a pathway is questionable; this suggests that the pathway may be incomplete. This criterion can be used to evaluate the biological relevance of the input protein list and to estimate whether the database contains enough interaction data to build the pathway from the input. Indeed, even if the input proteins do perform a common function, little may be known about their interaction with each other and, therefore, pathway building becomes impossible.

The existence of scaffolding interactions implies that physical interactions in a pathway can be divided into two classes: structural interactions that hold pathway components together, and interactions participating in the information flow. Some structural interactions may not participate directly in the information flow and can be omitted from the pathway diagram for simplification before proceeding with pathway kinetic simulation.

(3) Fast decay of pathway cross talk mediated by binding interactions [22]. The information does not propagate well through physical interactions alone. Additional mechanisms such as protein modification and translocation must be involved for efficient signal propagation.

Both rules (2) and (3) can provide a way to identify the borders of a physical interaction cluster that form a pathway and also can help to remove some erroneous proteins from the input list. The errors in the input protein list may arise due to the experimental noise or errors in the BAN database. It is safe to assume that these errors will always exist in molecular biology. Therefore, they have to be dealt with, rather than ignored. Refining the input protein list is essentially an attempt to make the protein clusters more well defined by increasing the clustering coefficient between proteins before proceeding with pathway building. This process can be described as slightly shifting a protein cluster around the original input list in the physical interaction network in order to increase cluster density relative to the surrounding network. The algorithm optimizing the network clustering is yet to be created.

(4) Positive feedback loops allow self-activation of a pathway [19,23]. An abundance of positive feedback loops among input proteins can serve as additional evidence for pathway existence and quality of available interaction data. In contrast, an abundance of negative feedback loops may be used as evidence for invalidating the input gene list. Since every regulatory loop must be explained through physical interaction, the rule can also help in validating and classifying physical interactions in the pathway. The physical interaction mediating the feedback loop should be classified as one participating in information flow rather than in pathway scaffolding. If the physical interaction mediates a negative feedback loop, it can be considered to be erroneous. Negative feedback loops also can be used to determine the end nodes in a pathway, because they indicate that information has reached its final destination and the pathway can be shut down.

(5) Feed-forward loops provide noise tolerance for a pathway [24,25]. This type of loop can be used in the same way as feedback loops to evaluate pathway quality and validate the input gene list. The existence of positive feed-forward loops should indicate a noise-tolerant pathway, while many negative feed-forward loops should invalidate the input protein list.

(6) Cross-pathway inhibition [19,24,26] is a type of regulatory relation that allows an activated pathway to compete and win out over other pathways in the cell and to refocus cellular resources to respond to stimuli that have activated the pathway. This phenomenon can provide additional criteria to determine a pathway boundary. Taking into account rules (2) and (6), the ideal pathway boundary can be summarized as having fewer physical interaction links and an increased number of negative regulatory links aimed at inhibiting other pathways.

(7) Consistency of the regulation effects through the regulatory path. If regulatory relations in the BAN are annotated with an effect sign, this annotation can be used to calculate the net regulatory effect at the end of every regulatory path in a pathway. The absence of consistency between the net effect along the path and the effect of the regulatory relation shortcutting the origin of a path to its end can be used as evidence to invalidate the protein list input for pathway building or for filtering out inconsistent BAN relationships from a pathway.

REFERENCES

1. Yuryev A, Mulyukov Z, Kotelnikova E, Maslov S, Egorov S, Nikitin A, Daraselia N, Mazo I. Automatic pathway building in biological association networks. BMC Bioinformatics 2006;7:171.
2. Hoffmann A, Levchenko A, Scott ML, Baltimore D. The IkappaB-NF-kappaB signaling module: temporal control and selective gene activation. Science 2002;298(5596):1241–1245.
3. Oda K, Kitano H. A comprehensive map of the toll-like receptor signaling network. Mol Syst Biol 2006;2:2006.0015.
4. Oda K, Matsuoka Y, Funahashi A, Kitano H. A comprehensive pathway map of epidermal growth factor receptor signaling. Mol Syst Biol 2005;1:2005.0010.
5. Crescenzi P, Kann V, Halldórsson M, Karpinski M, Woeginger G. Minimum Feedback Arc Set. A compendium of NP optimization problems. http://www.nada.kth.se/~viggo/wwwcompendium/node20.html. Last modified March 20, 2000.
6. Kann V. On the approximability of NP-complete optimization problems. PhD thesis. Department of Numerical Analysis and Computing Science, Royal Institute of Technology, Stockholm. 1992.
7. Dijkstra EW. A note on two problems in connection with graphs. Numerische Mathematik 1959;1:S269–271.
8. Castel D, Pitaval A, Debily MA, Gidrol X. Cell microarrays in drug discovery. Drug Discov Today 2006;11(13/14):616–622.
9. Rausch O. Use of high-content analysis for compound screening and target selection. IDrugs 2005;8(7):573–577.
10. Minor LK. Assays to measure the activation of membrane tyrosine kinase receptors: focus on cellular methods. Curr Opin Drug Discov Dev 2003;6(5):760–765.

11. Herschman HR. Noninvasive imaging of reporter gene expression in living subjects. Adv Cancer Res 2004;92:29–80.
12. Loo DT, Rillema JR. Measurement of cell death. Methods Cell Biol 1998;57:251–264.
13. Caldwell JS. Cancer cell-based genomic and small molecule screens. Adv Cancer Res 2007;96:145–173.
14. Ashburner M, Ball CA, Blake JA, Botstein D, Butler H, et al. Gene Ontology: tool for the unification of biology. Nat Genet 2000;25(1):25–29.
15. Daraselia N, Yuryev A, Egorov S, Mazo I, Ispolatov I. Automatic extraction of gene ontology annotation and its correlation with clusters in protein networks. BMC Bioinformatics 2007;8:243.
16. Abdollahi A, Schwager C, Kleeff J, Esposito I, Domhan S, Peschke P, Hauser K, Hahnfeldt P, Hlatky L, Debus J, Peters JM, Friess H, Folkman J, Huber PE. Transcriptional network governing the angiogenic switch in human pancreatic cancer. Proc Natl Acad Sci U S A 2007;104(31):12890–12895.
17. Lim J, Hao T, Shaw C, Patel AJ, Szabó G, Rual JF, Fisk CJ, Li N, Smolyar A, Hill DE, Barabási AL, Vidal M, Zoghbi HY. A protein–protein interaction network for human inherited ataxias and disorders of Purkinje cell degeneration. Cell 2006;125:801–814.
18. MAQC consortium. The MicroArray Quality Control (MAQC) project shows inter- and intraplatform reproducibility of gene expression measurements. Nat Biotechnol 2006;24(9):1151–1161.
19. Morrison DK, Davis RJ. Regulation of map kinase signaling modules by scaffold proteins in mammals. Ann Rev Cell Dev Biol 2003;19(1):91–118.
20. Levchenko A, Bruck J, Sternberg PW. Scaffold proteins may biphasically affect the levels of mitogen-activated protein kinase signaling and reduce its threshold properties. Proc Natl Acad Sci U S A 2000;97(11):5818–5823.
21. Klipp E, Liebermeister W. Mathematical modeling of intracellular signaling pathways. BMC Neurosci 2006;7(Suppl 1):S10.
22. Maslov S, Ispolatov I. Propagation of large concentration changes in reversible protein-binding networks. Proc Natl Acad Sci U S A 2007;104(34):13655–13660.
23. Yeung K, Janosch P, McFerran B, Rose DW, Mischak H, Sedivy JM, Kolch W. Mechanism of suppression of the Raf/MEK/extracellular signal-regulated kinase pathway by the raf kinase inhibitor protein. Mol Cell Biol 2000;20:3079–3085.
24. Acar M, Becskei A, van Oudenaarden A. Enhancement of cellular memory by reducing stochastic transitions. Nature 2005;435(7039):228–232.
25. Brandman O, Ferrell JE, Jr., Li R, Meyer T. Interlinked fast and slow positive feedback loops drive reliable cell decisions. Science 2005;310(5747):496–498.
26. Dang VD, Bohn C, Bolotin-Fukuhara M, Daignan-Fornier B. The CCAAT box-binding factor stimulates ammonium assimilation in Saccharomyces cerevisiae, defining a new cross-pathway regulation between nitrogen and carbon metabolisms. J Bacteriol 1996;178(7):1842–1849.

4

ALGORITHMIC BASIS FOR PATHWAY VISUALIZATION

SERGEY SIMAKOV, IAROSLAV ISPOLATOV, SERGEI MASLOV, AND ALEXANDER NIKITIN

Table of Contents

Pathway Analysis for Drug Discovery, Edited by Anton Yuryev

4.1 INTRODUCTION

System biology deals with a large amount of data, often with a very complex structure. Visualization of these data is very important in many scientific applications to make it available to be reviewed conveniently by experts. In most cases, the problem can be reduced to the placement of graph nodes at a plane surface using specific aesthetic criteria and an *a priori* assumed graph structure. Here, we summarize the most popular approaches for two-dimensional graph layout, provide some details for their implementation, propose several useful modifications to standard algorithms, and give examples of their application. An extended review of these approaches to graph drawing may be found in the literature [1–3]. We first describe force-based (direct force) and energy-based algorithms and their application to reveal hubs and clusters in the network. We also propose several modifications to the standard methods that both increase layout speed and improve quality of network clustering. The main idea behind these methods lies in ascribing certain physical properties to the nodes and edges of the graph and looking for the equilibrium state of the system. It uses both repulsive and attractive forces between nodes that may have no direct physical analogy. Energy-based algorithms are similar to direct-force algorithms, as the equilibrium state of the system corresponds to a minimum of energy. These methods provide a good basis for arbitrary graph placement but require relatively large computational resources. Some modifications of these algorithms utilize the LinLog [4] force model and the energy-based GEM algorithm [5]. The first method helps to reveal clustering properties of the graph without any additional assumptions. The second method exploits randomization and a specific technique for detecting and decreasing oscillations and rotations of the graph during iterations. The second algorithm is very fast, while the first provides a qualitative appearance. A new modification combining these approaches is proposed in the conclusion of the first part of this chapter.

We also discuss the space-limited layout, which is a novel extension of the direct-force method. This technique is used to lay out nodes in a biological pathway using cellular localization. The nodes of the graph are assigned to the specific regions of the plane surface where they must be arranged. The application of this method is obvious when the placement of proteins relative to the cellular elements is considered in order to see how proteins in a pathway are distributed in relationship to the major direction of information flow from the plasma membrane to the nuclei, as well as among other cell organelles. Two modifications are proposed based on the visual appearance of the membrane: belt-shaped and ring-shaped. The geometry of the other cellular elements depends on the type of the membrane configuration and is discussed in this part. Several graph drawings are presented at the end of this section. They show pleasant layouts that are straightforward and easy to analyze.

The fourth section describes the so-called symmetrical algorithm. The idea is that the root node is selected and incident nodes are placed relative to it.

These nodes are used as roots for the placement of subsequent incidentals, and so on. The result is a tree-like drawing that is useful both for trees and for almost acyclic directed and undirected graphs. Because the original algorithm implementation utilizes recursion, some modification is required to adapt it for large graphs. Hereafter, some implementations of well-studied orthogonal drawings are considered. The general idea of these methods is to place nodes of the graph at the points of 2D orthogonal mesh, provided that square bends of the edges are allowed. These tasks are often reduced to the well-known cost-flow minimization problem with subsequent refinement and compaction. A great number of methods have been proposed for these tasks. A brief overview is provided, and one of the possible combinations of constituent algorithms is proposed in this section. In addition, we describe hierarchical drawing. This approach assumes the presence of hierarchical structure in the data that composes a graph. Common stages of this algorithm are described and include the following: assigning nodes to layers, decreasing edge crossing by node permutation within layers, and final assignment of coordinates.

The section that follows describes an effective method of acyclic core graph construction. It is based on a probabilistic approach, exploiting simulated annealing. This method allows the discovery of a not necessarily exact but sufficiently near-optimal solution. In particular, it is very important as part of the hierarchical drawing of large graphs, as it increases overall performance and imposes fewer memory restrictions.

At the end of this chapter, we discuss the collapsing network method as an approach to reduce the complexity of a large biological graph and present it as a smaller graph that connects more global biological concepts, including components of the original biological network. We describe several strategies that exploit the hierarchy of biological knowledge to collapse a network. The main idea is to collapse some parts of the graph into subgraphs using heuristic criteria. The most obvious example of such a criterion may be subgraphs representing complete graphs. A more complex approach may utilize the Markov Cluster Algorithm (MCL) or others to collapse strongly connected components of the graph into subgraphs. A completely different approach may exploit specific biological information contained in the graph. For example, hierarchical protein classification stored in the database may be used for collapsing, or a set of smaller sub-pathways already stored in a database may be part of the larger pathway and thus may be collapsed. After collapsing, a force-directed method may be applied to lay out the whole graph and subgraph independently. These approaches highly simplify the drawings of the large graphs, while focusing on their biological structure. At the same time, it provides much better performance, as weakly connected components are processed regardless of the presence of one another.

Let $G = (V,E)$ be a graph (not necessarily connected) where V = a set of nodes (vertices) and E = a set of selected node pairs, i.e. edges (links), and N = number of vertices and M = number of nodes. A plane drawing for this graph is required. Thus, an assignment of a vector $\mathbf{r}_u \in R^2$ to every node $u \in E$ of the graph is needed. Let the nodes of the graph represent proteins, molecules, diseases, treatments, functional classes, cell processes, etc., and the edges or links represent chemical reactions, regulations, bindings, etc. Using this analogy, we can think of the graph as a biological pathway and vice versa. Most links of the pathway may have controls—specific nodes assisting in the identification of the biological meaning of the link and its geometric shape. Controls having two incidental nodes require specific processing. The other controls are processed as common nodes.

4.2 FORCE-BASED AND ENERGY MINIMIZATION ALGORITHMS

4.2.1 Analysis

Numerous algorithms of graph layout exploit the N-body simulation method. Nodes on the graph are treated as a set of interacting particles. The interaction law is described by forces acting among the particles. Usually, these forces are divided into the forces of repulsion and attraction. They depend on the distance between particles. Thus, these force-based methods are generally known as force-directed (direct force) methods. Attractive forces are sometimes associated with spring interaction along the graph edges, while repulsive forces are associated with electrostatic interaction between every pair of nodes. The task is to find the stationary node position corresponding to the equilibrium state of the system. For every fixed node $\bar{u}$ ($\bar{u} \in V$), we have:

$$\sum_{v \in V} \mathsf{F}_{rep}(\mathsf{r}_{\bar{u}} - \mathsf{r}_v) + \sum_{(\bar{u},v) \in E} \mathsf{F}_{attr}(\mathsf{r}_{\bar{u}} - \mathsf{r}_v) = 0 \tag{4.1}$$

It should be mentioned that so-called energy minimization algorithms are similar to force-based ones. But sometimes these approaches are referred to as different types of algorithms. Both approaches are focused on the equilibrium state of the set of nodes. The result of force-based algorithms is characterized by zero net force, while the result of energy-based algorithms is characterized by the state corresponding to the minimum energy.

$$U_{min} = \min\left[\sum_{\{u,v\} \in V} U_{rep}(\mathsf{r}_u - \mathsf{r}_v) - \sum_{(u,v) \in E} U_{attr}(\mathsf{r}_u - \mathsf{r}_v)\right] \tag{4.2}$$

where U_{rep} and U_{attr} = absolute values of the repulsion and attraction potentials.

In both algorithms, a stationary equilibrium state is the goal. The force-based algorithm can be easily transformed to the energy-based one. From a mathematical point of view, the force-based algorithm implementation consists of a solution of a nonlinear set of algebraic equations. The energy-based algorithm implementation performs functional minimization. Both of these tasks can be derived from one another. The following statement is derived from equation (4.2).

$$\sum_{v \in V} \nabla U_{rep}(\mathbf{r}_{\bar{u}} - \mathbf{r}_v) - \sum_{(\bar{u},v) \in E} \nabla U_{attr}(\mathbf{r}_{\bar{u}} - \mathbf{r}_v) = \mathbf{0} \qquad (4.3)$$

It is similar to the problem in equation (4.1). Because the two algorithms share these similarities, we will mainly discuss force-directed algorithms. It also allows us to combine the descriptions of LinLog and GEM, since they were stated in different terms.

Force-directed methods have several strong advantages and are valid for an arbitrary undirected graph that has straight-line edges. It produces visually attractive graph placements. Algorithm implementation is quite straightforward because its theory is an obvious concept. An algorithm can be used in the same manner in both 2D and 3D Euclidean space. However, numerous studies reveal drawbacks that are almost as strong as the advantages. They can be divided into two groups. The first group follows from physics. It is commonly known that the equilibrium state is not necessary to be stationary. Spring frictionless systems can produce oscillations. The system spinning around a common center represented as a single graph is also in an equilibrium state. The second drawback is related to the potential wells that have occurred as a result of the formulation of specific force laws. A set of nodes may have none, one, or multiple equilibrium states depending on the force law. Other drawbacks are related to the numerical implementation of a particular method. They can be referred as numerical drawbacks. For large graphs (several thousands nodes and tens of thousands of edges), the force-based algorithms require many computational resources.

Another aspect of these algorithms is that the interruption condition should be explicitly stated because the iteration procedure is used in many implementations. In our experience, little or no node movement during several iterations cannot be used as a criterion for the system to be near the equilibrium state. Therefore, it cannot be used as an interruption condition. Convergence is also important as a faster converging iterative procedure that decreases the number of iterations. Potential wells may occur during the iterative process, especially in the case of large graphs. These wells may arise even if force laws do not allow local minimum states. For example, the initial placement of the nodes can create the wells: a node placed inside a graph loop may never leave it if attractive forces toward it are incidentals and are relatively small. In order to overcome the aforementioned difficulties, we developed different approaches. Our algorithm exploits some useful ideas and provides several extensions specific to biological pathways.

We first consider how physical drawbacks can be resolved. To achieve a stationary equilibrium state, friction forces can be added to the system. A variation of node mobility can also be very useful. It allows the reduction of the length of a node track for every iteration. The track length becomes shorter as the temperature decreases, and a correctly chosen cooling schedule allows the improvement of the results. The presence of potential wells and multiple equilibrium states is hard to predict, even in the case of 10 nodes. The possibility always exists for a convergence to the local equilibrium state instead of global equilibrium. Certain algorithms can deal with this problem, such as Sim [6], Frick et al. [5], and Davidson and Harel [7]. They exploit the idea of simulated annealing by not allowing monotonic decrease in system energy between two successive iterations. Other methods must accept that local minimum equilibrium is satisfactory and may do nothing to overcome it.

When formulating force laws, it is important to remember that at least one equilibrium state must exist. Generally, this is a minor problem, as attractive forces are frequently proportional to the distance between the nodes, and repulsive forces are inversely proportional to this distance. Thus, the sum of decreasing and increasing functions definitely has at least one minimum. The key point of numerical implementation is speed of calculation. Interactive speed is highly desirable because a large graph can be processed within seconds. This speed can be achieved either by decreasing algorithm complexity or by increasing convergence speed. The first approach usually results in the increased complexity of algorithm implementation and in the increased sophistication of the fitting parameters (such as force intensity, power laws, etc.). The second approach seems to be more useful. It is worth mentioning that the force laws for repulsion and attraction may have no direct physical analogy. This fact can be used for faster converging algorithm construction. Several models were proposed [8,9] that can be formulated in the same terms. (See Noack [4] for a more general approach.)

4.2.2 Algorithm Implementation

We used the following relations for the repulsion and attraction forces exerted from node v to node u directed along the vector $\mathbf{r}_v - \mathbf{r}_u$.

$$F_{rep} = \begin{cases} 20R/(L-0.85L_{av})^2, & L \geq L_{av} \\ 20R/(0.15L_{av})^2, & L < L_{av} \end{cases} \tag{4.4}$$

$$F_{attr} = \begin{cases} -10(L/L_{av}-1)^{3/2}, & L \geq L_{av} \\ 10(1-L/L_{av}), & L < L_{av} \end{cases} \tag{4.5}$$

where $R = d_u d_v$, $L_{av} = 2R/(d_u + d_v)$, $L = \|\mathbf{r}_u - \mathbf{r}_v\|$, d_u, d_v = average diameters of the vertices u and v, respectively. A force formulation as in the previous example has the following characteristics.

First, node dimensions are considered to prevent their adhesion. Node diameter is introduced as the average of the height and width of the bounding box of the node. The distance L_{av} determines the scale of the stable state. As follows from equation (4.4), L_{av} is used to avoid infinite values of repulsion force for the short-distance nodes. In addition, due to the second part of equation (4.5), the short-distance states become instable when attraction turns to repulsion. Thus, even if at a given step, we have $\mathbf{r}_u = \mathbf{r}_v$, this configuration is hardly observed in the final layout. In order to increase stability and decrease oscillations and rotations, a global temperature is introduced. The temperature value determines a maximum distance for moving a node in the iteration step. This maximum $L_{\max}$ is constantly decreased during the iterations, and the value of the resulting node displacement $\mathbf{r}_u$ of the node u is determined by the following:

$$\|\mathbf{r}_u\| = \frac{\|d\mathbf{r}_u\|^{0.2} L_{\max}}{\max\limits_{v \in E}(\|d\mathbf{r}_v\|)} \tag{4.6}$$

where $d\mathbf{r}_u$ is the node u displacement induced by the forces.

For biological pathways, convergence can be improved by dividing the iteration procedure into several stages. Pathways often include chains—a set of double-connected nodes ending with a single incident node and having a node connected with an n-connected node ($n > 2$). These chains may cause disturbances that slow down the layout of the rest of a graph near equilibrium. The collapsing of chains before the first stage highly improves both the layout speed and the graph aesthetics. Since the skeleton layout of the graph is near equilibrium, the chains are rolled out along a straight line and the second stage begins. The other feature of biological pathways is that one pair of nodes can be connected by multiple links (multilink) representing different types of biological interactions. In the final layout, it is of no importance how many links connect one node pair. However, processing all links from the multilink slows down the algorithm. It also results in a tighter but unnecessary positioning of the pair. Therefore, a multilink can be safely replaced by one single link during the layout with the appropriate procedure of backward substitution at the end of the algorithm run. One possible solution is to split controls of the multilink into a fan. Special attention should be paid to graph connectivity. The problem of finding unconnected components is addressed previously [10,11]. After finding connected components, each subgraph can be processed and scaled separately. Scaling is needed at this point, because algorithms provide just the relative node positions and because the actual layout may be limited both below and above. For example, every node must be shown separately, or the whole graph must fit into the screen. The last step is the question of the placement of unconnected components after the main algorithm [12]. Polyomino covering is generated for every unconnected component. A polyomino is a shape made up of several rectangles joined by complete edges; it covers all nodes and edges of the component. Next, a drawing plane is tightly filled out with these polyominoes, and unconnected components are placed at the appropriate polyomino's locations.

Let us summarize the steps included in a direct-force algorithm:

1. Hide controls and collapse multilinks.
2. Identify connected components.
3. Roll up chains.
4. Perform stage 1 (a directed-force algorithm for the graph skeleton).
5. Roll out chains.
6. Perform stage 2 (a directed-force algorithm for the entire graph).
7. Restore controls and split same end links.
8. Scale connected components separately.
9. Calculate a polyomino for every connected component and assign its position.

Stages 1, 2, 8, and 9 are common for most of the layouts considered in this chapter and may be used for a wide range of other implementations. Stages 1 and 2 are used for preprocessing and stages 8 and 9 for postprocessing refinement. The result of the algorithm implemented in Pathway Studio software is depicted in Figure 4.1.

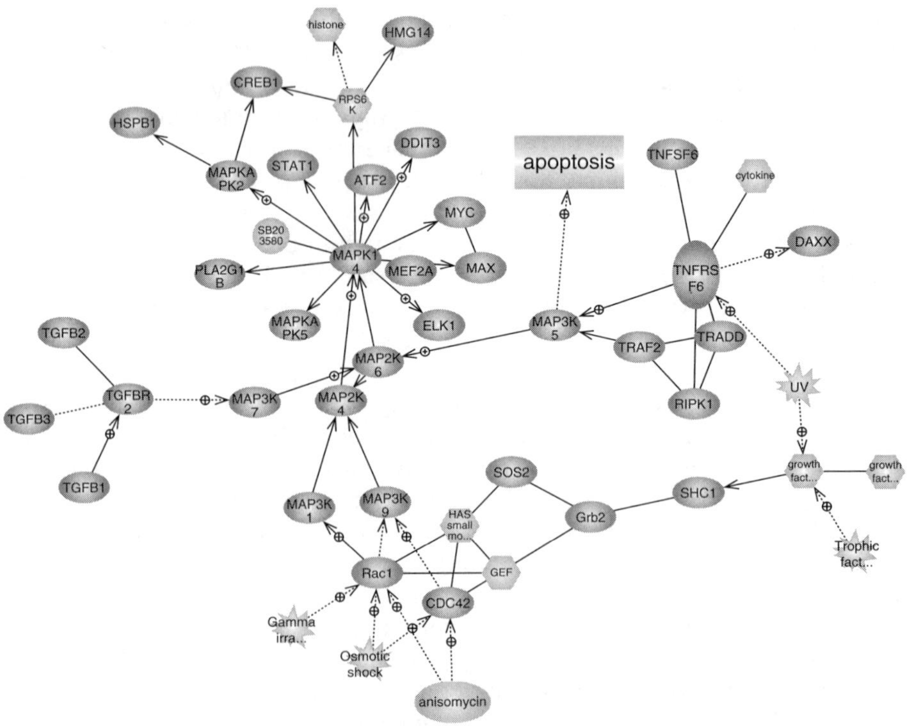

Figure 4.1 Direct-force layout.

4.2.3 Fast Clustering Method

Force-based and energy-based algorithms are very flexible. A number of modifications that satisfy different aesthetic criteria are possible. One such criterion is clustering, i.e. depicting strongly connected components of the connected graph. Several approaches reveal clustering such as the Markov Cluster Algorithm (MCL), which is thoroughly studied by Enright et al. [13]. It allows the extraction of strongly connected components, but an additional procedure for assignment of the coordinate is required. For example, a force-based model may be constructed so that the attraction of the nodes from the same strongly connected component should be greater than that of the nodes from different connected components. Additional repulsion should also be added to the nodes from different components. The performance of this procedure is the same as the performance of the force-based method itself, but the computational cost for MCL must be added. The resulting performance is quite complicated. Therefore, we introduce a fast algorithm that provides the best possible clustering property, the fast clustering method (FCM). Its main idea is to combine algorithms, giving the best clustering but a slow performance (LinLog) [4], and an algorithm providing an enhanced performance but with no attention to clustering (GEM) [5]. FCM is, essentially, a hybrid between GEM and LinLog that incorporates their best properties: network clustering and low computational cost. It is important to note that no special pre- or postprocessing is required to obtain clustering with FCM. It is based simply on selecting the right force for the algorithm.

The LinLog algorithm is a specific energy-based algorithm with the energy function formulated as equation (4.7). It was proven that the shape of this function provides the best clustering in a wide range of energy models [4]. Remarkably, no additional assumptions are needed, such as desired edge length, for example. The key feature of the GEM algorithm is the use of a randomization technique that allows moving the graph out from potential wells that correspond to the full energy local minimum. The local temperature is also used to decrease oscillations and rotations of the graph nodes during iterations. The temperature is defined as node mobility for every iteration. It is decreased throughout calculations as the system approaches the equilibrium. As reported [5], this algorithm is very fast but provides no clustering as many direct-force algorithms do. The following energy model was proposed by Noack [4].

$$U_{LinLog} = \sum_{\{u,v\}\in E} \|\mathbf{r}_u - \mathbf{r}_v\| - \sum_{\{u,v\}\in V} \ln\|\mathbf{r}_u - \mathbf{r}_v\| \tag{4.7}$$

That formula can be rewritten in terms of the following force-based model:

$$\nabla U_{LinLog} = \sum_{v\in E} \frac{\mathbf{r}_{\bar{u}} - \mathbf{r}_v}{\|\mathbf{r}_{\bar{u}} - \mathbf{r}_v\|} - \sum_{\{\bar{u},v\}\in V^{(2)}} \frac{\mathbf{r}_{\bar{u}} - \mathbf{r}_v}{\|\mathbf{r}_{\bar{u}} - \mathbf{r}_v\|^2} = \mathbf{0} \tag{4.8}$$

After this transformation, the model can be used in the GEM algorithm. The first summation in equation (4.8) is responsible for repulsion, and the second is responsible for attraction. Special attention should be paid to this formula, as an infinite force value is possible for the short-distance nodes. We add a small random vector a($\|a\| < 10^{-2}$) to the node position whenever $\|\mathbf{r}_u - \mathbf{r}_v\| < 5 \cdot 10^{-3}$ exists to avoid division by zero. This addition is another way of removing the singularity similar to the second case in the equation (4.4). The results obtained with GEM and FCM are presented in Figures 4.3 and 4.4. The input for both algorithms was a random initial placement of the nodes on a graph, as shown in Figure 4.2. The input graph consists of seven strongly connected top-level components. Every component consists of seven strongly connected subcomponents formed by seven nodes that are linked by 21 edges. Subcomponents of every component are linked by 100 edges. Two components are linked if a path exists from the node of one subcomponent to another and only the nodes from these two subcomponents are visited along the path. In addition, components are linked by 100 edges. Thus, we have 343 nodes that are linked by 1,829 edges.

As follows from Figure 4.3, the GEM algorithm is also capable of revealing clustering that is due to the weak connectivity of the top-level components. The FCM algorithm reveals more distinct clustering that can be adjusted to the desired scale (see Figure 4.4).

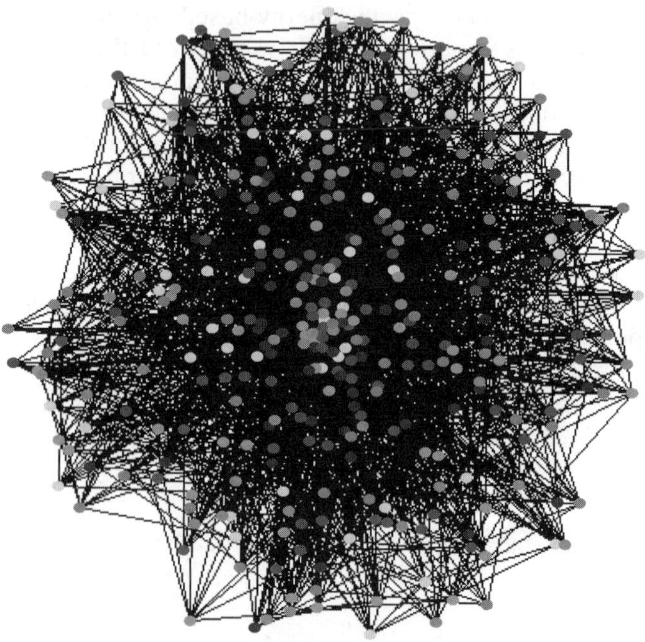

Figure 4.2 Initial random placement.

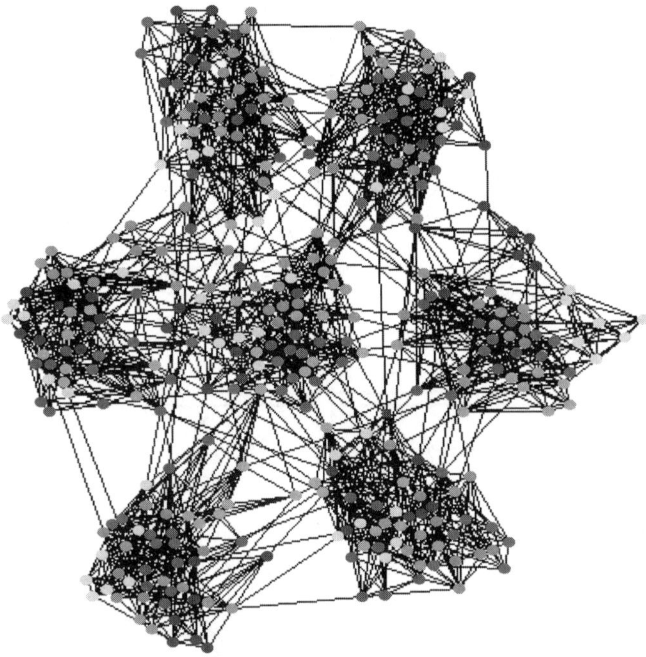

Figure 4.3 GEM layout.

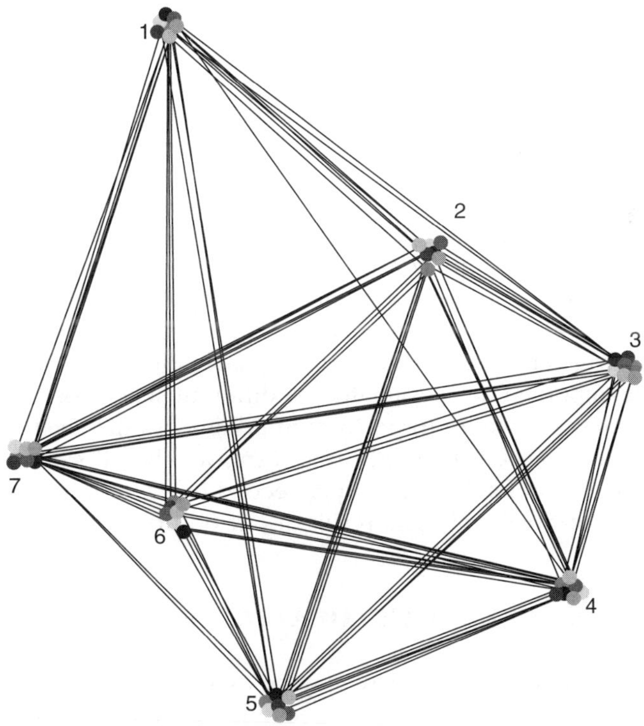

Figure 4.4 FCM layout.

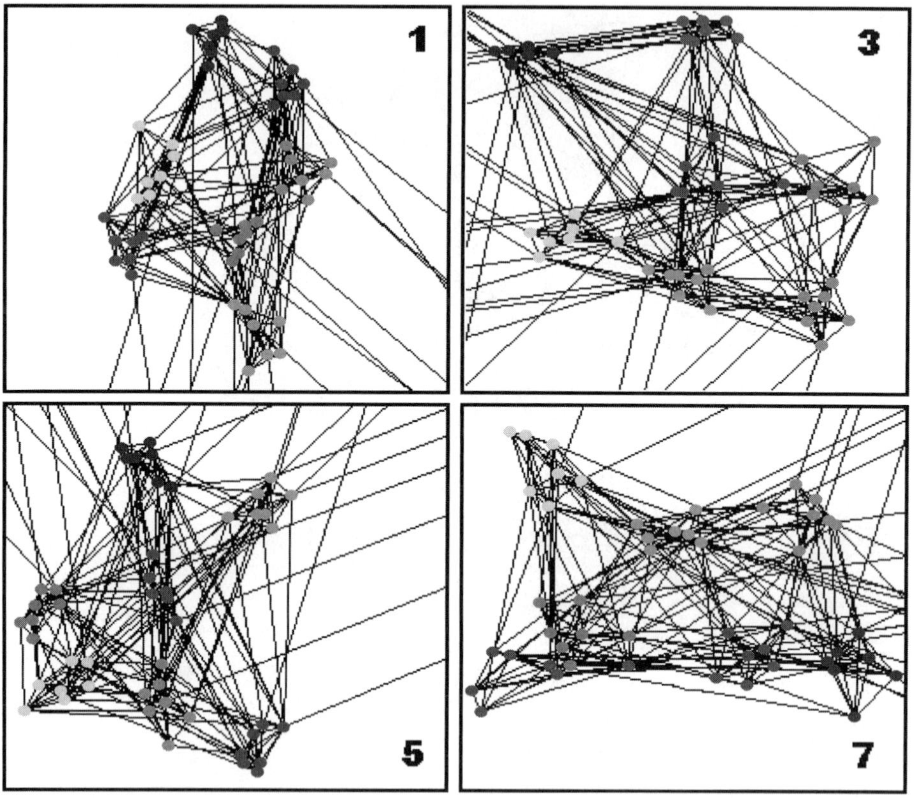

Figure 4.5 Multidimensional clustering with FCM.

The proposed FCM algorithm has one interesting property that we observed during algorithm implementation: it allows the depiction of multidimensional clustering. If we have a graph consisting of several subgraphs that in turn have rarely linked strongly connected components, applying FCM to this graph reveals clustering at every subgraph level. The initial node placement of this graph may be arbitrary. This feature is illustrated in Figure 4.5, where the nodes of every subcomponent are designated by different shades of gray.

It appears that the ability to perform multidimensional clustering remains for any number of nested subgraphs that form weakly self-similar clusters. A more thorough theoretical discussion is needed to address this question. We describe this property simply as a result of our observations.

4.3 LAYOUT BY CELLULAR LOCALIZATION

Every node of a biological pathway may be ascribed to a specific part of the cell (nucleus, cytoplasm, membrane, mitochondria, etc.) or extracellular space.

The depiction of the subcellular localization of nodes in a pathway, while maintaining relations between the nodes of the pathway, has important biological meaning. This layout type can be classified as space-limited. The general idea for this layout is similar to the direct-force layout or energy minimization technique. From a physical point of view, it is supposed that the space is uniform and the nodes are the only sources of attraction and repulsion.

Let us assume that there are several zones in space interacting with the nodes of the pathway. Each node is ascribed to or belonging to a specific zone and is owned by this zone. The node is attracted to its owner zone and is pushed away from the foreign zones. The forces exerted from the zones should be appropriately balanced in such a way that repulsion from the foreign zone and interaction between other nodes should not push the node away from its owner. Pathway zones may be classified according to their geometric primitives: elliptic

$$\frac{(x-x_0)^2}{a^2}+\frac{(y-y_0)^2}{b^2}\le 1 \tag{4.9}$$

where a,b = major and minor semiaxes; rectangular (belt-shaped)

$$|x-x_0|\le W/2, \quad |y-y_0|\le H/2, \tag{4.10}$$

where W,H = zone width and height; circular (ring-shaped)

$$\alpha^2\frac{(x-x_0)^2}{a^2}+\frac{(y-y_0)^2}{b^2}\le 1, \quad 0<\alpha<1, \tag{4.11}$$

where a,b = major and minor semiaxes for the outer boundary, α = ratio of the inner and outer boundary dimensions, and (x_0,y_0) = zone center coordinates.

The geometric type of the membrane determines the types for other zones. Two types of membrane can be considered: ring-shaped and belt-shaped. In the first case, the cytoplasm and extracellular space must have a circular shape concentric to the membrane. In the second case, the cytoplasm and extracellular space must have a rectangular shape. Other zones have an elliptic shape in all cases and are placed inside or below the membrane according to the first and second cases mentioned previously. The node–node interaction is described by equations (4.4) and (4.5). We also need a force-based model for node–zone interaction. A new reference system is introduced before calculating node–zone forces that is a parallel translation and scaling along the axes. For the elliptic and the ring shapes:

$$\tilde{x}=(x-x_0)/a, \quad \tilde{y}=(y-y_0)/b \tag{4.12}$$

and for the rectangular shape:

$$\tilde{x} = (x - x_0)/W, \quad \tilde{y} = (y - y_0)/H. \tag{4.13}$$

Thus, the ellipse and elliptic ring are transformed to the circle of radius 1 and a pair of concentric circles of radii 1 – α and 1. The rectangle is transformed to the square of side 1. $L = \sqrt{\tilde{x}^2 + \tilde{y}^2}$ is a dimensionless distance between the node and zone center. Using the described reference system, the following force-based models may be used. For the elliptic zone:

$$F_{elliptic} = 80L^6 \tag{4.14}$$

For the ring-shaped zone:

$$F_{ring} = sign(L - 1 + \alpha)\left(\frac{L - 1 + \alpha/2}{0.35\alpha}\right)^8 \tag{4.15}$$

For the belt-shaped zone:

$$F_{belt} = \begin{cases} 6L^8, L > 1 \\ 6L, L \leq 1 \end{cases} \tag{4.16}$$

In all cases, the positive force direction is toward the center. After calculation of the forces, an inverse transformation to the initial reference system is required. Equations (4.4), (4.5), and (4.14) to (4.16) give us the full-force model for layout by cellular localization. The forces given by equations (4.14) to (4.16) must provide sufficient intensity to hold out their own nodes.

If random initial placement of the nodes is used, zone attraction forces must predominate over other forces, especially for the nodes outside of the zone. Node–node interaction may be switched off during the first stage of the algorithm. Switching off this interaction allows the nodes to take place inside the appropriate zone. Specific attention should be paid to the nodes owned by the ring-shaped zones. Node repulsion along with attraction to the same zone may be the reasons for the local minimum energy well occurrence. It may prevent the pathway layout from reaching a global minimum state. An additional stage is introduced to the algorithm to avoid this problem. Repulsion of the nodes owned by the same ring-shaped zone is switched off during this stage, which allows the nodes to be correctly positioned with respect to each other. After this stage, repulsion is switched on and the nodes occupy their final places.

We also use an additional force for node–zone interaction, which is the zone reaction to the net force pushing out its own node from the zone. This force is calculated after all other forces as the difference between the net force and the maximum possible force along the same direction, leaving the node inside the owner zone. If a magnitude of the zone reaction force is less than a magnitude of the net force zone, reaction force is not applied. This procedure is analogous to the cutoff of the coordinate whenever a node tends to leave zone boundaries. The problems of resizing zones automatically and their initial placement happen when no previous information was specified for these parameters. A minimum zone area S^Z may be calculated as the total area of the bound boxes of its nodes S_n^Z added with some extra space S_{extra}^Z and the area of the zones is placed inside S_{sub}^Z:

$$S^Z = S_n^Z + S_{extra}^Z + S_{sub}^Z, \quad S_{extra}^Z = \beta S_n^Z, \tag{4.17}$$

where β = the coefficient depending on the nodes' number and graph density. Before calculating the zone area, the areas of all the internal zones must be determined and S_{sub}^Z calculated. Final elliptic- and ring-shaped zone dimensions may be calculated based on the viewport aspect ratio $\gamma = W_v/H_v(W_v =$ width, $H_v =$ height) by the following:

$$S_{elliptic}^Z = \pi ab, \quad S_{ring}^Z = \pi ab(1-\alpha^2), \quad a/b = \gamma. \tag{4.18}$$

The results of the layouts by the cellular localization algorithm implemented in Pathway Studio software with ring-shaped and belt-shaped membranes are presented in Figures 4.6 and 4.7.

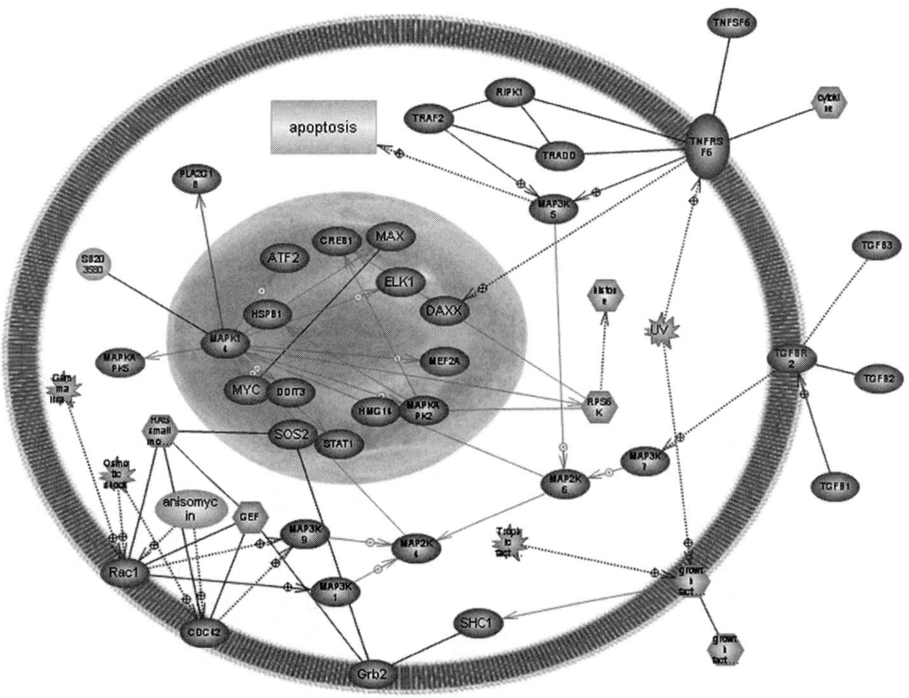

Figure 4.6 Layout by cellular localization with ring-shaped membrane.

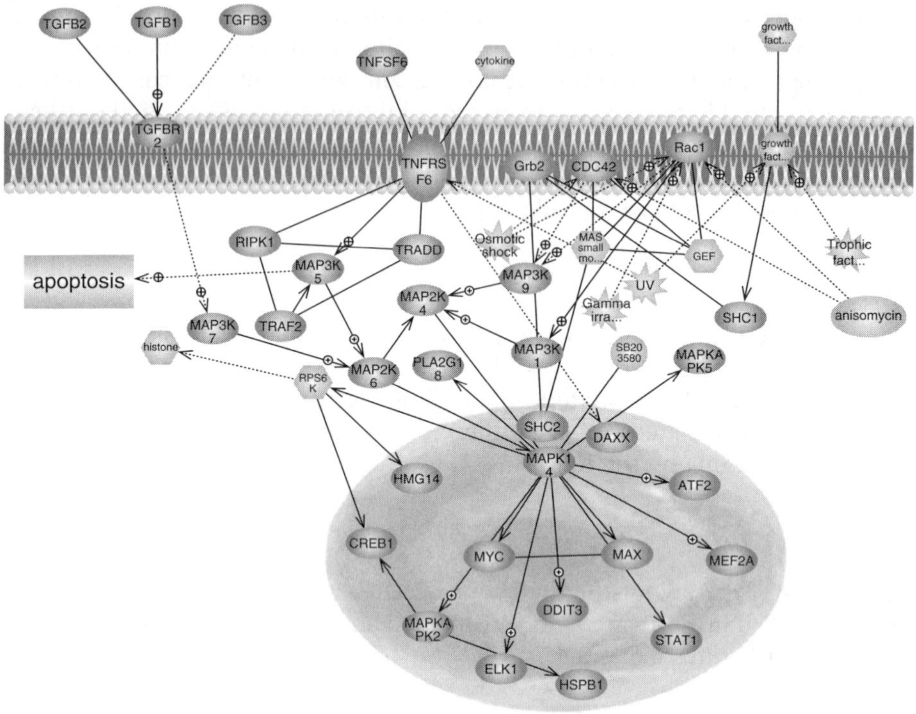

Figure 4.7 Layout by cellular localization with belt-shaped membrane.

4.4 SYMMETRICAL ALGORITHM

The symmetrical algorithm is a quite simple approach mostly intended for tree-like graphs. It is an extension of the level drawing approach [1]. This approach is also useful for acyclic directed graphs and thus may be applied to an arbitrary connected graph if a core acyclic graph must be computed. The details of acyclic directed graph construction will be discussed in subsequent sections. In the area of system biology, a symmetrical algorithm may be applied after pathway expansion, which results in a tree-like graph that can be nicely drawn using this approach.

The main aesthetics in this approach is to reveal the central symmetry of a graph. A central node is selected and the other incident nodes are placed around it along a circle. Then, the same algorithm starts for every placed incident node and its incident nodes are placed inside an angular segment. Several algorithms exploit this idea [1,14] to construct symmetrical drawings that are also known as radial drawings.

The procedure described previously is an expansion tree construction based on a core acyclic directed graph. It can be performed using breadth-first search (BFS).

One important aspect for software implementation is that a recursive procedure must be avoided if huge graphs with thousands of nodes and edges would be processed. It is known that every recursive algorithm may be implemented without recursive calls. In the case of a symmetrical layout, a list of currently placed incident nodes can be stored at every pass. The next pass uses this list for the same procedure. The algorithm continues until the list is not empty. The angle of a spanning segment is computed based on the size of the BFS sub-tree of the node.

A central node can be selected based on the maximum distance from the leaves [14], and the vertex of maximum degree can be chosen if no leaves are present in the graph (i.e. all vertices are at least of degree two).

The problem that cannot be solved in this approach is related to non-tree edges that may induce substantial complications for the overall drawing. A sample layout obtained with the symmetrical algorithm implemented in Pathway Studio software is shown in Figure 4.8. It is easy to observe several

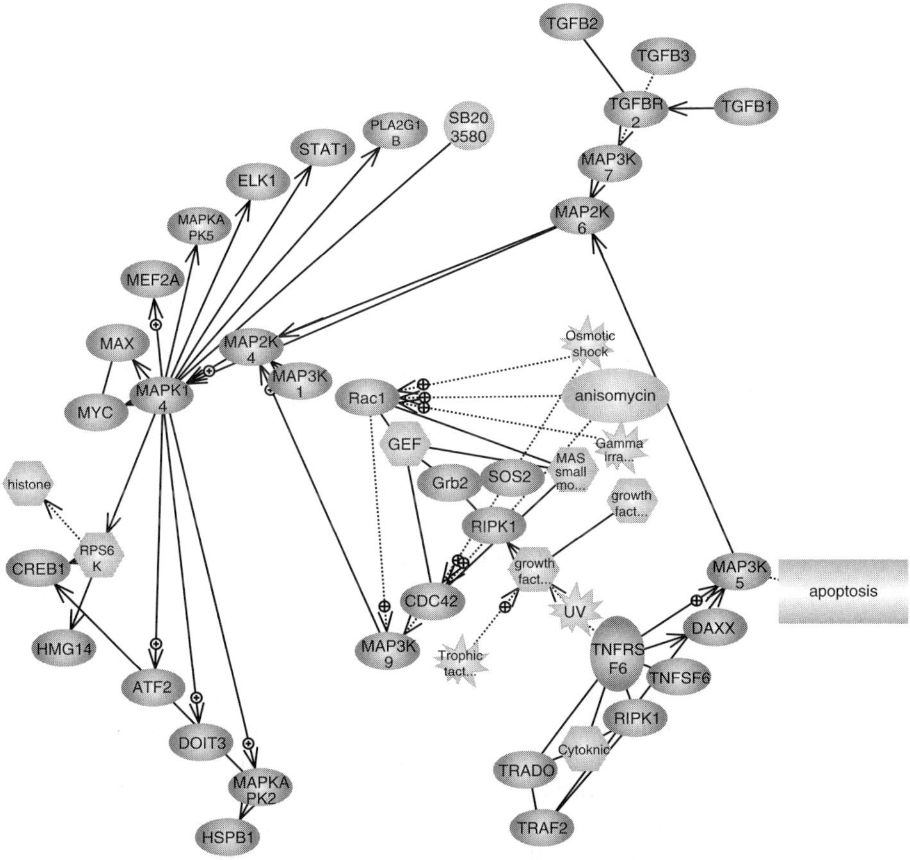

Figure 4.8 Symmetrical layout.

non-tree edges that are slightly decreased aesthetics. The result can be more complicated for more complex graphs.

4.5 ORTHOGONAL LAYOUT

4.5.1 Analysis

An orthogonal drawing is a classic plane representation of a graph in which each edge is drawn as a polyline consisting of orthogonal (horizontal and vertical) segments and each vertex is drawn as a rectangle or inside a rectangle (bound box) of calculated dimensions. These drawings are widely used in VLSI circuits modeling, visual CASE tools, CAD software, etc. Naturally, the use of these drawings also extends to the area of system biology. The most common aesthetic criteria are bend minimization and area minimization. In addition, the coordinates of nodes and bends may be limited to integers.

The common steps in the algorithm are similar for different implementations. They include planarization, orthogonalization, and compaction [15] or the Topology-Shape-Metrics approach [16] that is an equivalent. It was shown that only a planar graph of the degree no greater than four can be drawn according to the requirements imposed previously [17]. Usually, an input graph is not planar and has an arbitrary degree. Planar embedding construction is quite difficult for large graphs. A planar graph $G' = (V',E')$ of a degree at most four must be constructed from the input graph G during the first step. We will refer to G' as a supporting graph. Three tasks arise at this step: initial embedding for the input graph; substituting the vertices of a degree greater than four with the cycle of virtual vertices (clones) of a degree of at most three; and substituting each edge crossing with a virtual vertex.

The second and third tasks are quite straightforward, while the problem of initial embedding is quite complicated. It was reported that, without initial planar embedding, the problem of orthogonal drawing is *NP*-hard [18]. The problem of embedding a graph in a grid of a specified area is also *NP*-hard [19]. Any planar graph of a degree at most four can be drawn at $O(n^2)$ area [17] while minimizing the bend number by an $O(n^2 \log n)$ time algorithm [20]. $O(n)$ bend drawings are possible with an $O(n)$ time algorithm [21].

4.5.2 Topology

For initial graph embedding, we use a simplified direct-force algorithm with higher intensity repulsion and weaker attraction to smooth out the graph. All subsequent operations are applied to this embedding. Before this stage, the endpoint vertices (i.e. vertices of degree one) of the chains are linked by virtual edges to another node of the same chain or to a specific node of the supporting graph, preferably of degree two or three. Strictly speaking, an arbitrary graph G does not allow an orthogonal layout. It is only possible for the supporting graph G'. The dilemma lies in changing the rectangle dimensions (or bound boxes) of the vertices or in using a quasi-orthogonal drawing. A quasi-orthogonal drawing is almost an orthogonal drawing that allows some terminal segments at the edge of a graph to form an arbitrary angle with its incidental segments. For a graph of maximum degree four, a quasi-orthogonal drawing is equivalent to an orthogonal drawing.

4.5.3 Orthogonalization

The process of orthogonalization includes the task of bend minimization. A reduction of a graph to a flow network is widely used to complete this stage. Minimum cost flow is calculated for the network as the flow pattern corresponding to a specific orthogonal drawing [20]. A different technique is available for solving the minimum cost-flow problem, starting with the algorithms developed by Ford and Fulkerson [22] and Edmonds and Karp [23] until recent works that exploit the network simplex method [24], etc. A brief overview may be found in a technical report by Goldberg and Rao [25].

The other possible approach to the orthogonalization stage uses topological numbering and visibility representation. This stage strongly depends on initial embedding assigned during the first stage. First of all, starting (s) and terminal (t) nodes are selected in G'. These nodes also referred as the *source* (s) and the *sink* (t). The lowest and the topmost nodes can be chosen, for example, as s- and t-nodes. After that, all edges of the undirected graph G' must be directed from s to t. This problem is also known as st-numbering, st-orientation, or bipolar orientation and was first introduced by Lempel et al. [26]. st-numbering is a function $f_{st}(v)\colon\{v \in V'\} \to \{1, \ldots, N\}$ having the following properties:

$$\begin{cases} f_{st}(s) = 1,\ f_{st}(t) = N, \\ f_{st}(p) < f_{st}(u) < f_{st}(q),\ (p,u) \in E', (u,q) \in E', u \in V'/\{s,t\} \end{cases} \tag{4.19}$$

Several algorithms allow calculation of the numbering, inducing the required orientation [27–29]. The edge $(u,v) \in$ E′ is directed from u to v if $_{st}(u) < f_{st}(v)$ and from v to u, otherwise. It should be mentioned that only a biconnected graph allows st-numbering. Thus, an additional update of E' may be required to make E'' biconnected. Then, a dual graph $D' = \left(V'_D, E'_D\right)$ is constructed, based on the supporting graph G'. The nodes of the dual graph correspond to the faces of E'. A face is an area representing a polygon formed by the vertices and edges of the graph. The edges of the dual graph correspond to the edges common to the two faces in E'. In addition, the outer face is divided into two parts, s_D and t_D. The half face s_D is formed by the vertices of the external face perimeter lying on the path from s to t in a clockwise direction relative to any internal face. Biconnected E'' may be constructed here, that is, to avoid the condition when a vertex of the graph belongs to both s_D and t_D. Additional edges may be installed and the dual D' must be properly updated to D''. Furthermore, E'' is referred to as E' and D'' as D'.

For every face ϕ, a source s_ϕ and a sink t_ϕ are induced by the previously calculated st-orientation, and two different directed routes from s_ϕ to t_ϕ must be noted. The path from s_ϕ to t_ϕ is called the left side of the face if it goes along a clockwise direction relative to the face and is otherwise called the right side. Only the left and right faces are assigned to every edge $(u,v) \in E' : \phi_l(u,v), \phi_r(u,v)$ and every node $v \in V' : \phi_l(v), \phi_r(v)$. The face including incoming and outgoing edges of v in a clockwise direction is called the left face for the vertex $\phi_l(v)$ and the right face for the vertex $\phi_r(v)$, in the case of a clockwise direction. If an edge lies on the left side of a face, then the face is called $\phi_r(u,v)$ for this edge (u,v); it is called $\phi_l(u,v)$ otherwise. s_D and t_D are selected as source and sink for D'. Fast st-orientation may be assigned by specifying edge direction from ϕ_1 to ϕ_2, if for any edge $(u,v) \in E' : \phi_1 = \phi_l(u,v)$ and $\phi_2 = \phi_r(u,v)$. No explicit st-numbering is required in this case.

Now, we have directed graphs (digraphs) G''' and D''' derived from G' and D' by imposing the st-orientation described previously. The visibility representation of G' may be derived using optimal topologic numbering of G''' and D''' [30]. Optimal topologic numbering is a function $\overline{f}_{st}(v) : \{v \in V'\} \rightarrow \{1, \ldots, N\}$

$$\overline{f}_{st}(v) = l(s, v) \tag{4.20}$$

where $l(s,v)$ = the maximum distance along all directed routes from s to v. The length of every edge is equal to 1.

The visibility representation of a planar graph is a graph drawing in which each vertex is depicted as a straight-line horizontal segment and every edge is depicted as a straight-line vertical segment. The edge segment endpoints lie on the segments corresponding to the vertices incident to this face and do not intersect any other segments. This concept was first introduced by Otten and Van Wijk [31], and several approaches have been proposed since that time [30,32,33].

4.5.4 Metrics

The orthogonal drawing of G' and the quasi-orthogonal drawing of G can be implemented using visibility representation. Typical structures of visibility representation can be easily transformed into an orthogonal layout (see Figure 4.9) that results in an orthogonal layout for G'. Finally, additional compaction procedures may be applied [34].

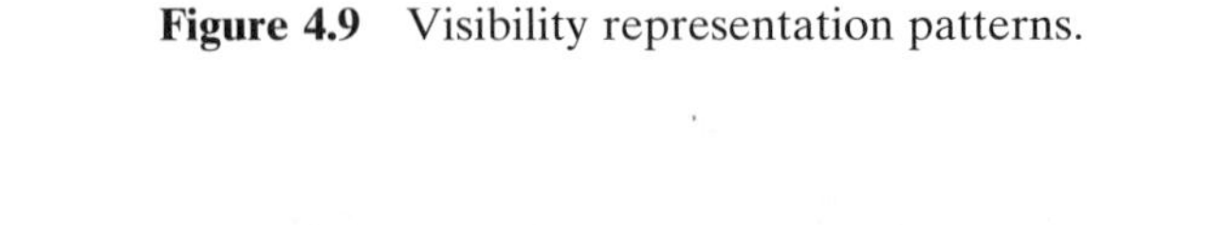

Figure 4.9 Visibility representation patterns.

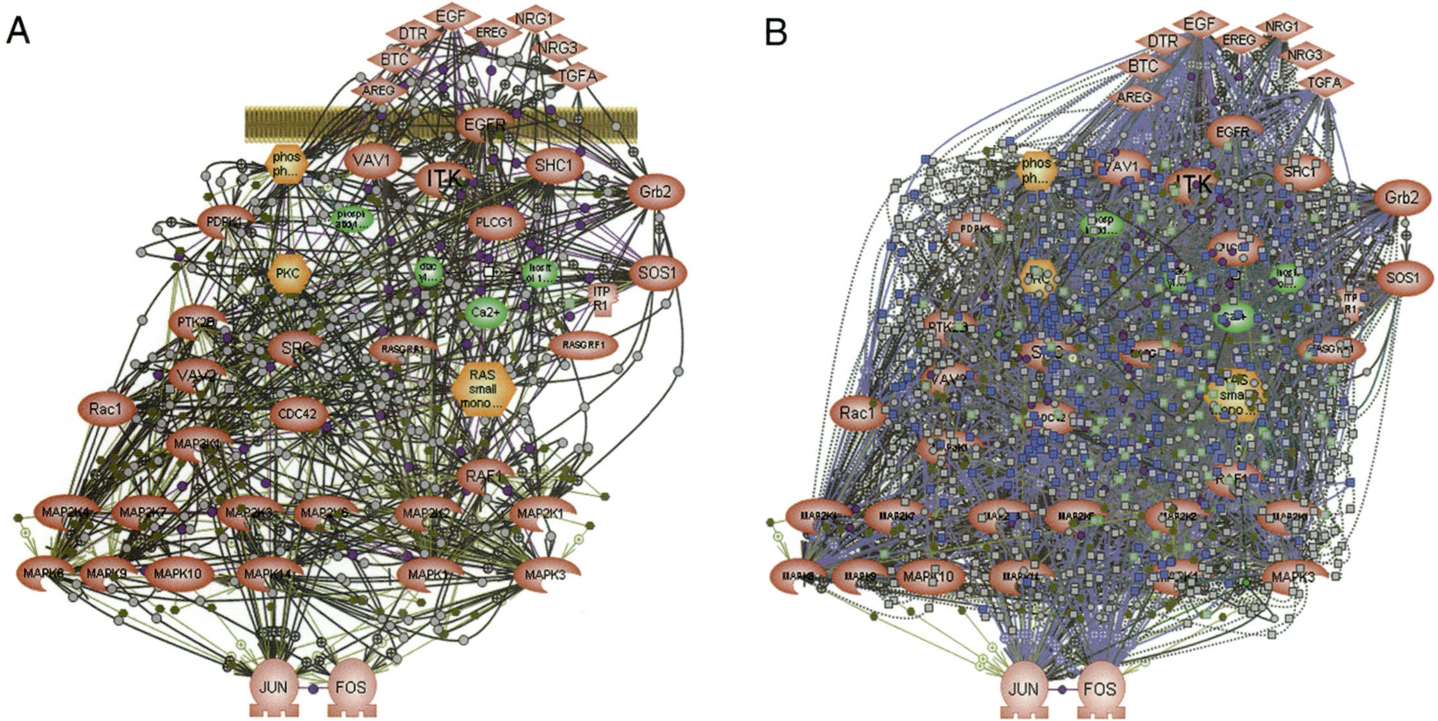

Figure 1.1 Canonical EGFR pathway shown in three ways: (A) physical interaction network; (B) physical interaction network combined with regulatory network; (C) as a pathway ready for kinetic modeling; and (D) kinetic model of MAP kinase cascade, which is a portion of EGFR pathway. One way to visualize the goal of pathway analysis is to imagine that it changes the contrast of the network image to make the main information flow more visible by hiding relations that are nonessential for depicting the principal information flow.

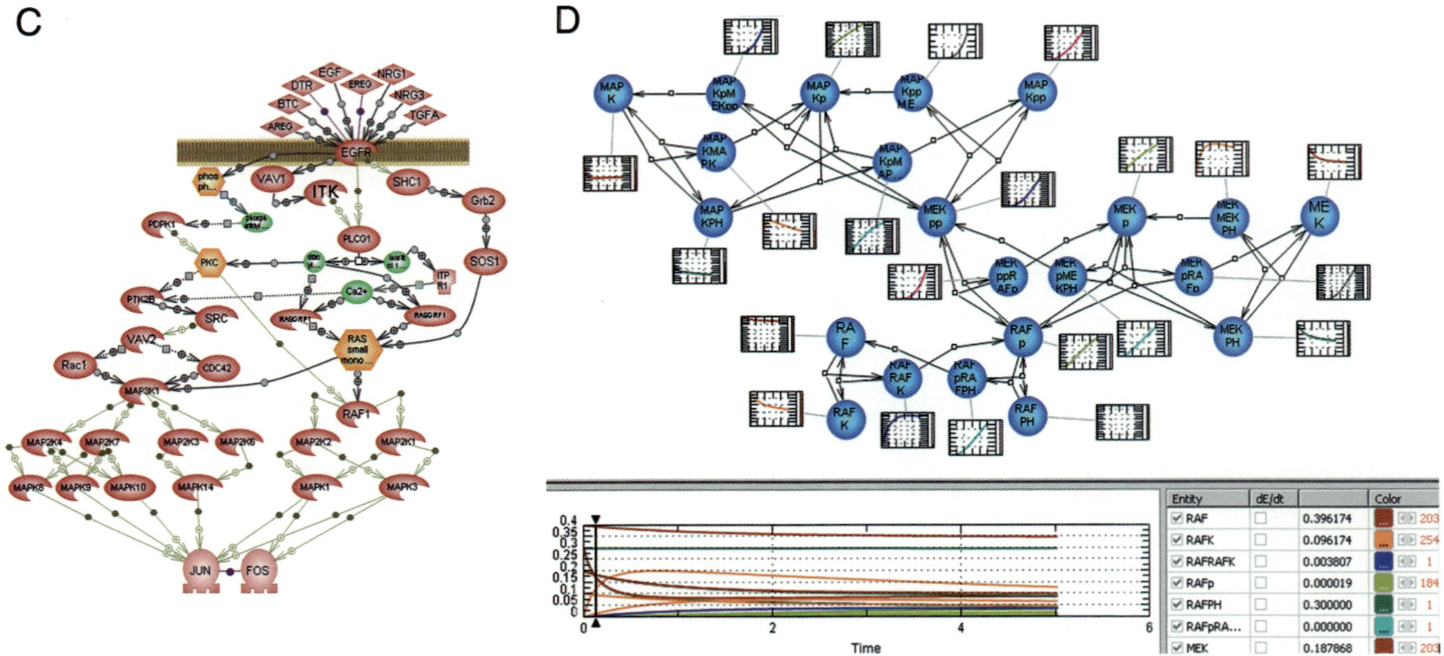

Figure 1.1 (*Continued*)

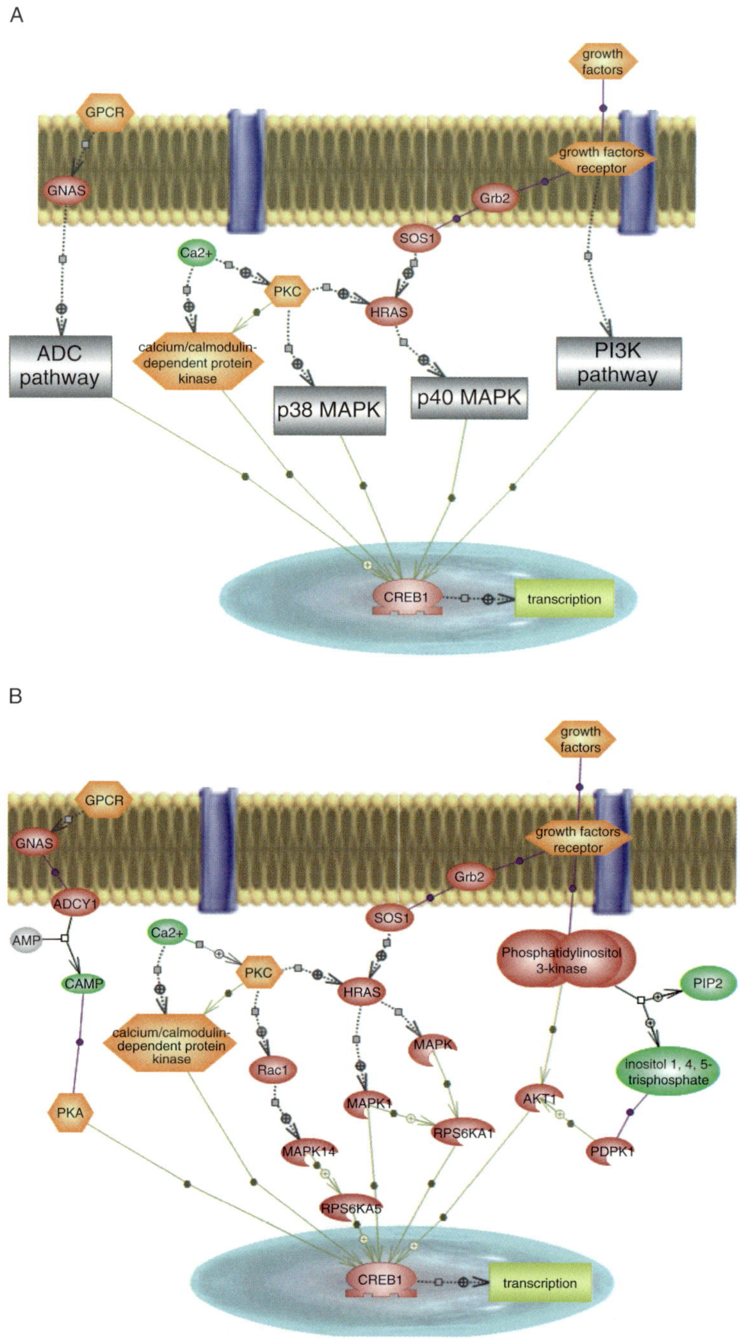

Figure 2.3 Different collapsing strategies of CREB activation pathway. (A) Using sub-pathways. (B) Using functional classes.

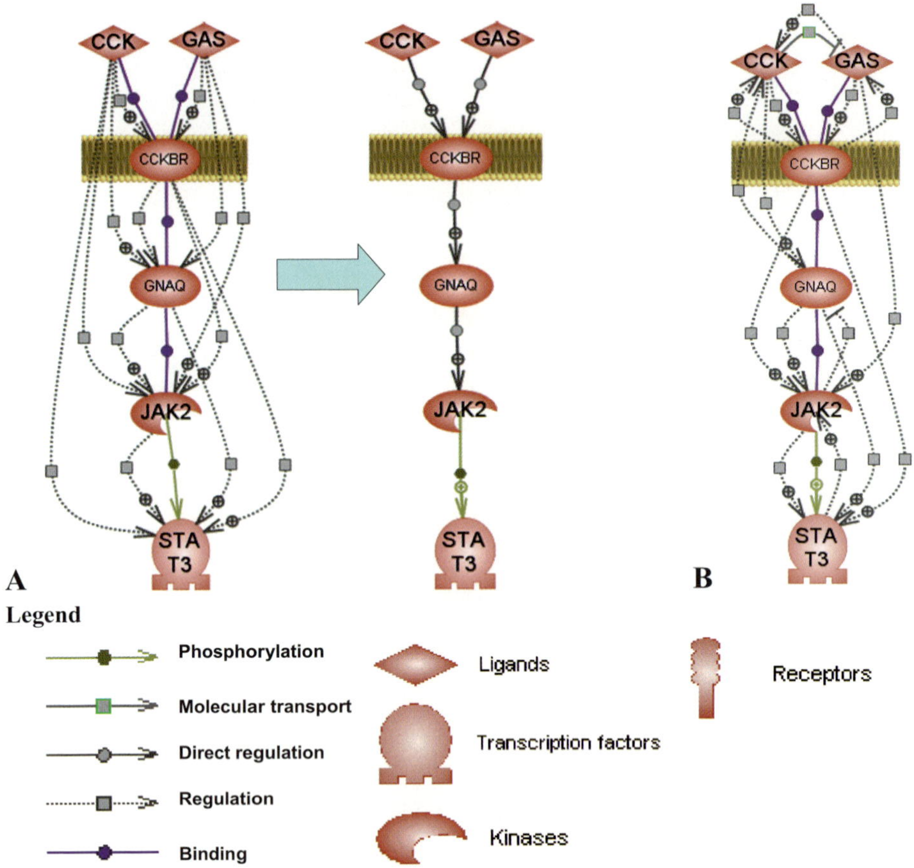

Figure 3.1 The basic principles of pathway building in a biological association network are similar to the manual compilation of experimental evidence for a consensus pathway. The order of components for information flow can be determined from the combination of regulatory interaction. Upstream molecules must have only outgoing relations to all downstream pathway components. After the order of components is determined from a regulatory network, the components must be connected by physical interactions that mediate signal propagation. (A) The relations are shown in an ideal BAN that contains a complete and accurate set of experimental interactions. (B) The relations between proteins in the same pathway are present in the ResNet 5 database. ResNet 5 contains biological association networks automatically extracted from scientific publications using MedScan, a natural language processing technology. Notice multiple feedback loops and incomplete regulatory network that complicate pathway calculation in a real-life BAN.

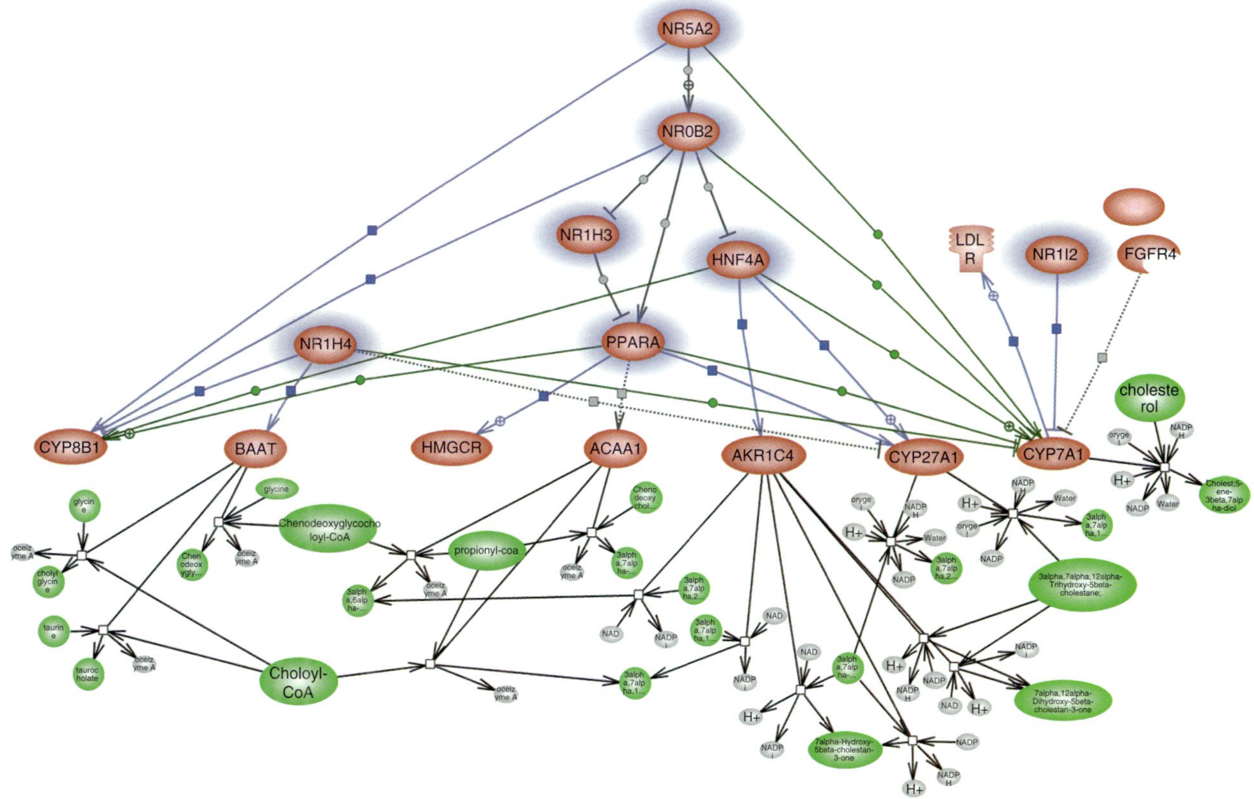

Figure 3.5 An example of a biological process pathway built from protein annotation with the Gene Ontology term "bile acid biosynthesis" (GO:0006699). Note that the GO annotation contains nuclear receptors (highlighted in blue) that regulate the gene expression of enzymes, catalyzing the biochemical pathway for bile acid biosynthesis. The pathway was built essentially as described by Caldwell [13]. Chemical reactions were added manually after the construction of the regulatory and physical interaction network.

COLOR PLATE

A

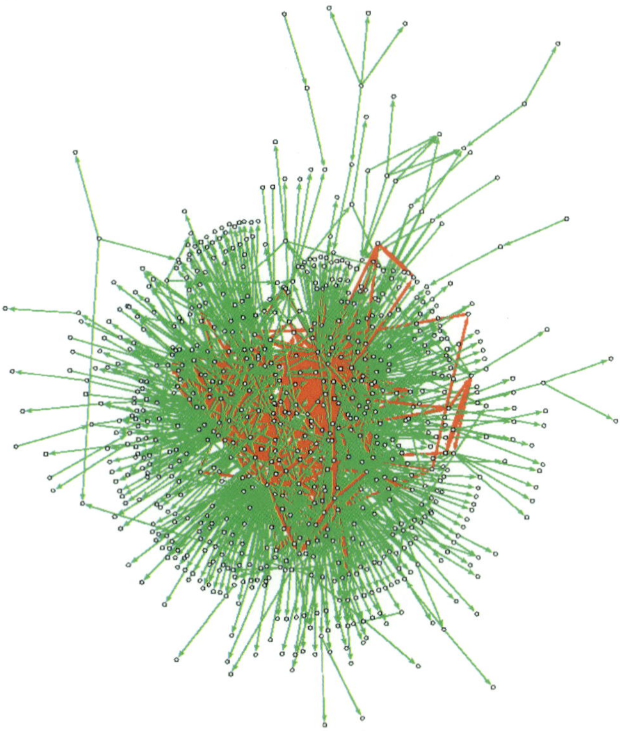

B

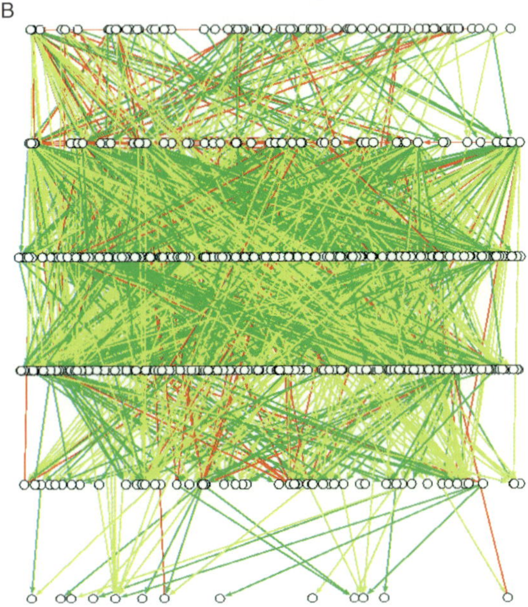

Figure 4.12 A part of the post-translational regulatory network in the human shown here includes 1,671 automatically and manually curated protein modification interactions (phosphorylation, proteolytic cleavage, etc.) between 732 proteins from our ResNet database [43]. Panel A contains the "hairball" visualization of the network structure emphasizing interconnections between individual pathways. Red edges lie within the strongly connected component of this network consisting of 107 proteins that could all be linked to each other by a path in both directions. This makes any two of these proteins to be simultaneously upstream and downstream from each other. In panel B, we optimally distribute proteins over a number of hierarchical levels. Red arrows represent 208 putative feedback links going from lower levels of the hierarchy to higher ones, while yellow ones represent 512 feed-forward links jumping over one or more hierarchical levels. Only proteins and links reachable from one of the 71 receptors placed at the top hierarchical level were included.

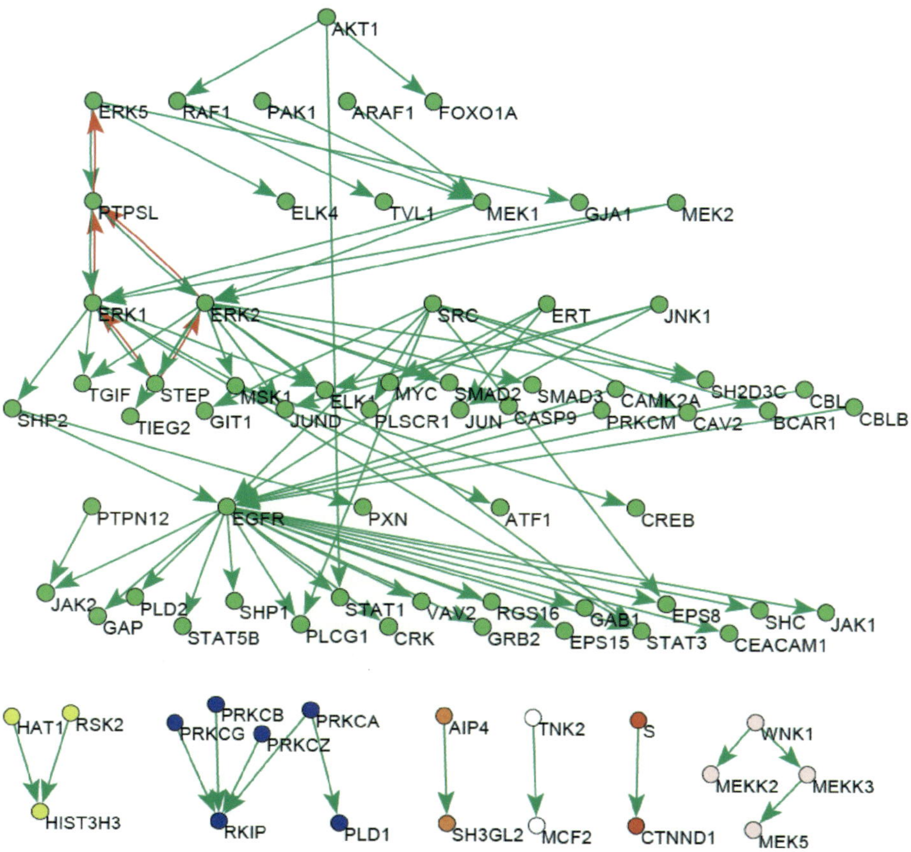

Figure 4.16 Hierarchical layout of EGFR1 from the HPRD pathway database. The counter-hierarchical links are shown in red.

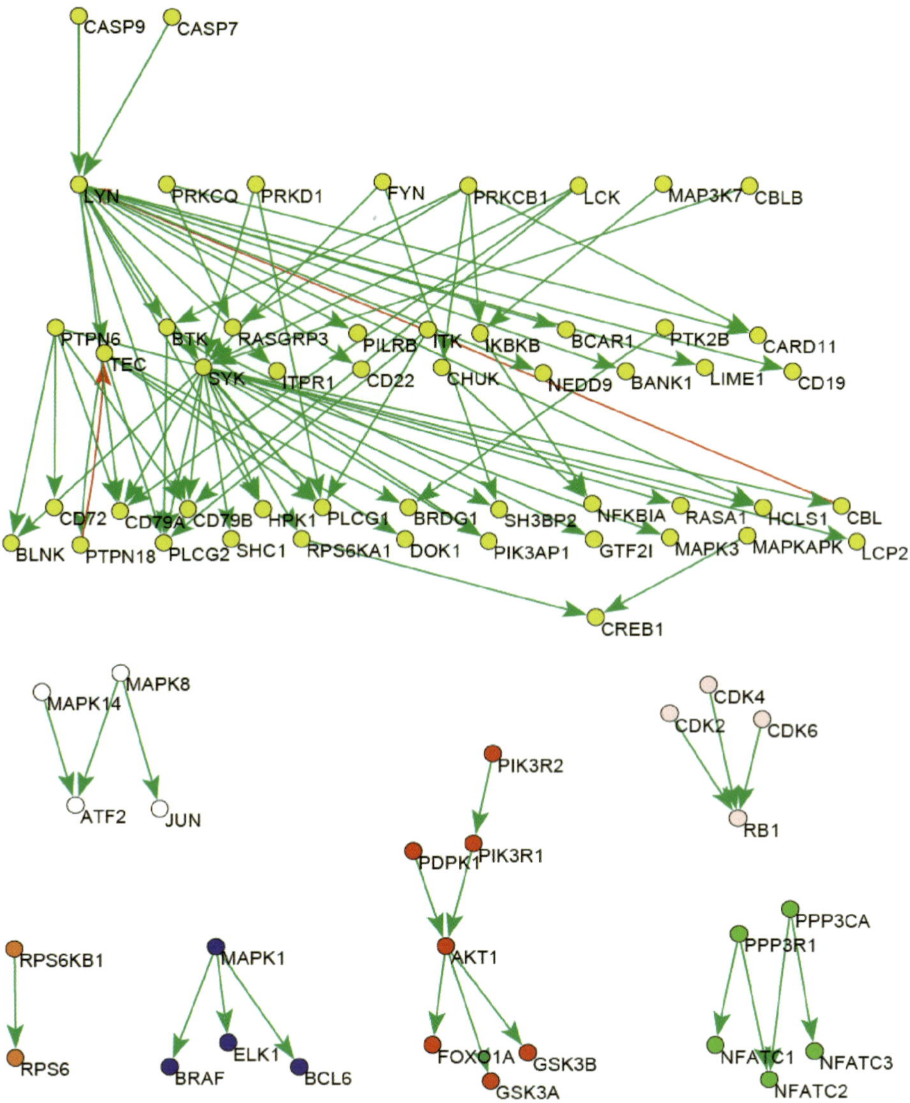

Figure 4.17 Hierarchical layout B-cell receptor pathways from the HPRD pathway database. The counter-hierarchical links are shown in red.

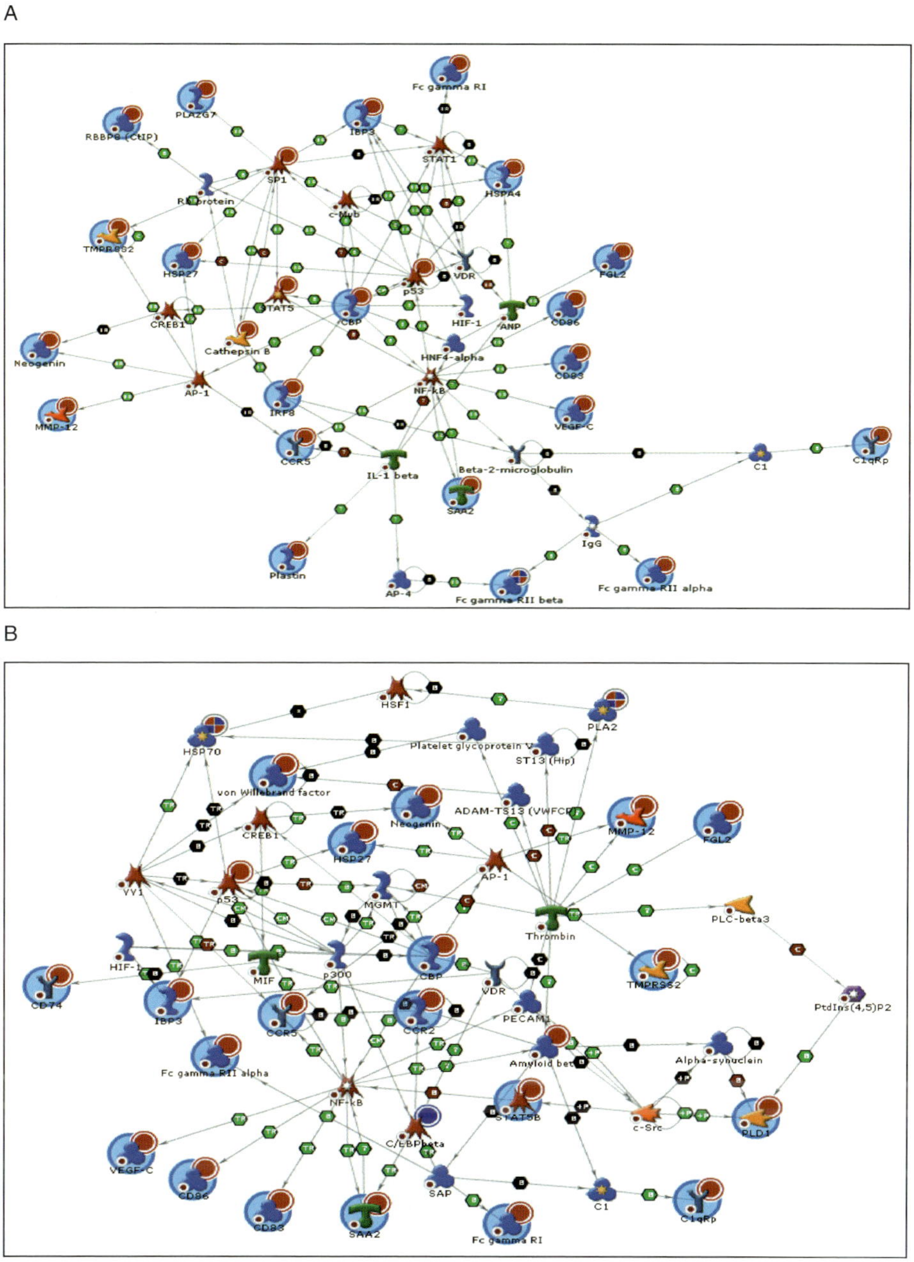

Figure 6.4 Biological networks of differentially expressed genes for HC allowing the identification of transcriptional master regulators. (A–C) Red (blue) dots in right corner of a gene indicate up-regulation (down-regulation) of a particular gene. Networks shown are representative networks used to identify transcriptional (master) regulators that control the gene expression changes under HC conditions (threshold of significance for networks $p < 0.01$). Metabolites are indicated as hexagons. (D) Concurrent network analysis of metabolomic and expression data reveals consistency between elevated level of eicosapentaenoic acid (red circle), an inhibitor of SREBP1 and underexpression of genes whose transcription is activated by SREBP1 (blue circles).

C

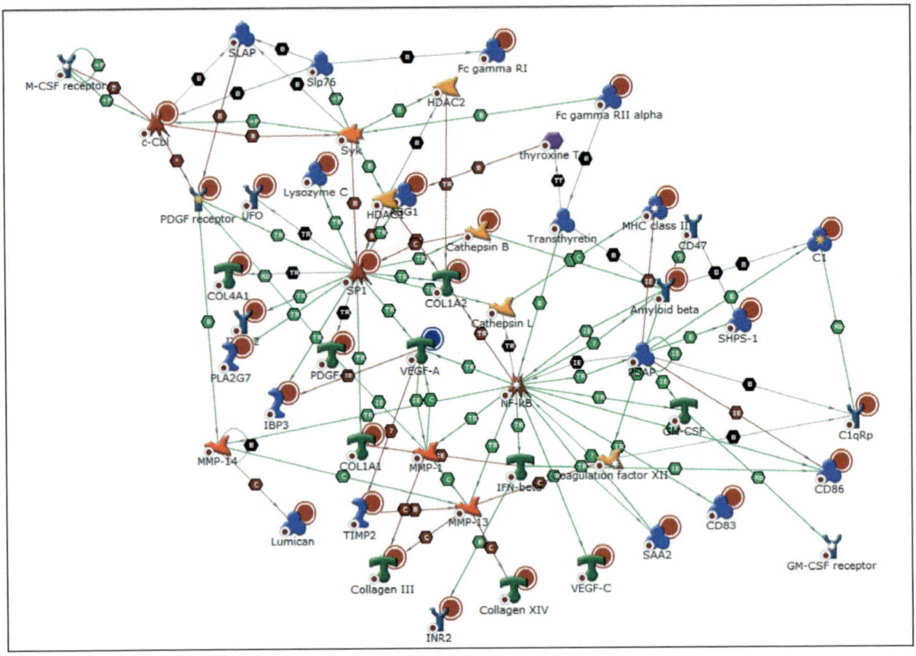

D

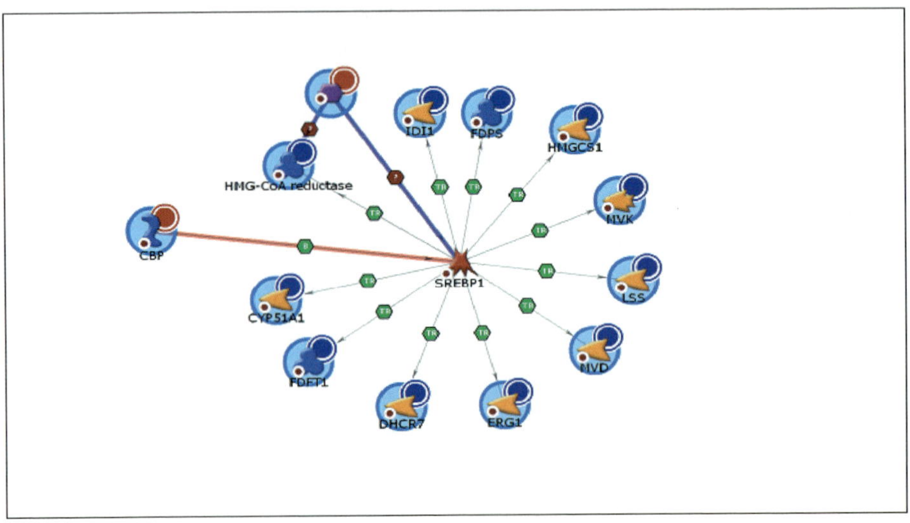

Figure 6.4 (*Continued*)

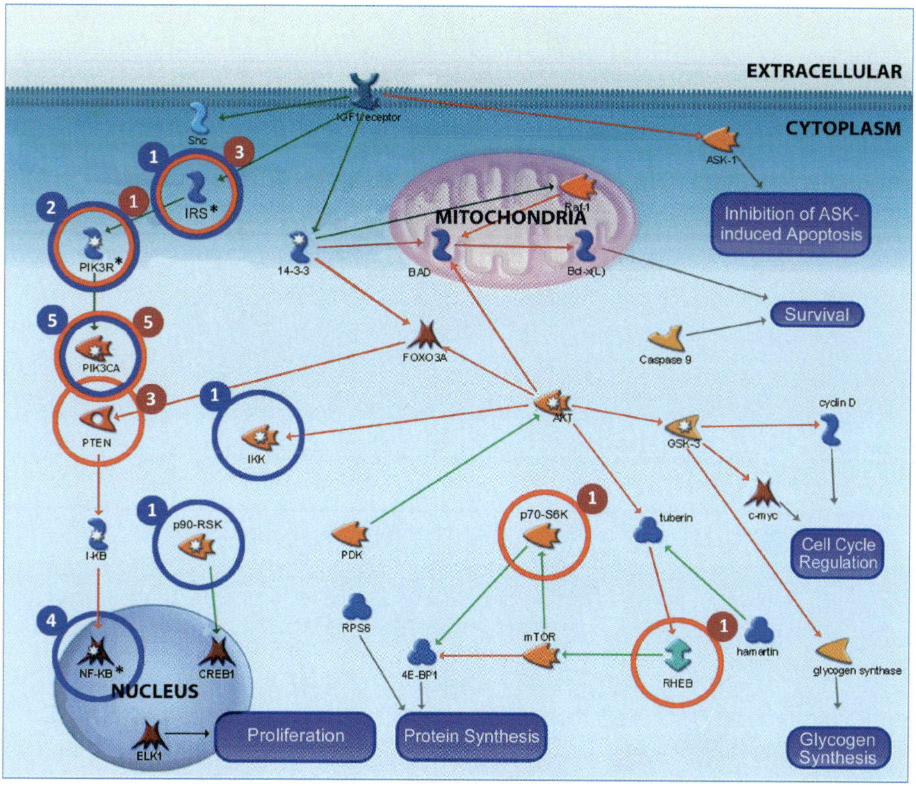

Figure 6.5 PI3K pathway mutations in breast and colorectal cancers. The identities and relationships of genes that function in PI3K signaling are indicated. Circled genes have somatic mutations in colorectal (red) and breast (blue) cancers. The number of tumors with somatic mutations in each mutated protein is indicated by the number adjacent to the circle. Asterisks indicate proteins with mutated isoforms that may play similar roles in the cell. These include insulin receptor substrates IRS2 and IRS4; phosphatidylinositol 3-kinase regulatory subunits PIK3R1, PIK3R4, and PIK3R5; and nuclear factor kappa-B regulators NFKB1, NFKBIA, and NFKBIE.

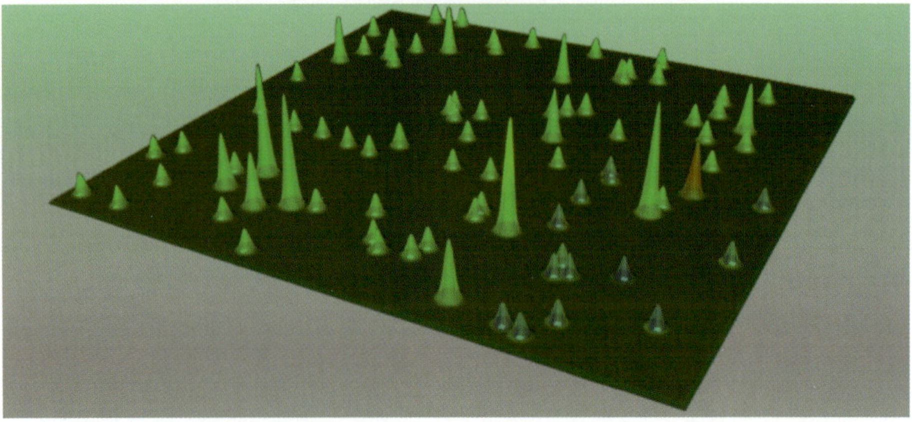

Figure 6.6 Cancer genome landscape. Non-silent somatic mutations are plotted in two-dimensional space representing chromosomal positions of RefSeq genes. (See text for full caption.)

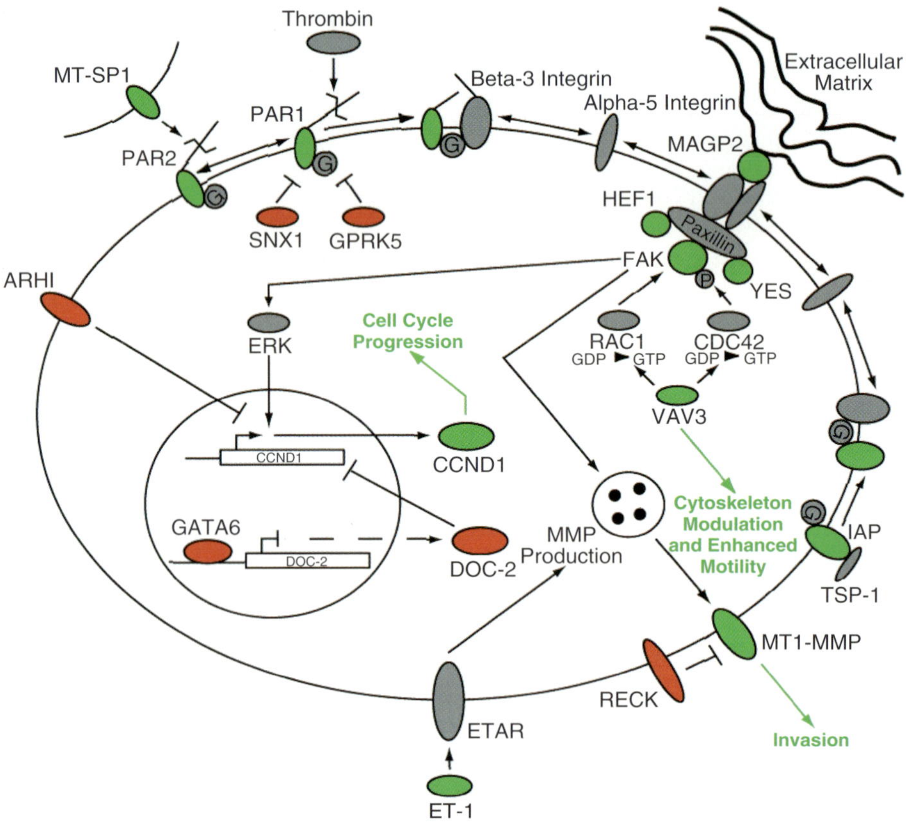

Figure 7.5 Schematic representation of potential signaling pathways involved in advanced serous ovarian cancer.

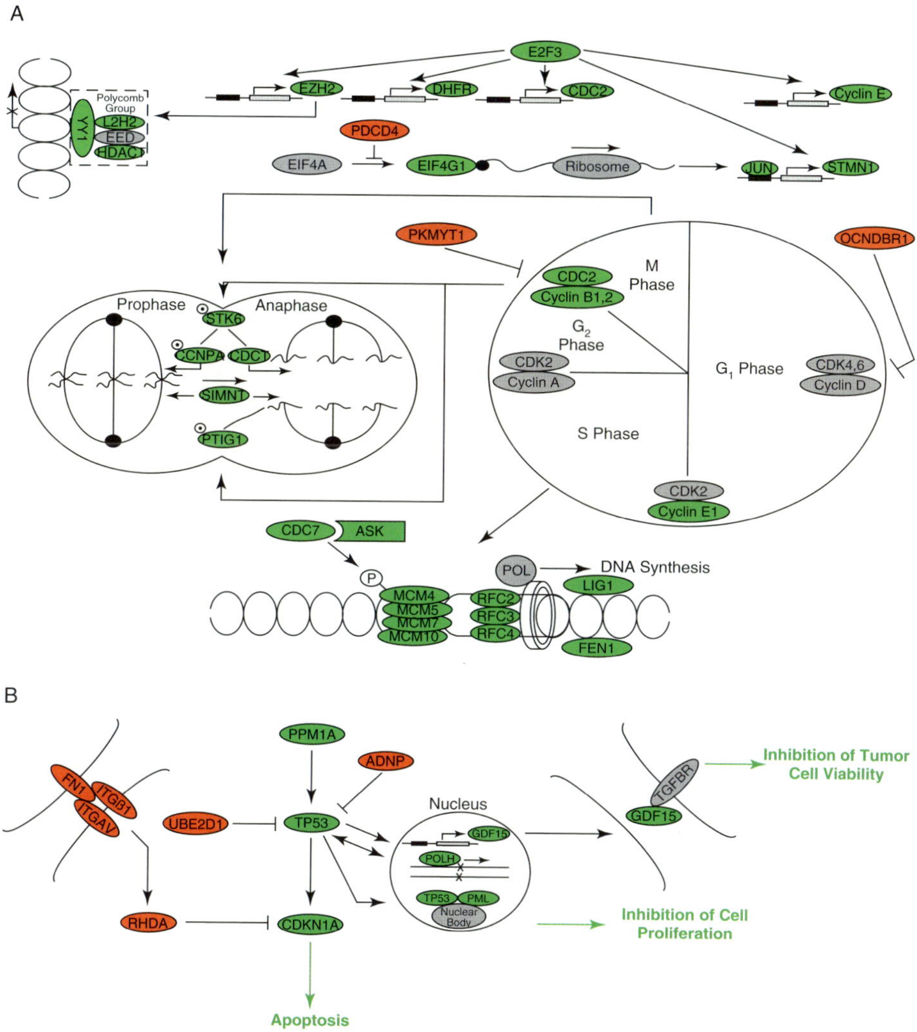

Figure 7.6 (A) Pathway analysis for proliferation and chromosomal instability in high-grade serous ovarian cancers versus OSE. (B) Pathway analysis of differentially regulated genes unique to LMP tumors versus OSE.

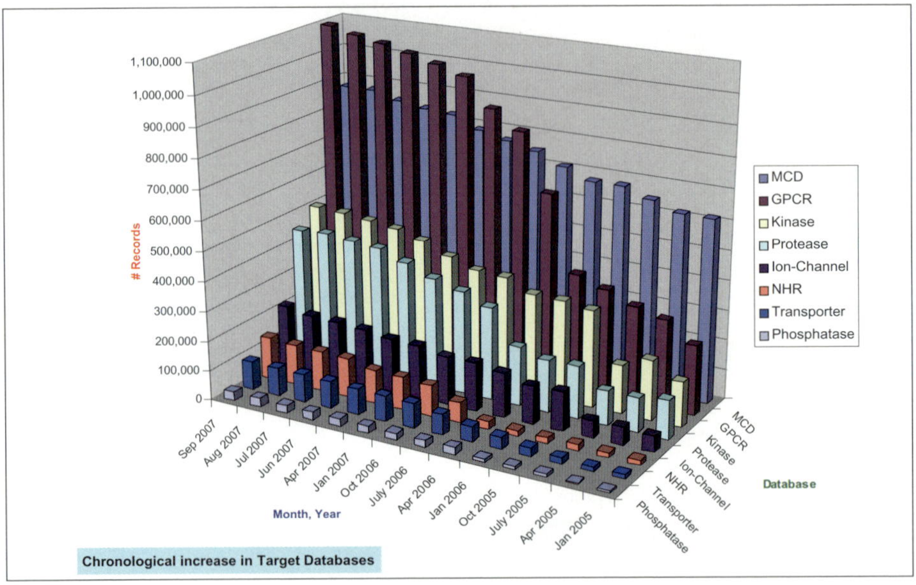

Figure 10.7 Rate of growth of content in Cheminformatics databases (Source: GVK BioSciences).

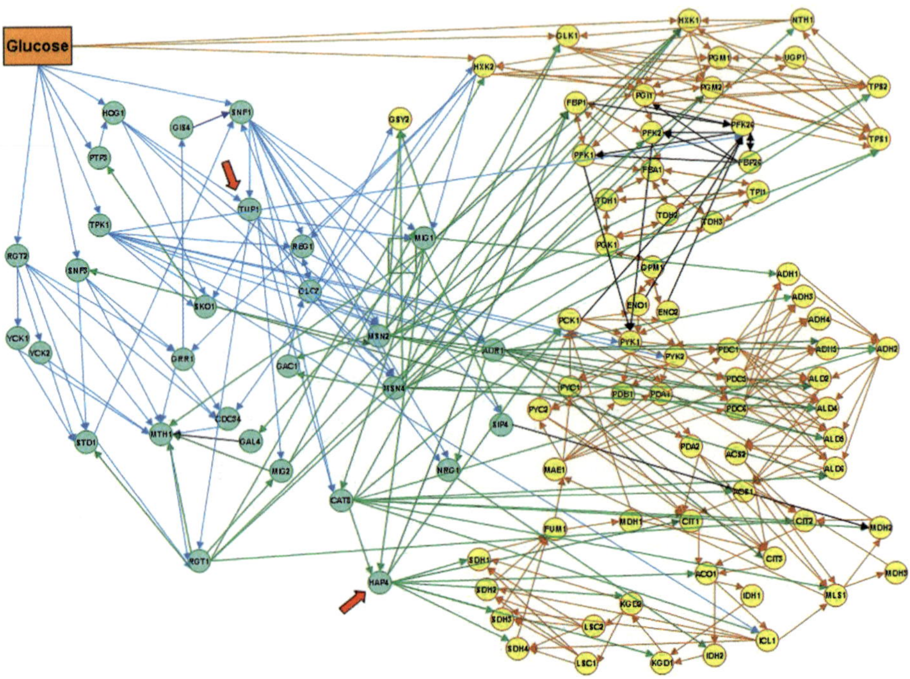

Figure 12.1 The glucose repressor/derepressor system. (See text for full caption.)

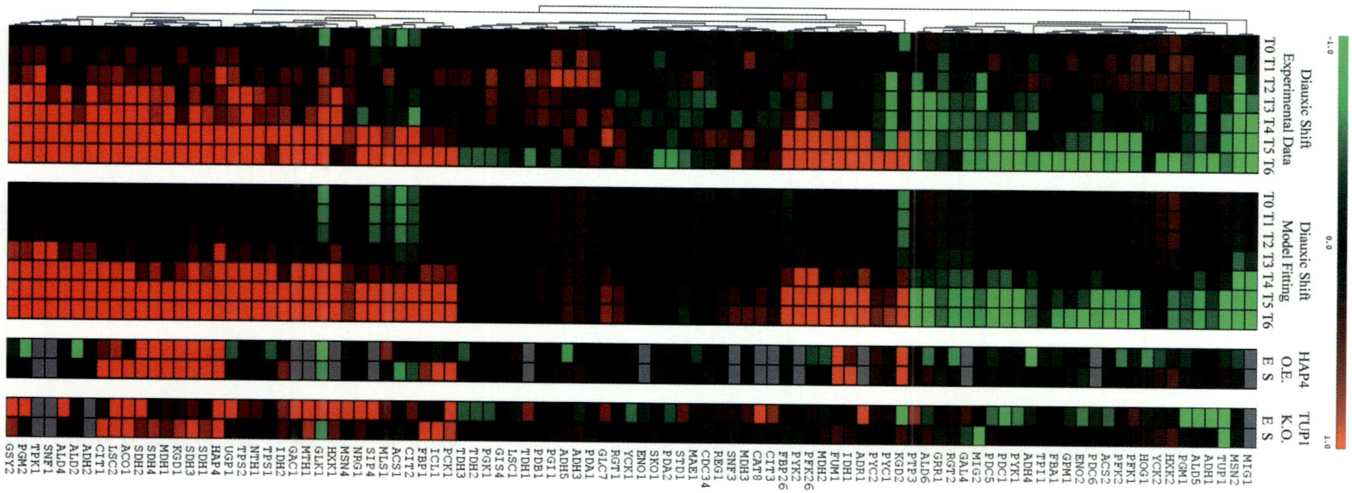

Figure 12.2 Model fitting and simulation results: Diauxic shift experimental data, simulated data (both at times T0 = 0, T1 = 150, T2 = 240, T3 = 360, T4 = 450, T5 = 570, T6 = 690 minutes), HAP4 overexpression data (O.E.) and TUP1 knockout data (K.O.). E = experimental, S = simulated. The gene expression ratios to a T0 experimental value corresponding to a similar independent experiment are log2-transformed. The resulting data were imported in the TIGR MeV MultiExperiment Viewer 3.1 software from the Institute for Genomic Research [37] to produce the images; hierarchical clustering of the genes/proteins according to their expression behavior (gene tree only) was performed using the Euclidean distance, with complete linkage clustering, 100 iterations, no resampling of samples, and all other parameters set to default values. Gray color = experimental data not available.

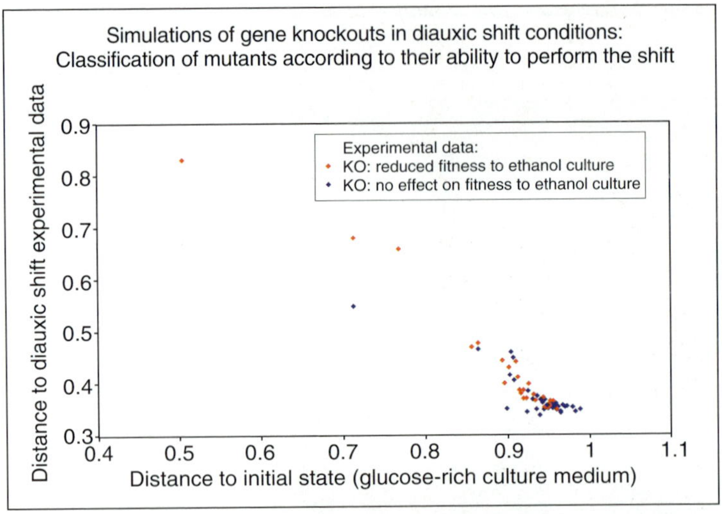

Figure 12.5 Classification of simulated KO mutants according to their fitness on ethanol-rich medium. Each dot represents a simulated gene KO. Gene deletion simulations were carried out under diauxic shift conditions. Most simulated NO mutants moving away from respiration to fermentation (initial state) are those with an experimentally confirmed deficiency in growth on ethanol. The distances have been expressed as ratios with respect to the distance between initial and diauxic shift state at time $T = 690$ minutes.

4.5.5 Summary and Results

We summarize orthogonal layout below in the following steps:

1. Planarization
   a. Initial embedding
      I. Linking chains endpoints to the supporting graph
      II. Using direct force with high-intensity repulsion
   b. Substituting vertices of degree n, ($n > 4$) with virtual cycles
   c. Substituting edge crossings with virtual vertices
2. Orthogonalization
   a. Dual graph construction
      I. Looking for faces in supporting graph
      II. Setting edges between faces
   b. *st*-orientation of the supporting graph
   c. Left and right faces assigning to the edges and vertices of the supporting graph
   d. *st*-orientation of the dual graph
   e. Supporting graph topologic numbering
   f. Dual graph topologic numbering
   g. Visibility representation
3. Metrics
   a. Coordinates assignment
   b. Compaction optimization

A sample of the orthogonal drawing layout is shown in Figure 4.10.

4.6 HIERARCHICAL ALGORITHM

The hierarchical algorithm is one of several solutions for fast arbitrary graph drawing [1,35]. It may be applied to any connected graph, but the results for the acyclic directed graphs are the most compelling. This approach also implies the presence of hierarchical structure in the data. An arbitrary connected graph should be reduced to a directed acrylic graph before the hierarchical algorithm is applied. The main feature of the algorithm is that all edges are turned along the same direction. For system biology, a hierarchical layout is the only way to establish nodes that are top regulators in a pathway and nodes that are major targets in the pathway, thus visualizing the information flow in the network.

The aesthetic criteria for this layout are the common direction of edges (i.e. from top to bottom or from left to right), drawing area minimization, uniform

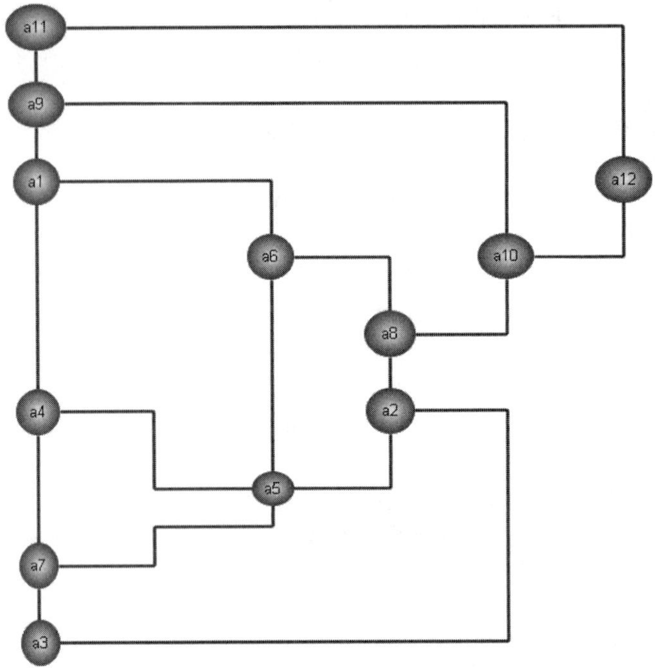

Figure 4.10 Orthogonal layout.

distribution of vertices in the drawing area, avoiding overly long edges, edge crossings minimizing, and straight-line edges [36]. All these criteria cannot be satisfied simultaneously; for example, a straight-line edge limitation contradicts the minimum area and uniform distribution. The steps of the hierarchical layout algorithm commonly include [37–42]: assigning nodes to layers so that all edges are turned in the same direction, decreasing edge crossings by node permutation within layers, and final coordinate assignment based on the edge length minimization.

Before the layout is started, the acyclic graph must be constructed from an input graph. This can be achieved by changing the direction of some edges and assigning a direction to the undirected edges. The procedure is similar to the st-orientation of the graph from the previous section. Source s and sink t nodes are selected, and depth-first search (DFS) is performed from s to t. Undirected and improperly directed edges for the DFS route are directed along this route and marked as non-tree edges, while the properly directed edges are marked as the tree edges. Another possible approach of acyclic core graph construction will be described later.

A node layering is similar to the topologic numbering (equation 4.20). Thus, for every vertex p a positive integer $\psi(p)$ should be ascribed in such a way that for every directed edge (u,v) (u = tail and v = head), $\psi(v) > \psi(u)$. The difference $L(u,v) = \psi(v) - \psi(u)$ is called topologic distance between *u* and *v*. The other approach is to reduce the task to the integer program [38]:

$$\min \sum_{(u,v)\in E} \psi(v)-\psi(u), \text{ where } \psi(v)-\psi(u)>\delta(u,v) \tag{4.21}$$

where $\delta(u,v)$ = minimum length function. The minimum cost flow or network simplex method may be used to solve the task (equation 4.21) in polynomial time. The first step completely determines the *y* coordinates of the vertices that are based on the function $\psi(p)$.

During the second step, vertices at every layer are permutated in such a way that the number of edge crossings is minimized. Thus, relative vertex positions are determined that impose limitations to the final horizontal coordinates assignment. A rigorous solution of this task is *NP*-hard and different heuristics are used for its approximate solution. One possible approach is local optimization based on vertex sorting. Three successive layers are selected and vertex positions of the highest and lowest layers are fixed. The vertex pairs of the middle layer are searched and the vertices are switched if the number of edge crossings is decreased. This approach is similar to the bubble sort method and is formulated as $O(N^2)$. The other approaches utilize the barycenter and median methods [38] that have $O(N)$ time complexity, reduction to planar graph, probabilistic approach, genetic algorithm, etc. [36].

Coordinates are assigned to the vertices at the third step of the algorithm. Vertical coordinates are fully determined by the node's layer index $\psi(p)$ at the first step of the algorithm. Horizontal coordinates must be assigned in such a way that the vertex relative positions determined during the second step are preserved. In addition, the total length of the edges can be minimized in this step using the integer program (equation 4.21).

Let us summarize hierarchical algorithm that includes the following steps:

1. Node layering
   a. Acyclic graph construction
   b. Topologic numbering computation
2. Decreasing edge crossings
3. Coordinate assignment

The result of the hierarchical algorithm implemented in Pathway Studio software is depicted in Figure 4.11.

4.7 ACYCLIC CORE GRAPH CONSTRUCTION

4.7.1 Analysis

Finding the dominant direction of information flow in densely interconnected regulatory or signaling networks is required in many applications in

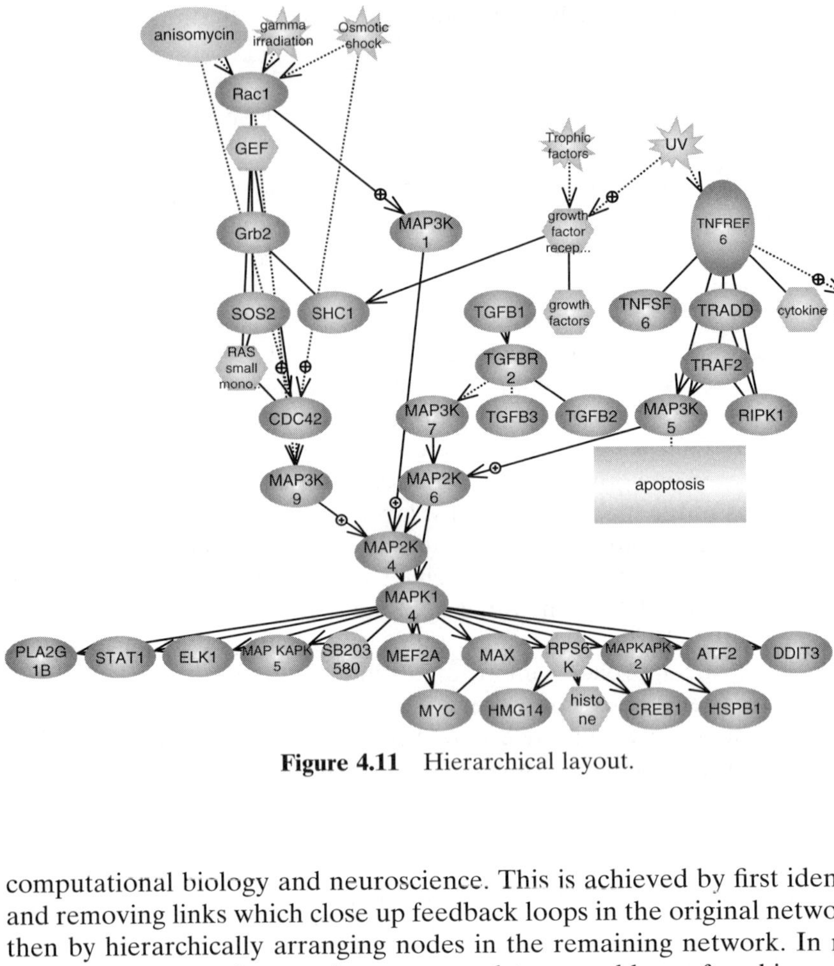

Figure 4.11 Hierarchical layout.

computational biology and neuroscience. This is achieved by first identifying and removing links which close up feedback loops in the original network and then by hierarchically arranging nodes in the remaining network. In mathematical language, these tasks correspond to a problem of making a graph acyclic by removing as few links as possible and thus altering the original graph in the least possible way. Practically in all applications, the exact solution to this problem requires an enumeration of all cycles and combinations of removed links, which, as an *NP*-hard problem, is computationally intractable even for modest-size networks.

We introduce and compare two algorithms: the probabilistic one based on a simulated annealing of a hierarchical layout of the network that minimizes the number of "backward" links going from lower to higher hierarchical levels and the deterministic, "greedy" algorithm that sequentially cuts the links that participate in the largest number of cycles. We find that the annealing algorithm outperforms the deterministic one in terms of speed, memory requirement, and the actual number of removed links. To further improve a visual perception of the layout, we perform an additional minimization of the length of hierarchical links while keeping the number of anti-hierarchical links at their minimum.

During the last several years, a substantial amount of information on large-scale structure of intracellular regulatory networks has been accumulated. However, the growth of our understanding of how these networks manage to function in a robust and specific manner was lagging behind the sheer rate of data acquisition. The fact that these networks are frequently visualized as a giant "hairball" (Figure 4.12) consisting of a multitude of edges, linking most constituent protein-nodes to each other, serves as a striking illustration of the complexity of the issue at hand.

To understand the functioning or even to efficiently visualize a densely interconnected directed network, it is desirable to determine the dominant direction of information flow and to identify links that go against this flow and thus close feedback loops. Ordering a network with respect to the dominant direction of flow can help to determine its previously unknown inputs and outputs, to track back to hidden sources of perturbations based on their observable downstream effects, etc. A simple-minded hierarchical layout of a densely interconnected network is often impossible due to the ubiquitous presence of feedback loops. Indeed, all nodes in a strongly connected component of a network are, by definition, simultaneously upstream and downstream of each other. However, if most feedback loops are closed by relatively few feedback-signaling links, the dominant direction of flow could still be reconstructed based on network topology alone. The identification and removal of these relatively infrequent feedback links would enable one to perform a hierarchical layout of the remaining acyclic network that still sufficiently resembles the original one.

We introduce a probabilistic algorithm for hierarchical layout that minimizes the number of counter-hierarchical links by using simulated annealing. This algorithm provides us with a new approach to one of the classic *NP*-hard problems, the search for the minimum feedback arc set [44], which enjoys an ever-present popularity and has a substantial number of approximate solutions (see, for example, [45,46] and references therein). To evaluate the performance of the annealing algorithm in identifying the minimal set of feedbacks, we compare it to a fairly straightforward, deterministic greedy algorithm that sequentially cuts the links that belong to the largest number of cycles. It turns out that the probabilistic algorithm outperforms the deterministic one in better minimizing the number of removed links and memory requirements, while maximizing the speed. A simple visual example is provided for the situation in which the deterministic greedy algorithm is nonoptimal. Following that, we discuss biological implications and applications of our findings, as well as how additional constraints such as *a priori* knowledge of the function and therefore the hierarchical position of certain nodes may affect algorithm performance.

4.7.2 Network Layout

Consider a graph of N vertices labeled as 1,2,3, . . ., N and L as directed links labeled by pairs of vertices they connect, $l_i \equiv (n_i, m_i)$. The goal is to distribute

A

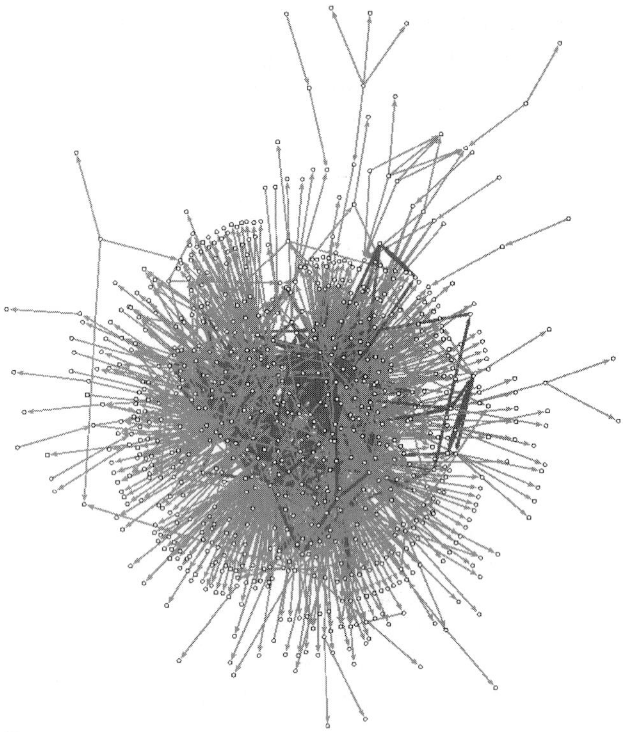

B

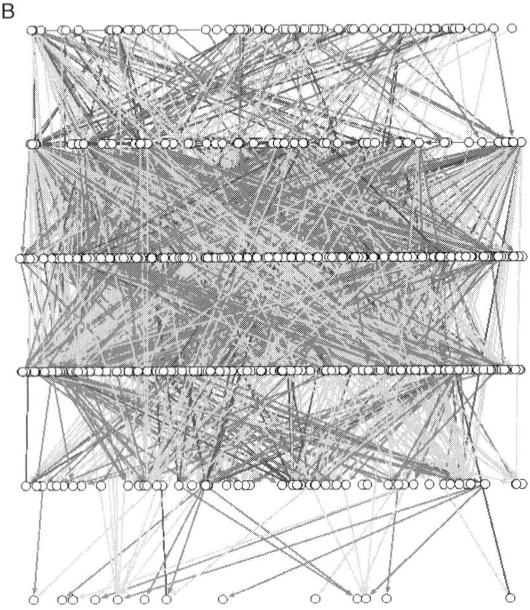

Figure 4.12 A part of the post-translational regulatory network in the human shown here includes 1,671 automatically and manually curated protein modification interactions (phosphorylation, proteolytic cleavage, etc.) between 732 proteins from our ResNet database [43]. Panel A contains the "hairball" visualization of the network structure emphasizing interconnections between individual pathways. Red edges lie within the strongly connected component of this network consisting of 107 proteins that could all be linked to each other by a path in both directions. This makes any two of these proteins to be simultaneously upstream and downstream from each other. In panel B, we optimally distribute proteins over a number of hierarchical levels. Red arrows represent 208 putative feedback links going from lower levels of the hierarchy to higher ones, while yellow ones represent 512 feed-forward links jumping over one or more hierarchical levels. Only proteins and links reachable from one of the 71 receptors placed at the top hierarchical level were included. See color insert.

the vertices on $M < N$ levels so that the number of links going against the hierarchy, or from a lower level to the same or higher one is minimal. If the number of levels M is sufficient (equal or larger than the longest simple graph path), this problem is equivalent to finding a minimum feedback arc set [44], or to removing as few as possible links to make the graph acyclic, or feedback-free.

A naive way to solve this problem exactly is to enumerate all cycles in a graph and then sample all possible combinations of links checking if they belong to all cycles. If one starts with enumerating individual links, then pairs of links, and so forth, until a removal of l links yield the first acyclic graph, this sampling would require checking the $\sum_{i=1}^{l}\binom{L}{i}$ combination of links. For the biologically relevant values of $L \sim 10^3$–10^4 and $l \sim 10$–10^2, this approach is clearly infeasible both in terms of the complete enumeration of cycles and minimal feedback link sets[1].

4.7.2.1 Simulated Annealing Network Ordering. The task of finding the minimum number of counter-hierarchical or feedback links can be interpreted as an optimization problem and tackled by probabilistic methods such as simulated annealing. Evidently, more than one way to define the optimization function exists, and after exploring several possibilities we converged on the following one:

- For a given network, a set of M levels is introduced ($M < N$, in reality, $M \ll N$ and is of the order of the graph diameter). Initially, all nodes are distributed on the levels randomly.
- For a particular distribution of nodes on levels, the number of links that oppose the hierarchy, that is, from a lower level to the same or higher

[1] From an obvious identity, $\sum_{i=1}^{L/2}\binom{L}{i} = 2^{L-1}$, it follows that even for fairly modest $L = 10^2$ and $l = L/2$ the number of such attempts is $\sim 10^{15}$.

one, is declared to be the energy E of the distribution, or the optimization function.

- A node and its new level are selected at random. A difference in energy ΔE that would occur if the node were moved to the new level is calculated. The node is moved to this new level with the probability $\min\{1,\exp(-\Delta E/T)\}$, where T is the temperature.
- After the network has been sampled a sufficient number of times (of the order of $N \times M$) so that each node has an opportunity to be moved to every level, the temperature is reduced by a specific factor, usually 0.9. Initially, the temperature is set sufficiently high, usually of the order of the average node degree L/N, to allow unobstructed level changes.
- When the temperature drops low enough to inhibit any level changes, the remaining ascending and same-level links are declared feedbacks and removed.
- The whole procedure can be repeated several times to check for consistency in the assignment of feedback links and to determine the solution with the lowest number of removed links.

A change of the level event and the associated energy difference is illustrated in Figure 4.13.

The number of levels M could be fixed by the requirements for a hierarchical layout. Otherwise, M could be determined self-consistently, by observing when the number of counter-hierarchical links stops decreasing, upon the increase in the number of levels. This is illustrated in Figure 4.14 in which a plot of the number of non-hierarchical links versus the number of levels is presented for the human protein phosphorylation network.

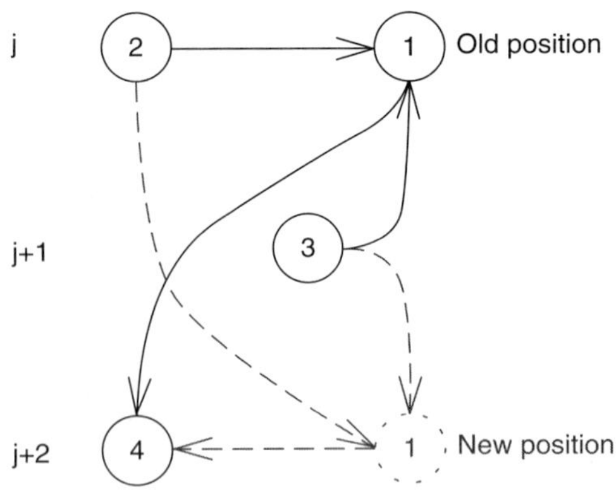

Figure 4.13 Node 1 with two incoming and one outgoing link is selected to move from its current position on level j to a new position on level $j + 2$. The associated energy difference is $\Delta E = -1 - 1 + 1 = -1$ where two -1 contributions come from making (2,1) and (3,1) links hierarchical and the single $+1$ contribution comes from turning the link (4,1) from hierarchical to non-hierarchical.

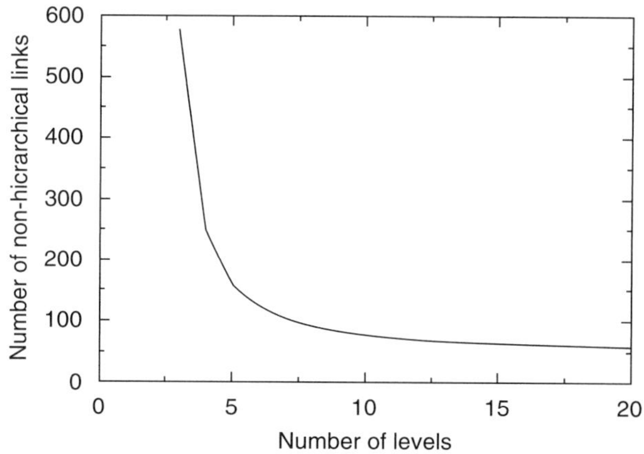

Figure 4.14 The number of non-hierarchical links versus the number of levels M in the annealing layout of the combined (a union of [47] and [43] datasets) protein phosphorylation network in a human cell. The network consists of $L = 2{,}880$ links and $N =$ 1,297 nodes (proteins). The nodes with zero in-degree and zero out-degree are always placed on the top and bottom levels, accordingly. The leftmost data point corresponds to the single intermediate level (three levels total), the number of non-hierarchical links clearly reaches its minimum of 59 links for $M > 17$, which apparently is the length of the largest simple path.

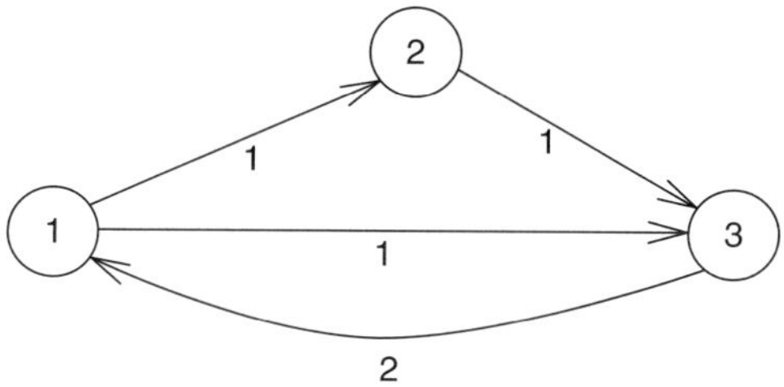

Figure 4.15 Removal of a single (3,1) link makes this 3-vertex graph acyclic.

The performance of the stochastic simulated annealing algorithm scales as $N \times M$. Despite the fairly large coefficient required for gradual multistep annealing, the algorithm readily performs layouts of protein networks of whole organisms, consisting of ~104 nodes and ~105 links. This whole organism layout is needed, for example, for a network backtracking to the source of a multigene differential expression pattern, in which feedback links need to be removed. The counter-hierarchical links identified by the annealing algorithm in biological signaling pathways usually are indeed feedback links in the biological sense. For example, in two of the most complex pathways in the HPRD pathway database (www.netpath.org), EGFR1 and B-cell receptor (Figures 4.16 and 4.17), we identified five and two counter-hierarchical links,

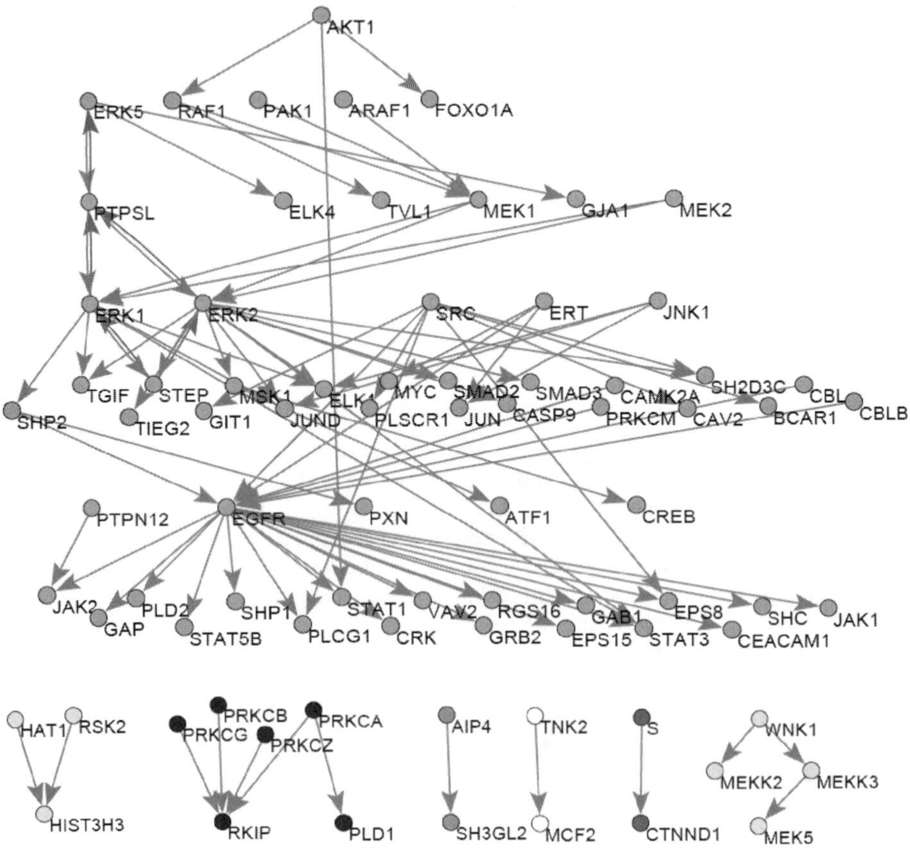

Figure 4.16 Hierarchical layout of EGFR1 from the HPRD pathway database. The counter-hierarchical links are shown in red. See color insert.

five of which correspond to Dephosphorylation and one to Ubiquitination, which are biological feedback mechanisms.

4.7.2.2 Greedy Algorithm. To illustrate advantages of the proposed annealing method, we compare it to a "greedy" algorithm which performs the "steepest descent" in the number of cycles. We implemented it in the following way:

- By enumerating all cycles in a graph, each link is assigned a score equal to the number of cycles of which it is a member.
- The link with the highest score is removed. When several links have the same highest score, a link to be removed is randomly selected among them.

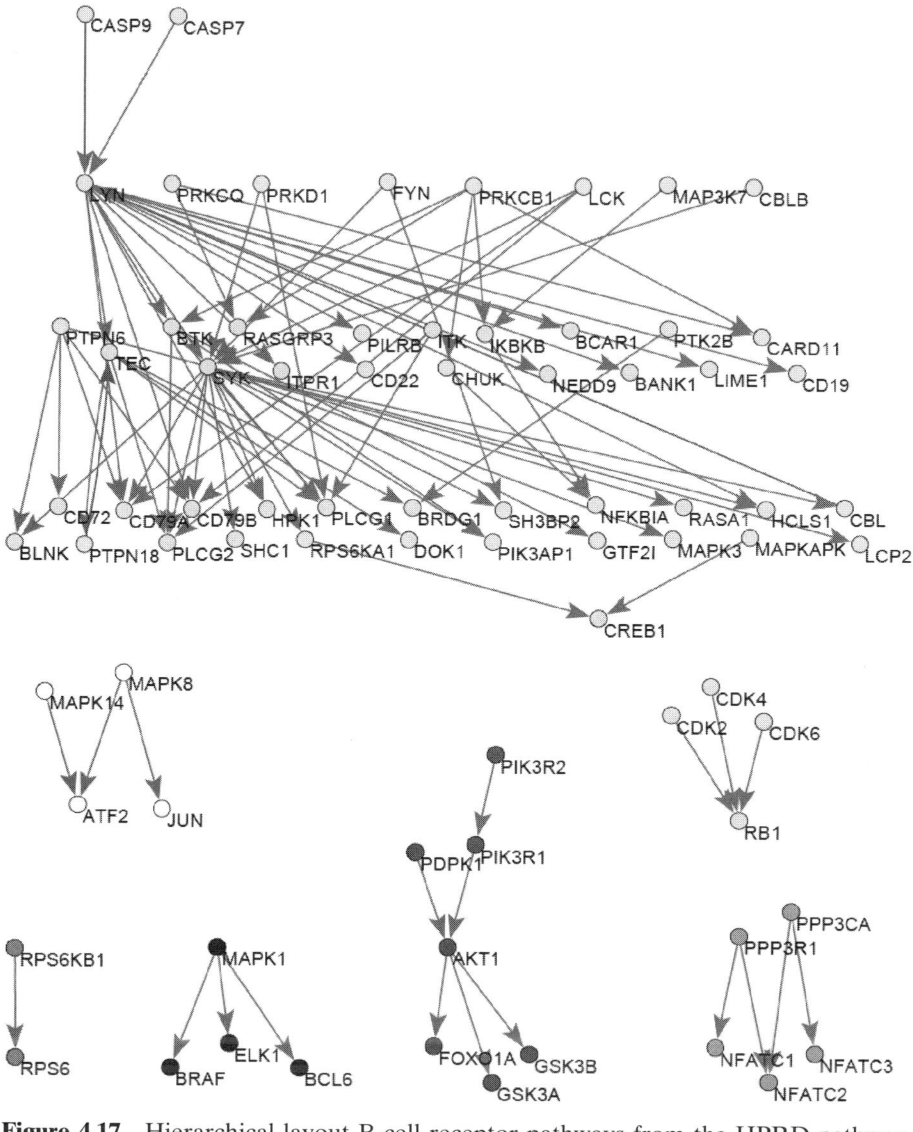

Figure 4.17 Hierarchical layout B-cell receptor pathways from the HPRD pathway database. The counter-hierarchical links are shown in red. See color insert.

- The score of each remaining link is reduced by the number of cycles that pass through this link and were cut in the previous step.
- The procedure of link removal and score reduction is repeated until no cycles remain (which means that scores of all links become zero).

Cycle enumeration can be implemented by following all paths that originate from a given vertex and recording only the cycles that come back to this

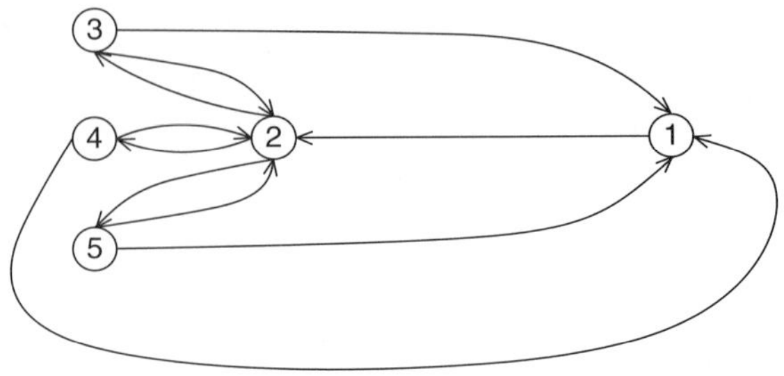

Figure 4.18 An example of a network in which the greedy algorithm fails to determine the optimal solution. The link (1,2) carries the highest score 3 and thus is cut first. However, three 2-node cycles {2,3}, {2,4}, and {2,5} remain to be eliminated, after which the number of removed links becomes 4. The optimal solution would be to cut only three links (2,3), (2,4), and (2,5) each carrying the score 2. This optimal solution has almost always been found by the annealing algorithm.

vertex. The procedure is repeated for each of the N graph vertices: evidently, each cycle of length C is counted C times and a proper normalization is performed. Naturally, the performance of the greedy algorithm is limited in terms of speed and memory requirement of the cycle enumeration step.

An example of network where the greedy algorithm performs flawlessly is shown in Figure 4.15. In this example, the link (3,1) carries a maximum score of 2. A removal of this link indeed makes the graph acyclic, while a removal of any other than the (3,1) link would require the subsequent removal of the second link to achieve the same goal. However, one would suspect that as any "steepest descent" method, the proposed greedy algorithm, performing a sometimes near-sighted, local one-step optimization, may miss the globally optimal solution. This is indeed often the case for bigger and more complex graphs; a fairly simple example of nonoptimal performance of the greedy algorithm is given in Figure 4.18.

4.8 COLLAPSING PROTEIN MAPS

Some recent reviews [48–51] reveal new trends in protein–protein network visualizing. It was proposed that a huge amount of data (thousands of nodes and edges and more) should be processed during pathway analysis. It greatly affects the throughput of many algorithms as they have considerable time complexity. In addition, rather cluttered drawings produced by well-known algorithms hardly allow visual analysis.

The main idea behind the collapsing protein map algorithm is to collapse some parts of the graph into subgraphs based on specific heuristic criteria. It was proposed to collapse subgraphs and process them as single components [48]. The first candidates for such collapsing procedure are cliques—the full subgraphs (i.e. a group of nodes completely connected to each other) and groups of nodes with the same interactions [48,49]. A similar idea of modular decomposition points out sub-complexes and shared components [50].

The idea may be extended and generalized for any algorithm that identifies network clustering. For example, LinLog can be used. A more complex approach may utilize the Markov Cluster Algorithm [13] that allows customization of the "strong connectivity" concept or other concepts to collapse strongly connected components of the graph into subgraphs.

The collapsing graph based on the network topology (e.g. network clusters) is one way to decrease computational cost and improve drawing readability. A more adequate approach should use hierarchical biological knowledge about protein functional classes and pathways to collapse a network map into a smaller set of nodes representing protein classes or sub-pathways. The clustering technique can be also applied here. Nodes in the sub-pathway can be connected by virtual links invisible to the user. It allows an increase in connectivity of the sub-pathways relative to the other graph. Certain clustering algorithms (LinLog, MCL, etc.) keep the vertices of the sub-pathway together. After collapsing, direct force or any other method may be applied to lay out the whole graph and subgraph independently. These approaches highly simplify the drawings of the large graphs, while focusing on their biological structure. At the same time, it should provide a much better performance as weakly connected components are processed independently.

REFERENCES

1. Eades P, Tamassia R. *Algorithms for drawing graphs: an annotated bibliography. Technical Report*, Brown University, Providence, RI, USA, 1988.
2. Garg A, Tamassia R. *Algorithms and Complexity* (Proc. CIAC′ 94), *volume 778 of Lecture Notes in Computer Science*, pp. 12–21. Rome: Springer-Verlag, 1994.
3. Di Battista G, Eades P, Tamassia R, Tollis IG. *Graph Drawing*. Upper Saddle River, NJ: Prentice Hall, 1999.
4. Noack A. Energy-based clustering of graphs with nonuniform degrees. In: *Graph Drawing*, edited by Healy P, Nikolov NS, pp. 309–320. Limerick, Ireland: Springer, 2006.
5. Frick A, Ludwig A, Mehldau H. A fast adaptive layout algorithm for undirected graphs. In: *Proceedings DIMACS Int. Work. Graph Drawing*, 1994 Oct–Dec, Berlin, Germany, edited by Tamassia R, Tollis IG, pp. 388–403. Berlin: Springer-Verlag, 1994.
6. Sim S. *Automatic graph drawing algorithms*. www.citeseer.ist.psu.edu/sim96automatic.html. Assessed December 17, 1996.
7. Davidson R, Harel D. Graphs nicely using simulated annealing. ACM Trans Graph 1996;15(4):301–331.
8. Kamada T, Kawai S. An algorithm for drawing general undirected graphs. Inf Processing Lett 1989;31(1):7–15.

9. Fruchterman TMJ, Reingold EM. Graph drawing by force-directed placement. Software Pract Experience 1991;21(11):1129–1164.
10. Shi W, West DB. Optimal algorithms for finding connected components of an unknown graph. In: *Computing and Combinatorics*, edited by Ding-Zhu Du, Ming Li, pp. 131–140. Xi'an: Springer, 1995.
11. Mehlhorn K, Näher S. *LEDA: A Platform for Combinatorial and Geometric Computing*. Cambridge: Cambridge University Press, 1999.
12. Freivalds K, Dogrusöz U, Kikusts P. Disconnected graph layout and the polyomino packing approach. In: *GD '01: Revised Papers from the 9th International Symposium on Graph Drawing*, pp. 378–391. London, UK: Springer-Verlag, 2002.
13. Enright AJ, Van Dongen S, Ouzounis CA. An efficient algorithm for large-scale detection of protein families. Nucleic Acids Res 2002;7(30):1575–1584.
14. Wills GJ. NicheWorks—interactive visualization of very large graphs. In: *GD '97: Proceedings of the 5th International Symposium on Graph Drawing*, pp. 403–414. London, UK: Springer-Verlag, 1997.
15. Bridgeman SS, Fanto J, Garg A, Tamassia R, Vismara L. Interactive-Giotto: an algorithm for interactive orthogonal graph drawing. Proc Graph Drawing 1997; 1353:303–308.
16. Brandes U, Güdemann M, Wagner D. Fully dynamic orthogonal graph layout for interactive systems. *Konstanzer Schriften in Mathematik und Informatik*, p. 111. Technical Report, Universität Konstanz, 2000.
17. Valiant LG. Universality considerations in VLSI circuits. IEEE Trans Comput 1981;30(2):135–140.
18. Garg A, Tamassia R. On the computational complexity of upward and rectilinear planarity testing. SIAM J Comput 2002;31(2):601–625.
19. Gonzalez TF, Serena D. Complexity of pairwise shortest path routing in the grid. Theory Comput Syst 2004;326(1–3):155–185.
20. Tamassia R. On embedding a graph in the grid with the minimum number of bends. SIAM J Comput 1987;16(3):421–444.
21. Tamassia R, Tollis IG. Planar grid embedding in linear time. IEEE Trans Circuits Syst 1989;36(9):1230–1234.
22. Cormen TH, Leiserson CE, Rivest RL, Stein C. Section 26.2: The Ford–Fulkerson method. In: *Introduction to Algorithms*, edited by Cormen TH, et al., pp. 651–664. Cambridge: MIT Press and McGraw–Hill, 2001.
23. Edmonds J, Karp RM. Theoretical improvements in algorithmic efficiency for network flow problems. J ACM 1972;19(2):248–264.
24. Löbel A. Solving large-scale real-world minimum-cost flow problems by a network simplex method. Technical Report SC 96–7, Konrad–Zuse–Zentrum für Informationstechnik Berlin (ZIB), 1996 Feb.
25. Goldberg A, Rao S. Length functions for flow computations. Technical Report 97–055; NEC Research Institute Inc., 1997.
26. Lempel A, Even S, Cederbaum I. An algorithm for planarity testing of graphs. In: *Proc. of Theory of Graphs*, International Symposium, pp. 215–232, 1966.
27. Even S, Tarjan RE. Computing an st-numbering. Theor Comput Sci 1976; 2(3):339–344.

28. Tarjan RE. Two streamlined depth-first search algorithms. Fundamentae Informatica 1986;9:85–94.
29. Tamassia R. On-line planar graph embedding. J Algorithms 1996;21:201–239.
30. Tamassia R, Tollis IG. A unified approach to visibility representations of planar graphs. Discrete Comput Geometry 1986;1:321–341.
31. Otten RHJM, Van Wijk JG. Graph representation in interactive layout design. In: *Proceedings of the IEEE International Symposium on Circuits and Systems*, pp. 914–918, 1978.
32. Lin C, Lu H, Sun I. Improved compact visibility representation of planar graph via Schnyder's realizer. SIAM J Discrete Math 2004;18(1):19–29.
33. Rosenstiehl P, Tarjan RE. Rectilinear planar layouts and bipolar orientations of planar graphs. Discrete Comput Geometry 1986;1:343–353.
34. Klau GW, Mutzel P. Optimal compaction of orthogonal grid drawings. In: *Proceedings of the 7th International Conference on Integer Programming and Combinatorial Optimization (IPCO–99), volume of 1610 Lecture Notes in Computer Science*, June 1999, edited by Cornuéjols G, Burkard RE, Woeginger GJ, Graz, pp. 304–319. Austria: Springer, 1999.
35. Sugiyama K, Tagawa S, Toda M. Methods for visual understanding of hierarchical systems. IEEE Trans Syst Man Cybern 1981;11(2):109–125.
36. Kasyanov VN. Graph applications in programming. Programming Comput Software 2001;27(3):146–164.
37. Kakoulis KG, Tollis IG. An algorithm for labeling edges of hierarchical drawings. In: *GD '97: Proceedings of the 5th International Symposium on Graph Drawing*, pp. 169–180. London, UK: Springer-Verlag, 1997.
38. Gansner ER, Koutsofios E, North SC, Vo KP. A technique for drawing directed graphs. IEEE Trans Software Eng 1993;19(3):214–230.
39. Kasyanov VN, Lisitsin IA. Hierarchical graph models and visual processing. In: *Proceedings of International Conference on Software: Theory and Practice. 16th IFIP World Computer Congress*, pp. 179–182. Beijing, 2000.
40. Kasyanov VN, Lisitsin IA. Support tools for hierarchical information visualization. In: *Human Computer Interaction: Communication, Cooperation and Application Design*, volume 2, pp. 117–121. London, UK: Lawrence Erlbaum Associates, 1999.
41. North SC, Woodhull G. Online hierarchical graph drawing. In: *GD '01: Revised Papers from the 9th International Symposium on Graph Drawing*, pp. 232–246. London, UK: Springer-Verlag, 2002.
42. Friedrich C, Schreiber F. Flexible layering in hierarchical drawings with nodes of arbitrary size. In: *ACSC '04: Proceedings of the 27th Australasian conference on Computer science*, pp. 369–376. Dunedin, New Zealand: Australian Computer Society, Inc., 2004.
43. Nikitin A, Egorov S, Daraselia N, Mazo I. Pathway studio—the analysis and navigation of molecular networks. Bioinformatics 2003;19(16):2155–2157.
44. Karp RM. Reducibility among combinatorial problems. Complexity of computer computations, Proceeding of a Symposium; March 20–22, edited by Miller RE and Thatcher JW, New York and London: Plenum Press, 1972, pp. 85–103.
45. Even G, Naor J, Schieber B, Sudan M. Approximating minimum feedback sets and multicuts in directed graphs. Algorithmica 1998;20:151–174.

46. Bender MA, Ron D. Testing properties of directed graphs: acyclicity and connectivity. Random Struct Algorithms 2002;20:184–205.
47. Peri S, Navarro JD, Amanchy R, Kristiansen TZ. Development of human protein reference database as an initial platform for approaching systems biology in humans. Genome Res 2003;13(10):2363–2371.
48. Ju BH, Han K. Complexity management in visualizing protein interaction network. Bioinformatics 2003;19(1):I177–I179.
49. Han K, Ju BH, Jung H. WebInterViewer: visualizing and analyzing molecular interaction networks. Nucleic Acids Res 2004;32:W89–W95.
50. Gagneur J, Krause R, Bouwmeester T, Casari G. Modular decomposition of protein–protein interaction networks. Genome Biology 2004;5(8):DOI 10.1186/gb-2004-5-8-r57.
51. Bosman DWJ, Blom E, Ogao PJ, et al. MOVE: a multi-level ontology-based visualization and exploration framework for genomic networks. In Silico Biol 2007;7(1):35–59.

5

PATHWAY ANALYSIS OF HIGH-THROUGHPUT EXPERIMENTAL DATA

ANDREY Y. SIVACHENKO

Table of Contents

5.1 INTRODUCTION

Recent advancements in biological sciences and emergence of new experimental techniques and platforms are commonly recognized as crucial ingredients for making a quantum leap in understanding the cellular processes at molecular level, elucidating disease mechanisms, and discovering new and efficient therapies. Large amounts of information have been accumulated and the rate of data generation continues growing. Even a simple inspection of NCBI database (http://www.ncbi.nlm.nih.gov) statistics shows that the amount of collected data increases exponentially over time, whether it is raw genomic

Pathway Analysis for Drug Discovery, Edited by Anton Yuryev

information (such as total number of DNA bases or total number of sequences recorded) or functional information such as the number of whole-genome microarray experiments. The very number of molecular biology databases themselves continues growing and more than a thousand are now listed [1]. The trend for exponential growth of information is in fact very broad and universal, and with the cost of obtaining data ever dropping there is no reason to believe that the pace is going to slow down anytime soon. It is thus an unexpected and disappointing situation that despite a number of striking successes, the overall rates of introducing new drugs and FDA approvals have dropped over the past decade. The slower trends for utilizing new mechanisms and for developing new drugs against novel targets as compared to re-engineering "old" drugs are especially discouraging [2–4].

In order to fully realize the "promise of genomic revolution," multiple challenges still have to be met by various aspects of the research and drug discovery processes both in academic and industrial settings. These challenges arise at all steps of the process: data acquisition, storage, and mechanistic database-level integration; presentation and visualization of large amounts of heterogeneous data, data analysis, extraction of complex information present in high-throughput (HT) datasets, and hypothesis generation. For instance, with rapid improvement of experimental techniques and emergence of novel ones, our databases have to keep pace with the increasing amount of data. To make the data useful and to provide the framework for data interpretation by the researchers, we must be able to keep these data organized, easily retrievable, and available for nontrivial querying. The data should be efficiently cross-linked and conveniently integrated with existing annotations. Finally, advanced data analysis techniques adequate to the volume, dimensionality, and hidden structure of the data must be applied to extract meaningful statistical and biological information.

It ought to be stressed that, naturally, there is no "consensus" approach to the data analysis and that it is, of course, unlikely in general that a single winning approach will ever exist. There are very different questions to be answered and different trade-offs offered by various methods. Identifying functional roles of genes and proteins in specific phenotypes, finding signature patterns (e.g. in gene expression) that strongly associate with specific phenotypes (diagnosis) and can predict clinical outcome (prognosis), elucidating molecular mechanisms and pathways—these are only a few examples of the broad range of applications. The properties of HT datasets are subject to active research and while a number of plausible and promising models were already developed, much better understanding and more powerful models are required. The complexity and inherent structure of the data are still poorly understood, even overlooked in some cases. Therefore, both further fundamental research and translational studies are of great importance (see for example Clarke et al. [5] for the extensive review from the perspective of translational science of the use of HT data in diagnostic and prognostic applications).

Among various analysis types, pathway analysis is an emerging paradigm in systems biology that aims at better integration and interpretation of HT data and deeper understanding of the underlying biological processes from the causal standpoint. It has been argued [6–10] that pathway analysis is also a critical ingredient for improving efficiency of drug development process, since HT experiments are necessary for drug action validation and understanding of disease mechanisms. These data have to be analyzed to evaluate drug efficacy as well as to identify drug targets, whereas pathway analysis helps to understand the biological function of responsive genes by identifying either known pathways or major regulators and targets involved.

Like other broad classes of approaches, pathway analysis is not defined by a specific collection of methods (although some techniques and models may be more common than others), but rather by its conceptual framework and goal. This goal is to elucidate molecular pathways: to reconstruct the sequence of molecular events, the actual flow of signals (in regulatory networks) and/or intermediates, nutrients or other chemicals (in metabolic networks) that might have resulted in the observed data and, ultimately, are responsible for the specific phenotype or disease. It should be stressed that pathway analysis nevertheless does not replace or supersede other methods. While it is true that we may want to know "the reason behind everything" (which is still a distant goal), applying pathway analysis can be impossible, impractical or simply not helpful for some purposes. For instance, pathway reconstruction may require much more data and processing power than sample classification, while such classification may be all that is needed for diagnostics applications, which should be fast and cost-efficient.

In this chapter, we will specifically review concepts and methods relevant for the pathway analysis of high-throughput datasets. The standard analysis techniques commonly used for the analysis of large-scale datasets on their own are well beyond the scope of this chapter, and so is the discussion of different types of available datasets and their relative advantages and drawbacks. In the following discussion, we will assume, for the sake of certainty, that the dataset is microarray-based expression profiling (differential expression) data. Since this is one of the older, more common, and more established types of HT experiments, the analysis methods discussed here were also initially developed with expression data in mind. In many cases, these methods can be directly used with or easily generalized to different HT datasets, such as protein expression data, genome-wide methylation assays, large-scale metabolic profiling screens, and so on. Hence, we invite the reader to think of references made here to microarray data as a primer intended to illustrate general concepts with concrete examples. We also assume that the reader is well familiar with standard microarray data analysis techniques, such as normalization, calculation of differential expression p-values, hierarchical clustering, etc., and do not review them here.

5.2 PATHWAYS, NETWORKS, AND NETWORK MODULES

The structure and functioning of a cell at the "pathway" level provide relevant and adequate description for understanding biological processes, addressing physiological differences between developmental stages, tissues, environmental responses or normal/disease conditions, and designing novel therapies. As simple and natural as this statement may sound, it should not be taken lightly. Cellular pathway is an intuitive concept that is difficult to define rigorously. Here, we will understand a pathway as a collection of molecular biochemical entities (enzymes, scaffold proteins, DNA and RNA molecules, metabolites, etc.) acting in concert to perform certain cellular function. Examples of such actions include sensing changes in the cell state or in the environment, providing a channel for the transduction of this information toward the cellular effectors or outside of the cell (intracellular signaling, intercellular signaling in multicellular eukaryotes or community sensing in bacteria), helping to synchronize different subunits and processes, providing cellular checkpoints, and executing a sequence of chemical reactions (e.g. metabolic pathway). At the molecular level, performing these tasks requires, and is enabled by, elementary biochemical events such as interactions, chemical modifications, enzyme activation or inhibition. Thus, these events should be considered an integral part of a pathway along with the constituting entities.

A pathway therefore can be formally thought of as a collection of "nodes," representing biochemical entities, connected by edges that represent interactions (e.g. protein–protein and protein–DNA binding), regulation events (e.g. transcriptional control), or modifications (such as phosphorylation of a substrate by a kinase). In the systemic context of a global bimolecular network—a graph consisting of all cellular molecular entities and all possible molecular events linking them—a pathway is a subgraph (subnetwork). However, not any subnetwork can be a pathway. According to our definition, it is the actual flow of information (signal) and/or chemical reaction flow in a specific condition at specific time that elevates part of a network to a pathway. Given this function-centric definition, a meaningful question is whether such functional pathways are "etched" somehow into the very structure of the global cellular network of all molecular events. In other words, one may ask if the global network is comprised of a number of well-defined, almost independent, and isolated modules. Such modules, if existed (and if conceivably small) would fit very well into a simplistic model of a pathway.[1] Notably, it would be possible in this case to predict pathways by simply analyzing the network structure. Alternatively, the global network could be completely "amorphous" and structureless if viewed statically, and existence and functioning of dedicated "pathways" could be a purely

[1] Note that alternative interpretations are possible depending on the kind of the network considered: for instance, a dense cluster, or "module," in a protein–protein binding network may be better interpreted as a putative protein complex.

dynamical effect driven by stoichiometry through fine-tuned co-expression, co-localization and co-activation of only required components of the cellular machinery.

With the emergence of global biomolecular network datasets available both from HT experiments and large-scale mining of peer-reviewed scientific literature, topological structure of biomolecular networks has been a subject of active research. Existing results indicate that neither of the two extreme cases outlined above is realized: biological networks were repeatedly shown to be scale-free and to lack a structure characterized by well-defined, (almost) independent, and isolated modules [11–16]. Instead, hierarchical organization of dense, highly connected structures is observed, with smaller blocks combining into larger super-blocks across all scales [17]. It is indeed possible to identify some modules (or "communities" using language borrowed from the theory of social networks) in biomolecular networks using various approaches [18–22]. These modules are somewhat better identifiable in protein–protein interaction networks (where they usually correspond to protein complexes), while there is currently lesser evidence for reliable extended modules in regulatory (signal transduction) networks.

Although the sensitivity of some of the methods used to identify network modules was questioned recently [23], and their results may be biased by different availability and coverage of known interaction and regulatory networks, it does appear that only weak partial modularity is imposed by structure of the global network alone. It is also interesting in this context that a pathway building approach was suggested that does not use any "clustering" or "modularity" criteria at all but relies on annotations [24]. In this approach, heuristic rules are applied to extract putative signaling pathways from the global network as paths connecting receptors and ligands to downstream effectors (transcription factors). It has been further demonstrated that pathways automatically reconstructed using this procedure exhibit large and significant (but still far from 100%) overlap with known, manually curated ones. To summarize, analysis of the structure of biomolecular networks, although undoubtedly helpful, cannot be used alone to predict pathways. Additional information, in the form of general (function, localization, etc.) and condition-specific (mRNA expression levels) annotations is required.

5.3 ROBUSTNESS AND TRANSCRIPTIONAL PLASTICITY

The observations outlined in the previous section are in line with current understanding of the organization of biological processes. Indeed, the original appeal of intuitive picture of sufficiently isolated and well-defined modules comes from a handful of "classical" textbook pathways. Apparently, they are the best-studied ones and the most common—in a sense that the signaling does follow these specific routes in the course of most fundamental and common cellular responses. However, these pathways are not truly isolated at the level of global network structure—and as the most basic integrating signal conduits onto which multiple external stimuli converge, they should not be. This view is supported by continued discoveries of novel interactions between and modulators of classical pathways; "noncanonical" signaling, including alternative sub-localization and activation as well as dosage compensation between homologues, is being described for many pathways, even such fundamental ones as MAPK [25].

This complexity and dynamical nature of the pathways reflects robustness, a fundamental characteristic of biological systems. Robustness is understood here as a "property that allows a system to maintain its functions against internal and external perturbations" (as defined by Kitano [26], where more references to the relevant work on the subject can also be found). Within this paradigm, it is argued that it is the high-level functions (such as energy and nutrient balance) and homeostasis of a cell or an organism as a whole that biomolecular machinery attempts to maintain. The stability and homeostasis of various subsystems (e.g. individual pathways) are not required and, in fact, are not maintained. The cell is "healthy" (from its own standpoint, even if it is a tumor cell) as long as it can maintain its state and/or proliferate, with its subsystems shifting through different steady states or unstable regimes as needed in response to different environmental cues.

A mechanism ensuring robustness of biological systems can be provided by transcriptional plasticity. This relatively novel notion was put forward as it was observed with the advent of large-scale transcription screening techniques that changes in mRNA expression profiles occur on a global scale rather than in a small group of dedicated genes when cells adjust to specific external or genetic perturbations [27]. It was further demonstrated in yeast model that despite exhibiting very different global gene expression patterns, mutants with different lesions at the same level in the MAPK pathway still possessed very similar invasive growth phenotype [28]. The latter result can be viewed as robust global (phenotypic) attractor state realized through very different molecular (transcriptional) configurations. The transcriptional plasticity itself may be considered a manifestation of redundant paralogs that provide functional backup for one another in case of mutation or external challenge [29]. Indeed, as it was shown by Kafri et al. [29], paralogs with the most efficient backup capability exhibited only partial overlap of their regulatory motifs. These paralogs were expressed dissimilarly in most growth conditions (due to the differences in their regulatory motifs), but were the most susceptible to transcriptional reprogramming and thus readily provided backup functionality. Finally, the question of the extent to which transcriptional response to various stimuli is "hardwired" through evolution was recently addressed by Stern et al. [30]. The authors engineered yeast cells to present them with a severe challenge that they never encountered before, and demonstrated that the cells adapted to the perturbation through global transcriptional reprogramming over a few generations. Interestingly, large fractions of the responding genes were non-reproducible in repeated experiments, thus suggesting individual rearrangement of transcriptional program passing through global shake-up and instability.

It should be also mentioned in this context, that the concept of transcriptional plasticity is actually related to the theory of regulatory divergence. This hypothesis proposes that protein (functional) divergence alone is insufficient to account for extensive differences observed between species, and that many adaptations may have arisen from changes in gene regulation rather than in gene function (see Fay and Wittkopp [31] for an excellent review). In other words, it was suggested that regulation rather than function may be an important target of evolution and that development of new regulation patterns has strong effect even if function of participating proteins is only slightly affected. In this context, transcriptional plasticity is a similar mechanism acting at the organismal level: while the evolution modifies regulation patterns globally to adapt species, these evolved patterns are flexible enough and can be modulated in an organism to adjust to local environmental changes.

5.4 OVERLAP METHODS

The most commonly used method in the pathway analysis of HT datasets is to process the data using any of the standard statistical methods and then to evaluate the obtained results in a biological context. The simplest procedure for such evaluation is to quantify the significance of the overlap between the set of differentially expressed genes (or other features) detected in the data and any number of externally and independently defined annotation groups, or classes. The logic behind this approach is straightforward and intuitive: assume that the total of N genes was measured in the HT experiment. Analysis of the dataset results in a gene list consisting of K genes—for instance, genes differentially expressed according to some criteria (e.g. differential expression p-value cutoff, combined p-value and fold-change cutoff, etc.). Let us now assume that genes are also annotated independently with a discrete set of group labels (or classes) L_i, so that each label is assigned to some number $n(L_i)$ of genes (in other words, these $n(L_i)$ genes belong to a group, or class, L_i). Note that the selection of differentially expressed genes effectively imposes additional, experiment-specific annotation label L_D, such that differentially expressed genes can be thought as "annotated" with this label and the number of genes so annotated is $n(L_D) = K$. A meaningful question is whether there is an association, for a gene, between being differentially expressed and belonging to any particular group L_i, i.e. whether the group L_i is enriched with differentially expressed genes.[2]

Let k_i be the number of genes common between K differentially expressed genes and $n(L_i)$ genes annotated with L_i. If there were no association what-

[2] In statistics, the described classical situation is represented by 2×2 contingency table between categorical indicator variables, L_d and L_i.

soever between L_i and L_D, this situation could be viewed as the two labels L_i and L_D being assigned to the subsets of $n(L_i)$ and $n(L_D)$ genes, respectively, selected completely randomly and independently from the original set of all N genes. Still, some number k_i of genes could be selected twice (i.e. assigned *both* labels) purely by random chance. However, if the observed k_i is sufficiently greater (smaller) than the value expected by chance, then the data suggest positive (negative) association between the labels, i.e. genes from group L_i have a tendency to be (to be not) differentially expressed. The degree of the confidence in favor of such association depends on N, K, $n(L_i)$ and k_i and can be quantified using statistical tools of contingency tables and Fisher (hypergeometric) test (for exact answer) or χ^2-test (for asymptotically accurate approximate answer). Regardless of specific statistical test, computational algorithm, or scoring method used (raw p-value of the overlap significance, z-score, etc.), we refer to the whole class of such approaches and software tools as "overlap" tests, since it is the significance of the overlap k_i between the two groups that is ultimately evaluated.

In practice, annotations used most often as labels L_i are GO ontology terms (especially functional categories and biological processes) and pathway annotations (KEGG, BioCarta, GenMAPP) [32–36]. It is the latter use that makes these general approaches and tools important in the context of the pathway analysis. On the positive side, the overlap methods are intuitive, simple and computationally inexpensive. Since the finite overlaps with known, fixed, and usually heavily curated annotations are sought, the results of such analysis are easily interpretable: an observation of statistically significant number of differentially expressed genes in a given pathway suggests the involvement of the pathway in the process or disease under investigation.

The overlap methods, however, suffer from some inherent weaknesses. The first principal problem lies in the fact that the HT dataset has to be fully analyzed and gene lists (differentially expressed genes, or clusters of co-expressed genes, etc.) must be preselected. Selection of a gene list from the set of all (measured) genes always involves an arbitrary cutoff, with different cutoff values resulting in different lists (for instance, genes with differential expression p-values <0.0001 may be selected into the gene list, or gene co-expression clusters can be defined by drawing a cutoff line at an arbitrarily chosen level in the hierarchical clustering tree, etc.). It has been observed, however (see Pavlidis et al. [37,38]), that when simple overlap tests are applied to such lists, the results (i.e. the significance scores for the over-representation of differentially expressed genes in specific annotation groups/pathways and the very identities of groups/pathways exhibiting such over-representation) are unstable and may depend strongly on the cutoff chosen at the earlier stage. Namely, suppose that genes with some differential expression p-value p_1 are selected into a list D_1. Let us now try to select genes with $p < p_2$ into another list of "differentially expressed" genes D_2 (clearly, if $p_1 < p_2$ then D_1 is simply a subset of D_2). If a specific pathway is significantly enriched with genes from D_2 but not from D_1, it is unclear how this information should be interpreted. The question of which of the cutoffs (and lists) is "better" cannot be answered in a satisfactory way using solely statistical approaches, since there is often no objective criterion for cutoff selection.

The second limitation of overlap tests comes from the fact that they rely on pre-existing annotations. Functional annotations (such as GO) are more universal, stable across multiple conditions and are indeed available for large share of the genes. However, as we have discussed earlier, a pathway is a dynamical notion and only a relatively small number of fundamental, canonical pathways are known to be stable enough to serve as generic annotations. The number of these "textbook," manually curated pathways is in the hundreds (note that even they, despite containing a "consensus" core, may be still "fuzzy," due to pathway cross talk, noncanonical signaling, etc). Comparing the number of available pathway annotations with 20,000+ genes, numbers of splice variants available for many genes, and the numbers of cell types/tissues where the same genes or their splice variants may play different roles, we conclude that most pathways remain unannotated so far, and that even if all pathways were universal and well defined, the sheer number of all pathways for all genes would probably preclude us from creating an extensive curated pathway annotation database, unless some automated methods are used. Moreover, even if we could build such comprehensive database, it would have to keep multiple closely overlapping variants of same pathways. It would be then very difficult to observe a significant overlap with a pathway or to be able to choose between redundant highly overlapping pathway variants, no matter what the experimental data are. Indeed, the larger the number of different pathways, the higher the chance of observing strong overlaps with a few of them purely by a random chance in any HT dataset. To fight increasing false discovery rate, multiple testing corrections have to be introduced, which can adjust properly the significance values and wipe out most of the results as insignificant. Finally, due to transcriptional plasticity and robustness of biological systems, such extensive set of well-defined pathways may not even exist in principle in a stable and universal form suitable for generic annotation. Hence, for the specific purpose of pathways analysis, overlap methods can only provide a fast answer with respect to involvement of a limited number of well-defined fundamental pathways.

5.5 GROUP SCORING METHODS

As we have just discussed, the results of overlap methods depend on the (arbitrary) choices made at preceding analysis steps. Another closely related problem of overlap methods is that additional evidence provided by the biological context is not used to adjust confidence levels for the experimental measurements because the analysis of the HT dataset is completed, and gene list is selected prior to functional analysis. Consider a situation, when some specific pathway P is truly affected in the disease, and most of the genes from that pathway are up-regulated, but the changes in their expression levels are relatively small. In this case, these small changes might not be discernible at the individual gene level due to inherent biological variation between samples and noise present in HT experimental data. As a result, when the dataset is analyzed solely on the level of individual genes by calculating their differential expression p-values, all genes from the pathway P may get poor individual p-values. None of them will end up in the preselected list of differentially expressed genes. Applying overlap analysis to this list will thus never turn P out as an affected pathway because the overlap is zero. Therefore, it would be a desirable and much more sensitive model, in which an event when a handful of *functionally related* genes (e.g. genes from the same pathway) exhibiting *consistent* expression pattern becomes significant, even if the individual changes are small.

This problem is solved by "group scoring" approach. Within group scoring methods, the data can be preprocessed (for instance, differential expression p-values can be computed), but, importantly, arbitrary cutoffs are not applied, no "final call" is made and no "list of interesting genes" is preselected. Instead, the experimental data available for *all* the genes are considered. Regardless of the specific type of data used in any particular method of this broad class (differential expression log-ratios, p-values, signal-to-noise ratios), all these data are compared across different pathways. The background (expected) distribution can be estimated, either analytically or by resampling, and consequently the method can determine whether the distribution of the data actually measured for the pathway P is significant. It is not absolutely necessary to compare raw distributions as well, as one may choose a suitable statistics and use it as a score function. For instance, average log-ratio for all genes constituting a pathway can be used and compared to the expected value of average log-ratio of a randomly selected set of genes. Importantly, this procedure is objective as it does not depend on any arbitrary cutoffs chosen in advance. One can say that it is expression of pathways (or other annotation classes) that is evaluated by group scoring methods rather than the expression of individual genes.

One early approach for group scoring was outlined [39], where distributions of gene expression log-ratios observed within groups of functionally related genes were compared to the total distributions of log-ratios observed on the whole arrays. A few different scoring functions were later explored [37,38] for the purpose of ranking and comparison of annotation classes. Gene Set Enrichment Analysis (GSEA) method [40,41] belongs to the same class of models and uses modified Kolmogorov–Smirnov test to evaluate the statistical significance of the differences between the distributions of expression values within predefined groups of genes (termed "gene sets") and the distribution of all the expression values measured on the microarray. GSEA uses its own custom-built collection of gene sets (but can work with any annotation classes in principle, such as GO), and its implementation can estimate the false discovery rate. Most importantly, group scoring methods can indeed detect subtle but consistent changes in gene expression at pathway level as it was the case, e.g. for the group of oxidative phosphorylation-related genes discovered by GSEA in the studies of human diabetes [40]. It should be noted, however, that due to the choice of Kolmogorov–Smirnov test as the statistical vehicle for calculation of the significance of difference between two distributions, GSEA can evaluate only relatively large groups of genes. It was also reported that original GSEA underperforms in some situations and multiple interesting extensions and modifications are being developed (e.g. [42–44]).

5.6 PATHWAY ANALYSIS IN BIOLOGICAL NETWORK CONTEXT

Group scoring methods discussed in the previous section are powerful, sensitive, and free from inconsistencies that arise when arbitrary cutoffs are used in HT data analysis. However, they still share a weakness with overlap methods: a requirement for external annotations, a set of predefined, curated pathways to be evaluated for signs of significant expression. Hence, group scoring methods are an excellent choice for assessing significance of relatively small number ("small" here stands for hundreds, even thousands) of well-defined canonical pathways (or rather their consensus "cores"), handpicked reference set of disease-specific pathways and so on.

Furthermore, pathway annotations used in overlap or group scoring methods are simply class labels that keep no information about causality and hierarchy of events, the interplay between different components of the pathway, order of their activation, etc. Results obtained with these methods do not provide any direct insight into the mechanism, or *how* the pathway works. Instead they just reveal the *identity* of the potentially affected pathway while the details of the functioning of the annotated pathways are assumed to be known *a priori*. To interpret the data at pathway level in the most general case, the physical structure that enables pathways—the biomolecular network—should be taken into account.

Since the structure of the network on its own does not unambiguously define pathways, integration of static networks of interaction, promoter-binding, protein modification, and other elementary molecular events with dynamic HT experimental data that provide a snapshot of the system under specific conditions is required for pathway reconstruction. While unbiased and accurate *de novo* inference of activated regulatory and biochemical cascades dynamically emerging in response to internal and external stimuli is still a distant goal, a number of feasible and meaningful approaches following this general direction were demonstrated. For instance, Gunsalus et al. [45] used the integration of co-expression data, phenotypic profiles, and interaction network as a basis for prediction of putative "molecular machines" involved in *C. elegans* early embryogenesis.

A group of methods relevant for pathways analysis can be characterized by using network knowledge *a priori* during the initial stages of HT data analysis. Thus, it was proposed to use in gene expression clustering a modified intergene distance measure depending on both expression profile correlation and the distance between the genes in the interaction network [46]. With such combined measure, a pair of genes is considered closer to each other at the same level of correlation between their expression profiles if these genes also physically interact. As a result, clusters of co-expressed genes obtained using such metrics tend to contain genes that are also neighbors in the network. This procedure was demonstrated to result in more compact clusters, which better matched specific pathways known to be affected in the experiment. In a related approach [47], a Bayesian likelihood model was developed for finding optimal clusters based on co-expression and interaction information. The conventional clustering based on correlations between gene expression profiles was used in this model to seed the partition into clusters, and then iterative expectation-maximization procedure was employed to optimize the clusters in the context of interaction network. Yet another technique was suggested [48] where multi-body correlations were computed using superparamagnetic clustering algorithm [49]; genes were sought that comprised a module in protein–protein interaction network and at the same time exhibited high correlation strength in gene co-expression network. Spectral decomposition of gene expression profiles with respect to eigenfunctions of the graph representing biomolecular network [50] is also a promising approach with potential applications in pathway analysis, as high-frequency components of expression profiles with respect to network topology (presumably noise) are suppressed, while retained components are more consistent with the underlying network structure and thus should better represent patterns of activation and suppression of individual genes at pathway level.

In contrast to approaches where clustering or other processing of gene expression profiles is augmented by *a priori* additional information provided by biological network, a class of more direct methods try to perform an opposite task: identify a compact and consistent connected module in the network using the expression information (or, in general, any other HT data). The concept of these methods, in principle, strictly follows the pathway analysis paradigm: while the network itself does not define pathways, it enables them by providing the "wiring" along which the signal can flow. Overlaying large-scale, condition-specific snapshot (such as expression dataset) onto the network should in principle make it possible to see how the pathways emerge in that specific state of the system. A classical model of this class was introduced in [51] (see also Ideker [52]), where a cumulative scoring function was suggested to quantify the significance of expression changes observed across a set of genes, thus making the model related in this respect to group scoring analysis methods discussed in the previous section. The principal difference from group scoring approach, however, was that an annealing algorithm was used to iteratively find subnetworks (termed "active subnetworks" [51]) in the protein interaction network that optimize the scoring function. Thus, in contrast to overlap or group scoring analysis methods, there are no fixed, predefined groups of genes in this approach, but rather the "most relevant" groups are sought and updated dynamically (defined as a set of proteins in the active subnetwork at the current step of the iterative optimization procedure). Importantly, the network imposes strong constraint in this approach, as only connected active subnetworks are sought, which in principle is more relevant for pathway prediction.

It is important that while taking into account *connections* present in the biomolecular network, many existing methods do not take into account the *connectivity*, namely, the number of connections made in the network by a given gene or protein. For instance, within the framework used by active subnetwork algorithm [51], the hubs (proteins with very large number of connections, which are always present in scale-free biomolecular networks) have larger probability to have a few of low p-value, highly expressed, or otherwise "interesting" genes among their neighbors by random chance. Hence, two otherwise independent subnetworks enriched with these "interesting genes" may accidentally touch the same hub. A naive algorithm that tries to find such "active subnetworks" through iterative optimization procedure can be prone to combining these two subnetworks connected through a hub into one. The Significant Area search algorithm [53] was developed to avoid this problem—it implements a similar logic of iterative search for a subnetwork that optimizes a scoring function (statistically sound score function based on Fisher's inverse χ^2 test is used), but also employs a greedy selection procedure based on local network topology to penalize connections through unspecific hubs. The problem of hubs was also recognized [54], where the scoring function of active subnetwork algorithm was modified to explicitly downplay connections that could be made by a random chance through hubs.

Network connections and connectivity are also explicitly taken into account in Network Enrichment Analysis (NEA) algorithm [55]. Given a regulatory network, this method first defines sets of targets of every regulator in the network as "groups," and then evaluates the significance of the difference between distributions of gene expression log-ratios within each such group of targets and the effective distribution constructed for the whole microarray. The latter step is similar to other group scoring approaches such as GSEA, but the effective baseline distribution of log-ratios on the microarray is built in such a way that it downplays the significance of hubs. Note that the regulators themselves can be activated non-transcriptionally and don't have to be differentially expressed to be reliably identified by this method. The "significant regulators" discovered with this algorithm, i.e. the regulators, for which their targets, as a group, exhibit overall significant deviation of expression levels from the background, can provide a first step in a bottom-up reconstruction of extended pathways consisting of a few levels of signal transduction. This method can be also viewed as "local" network analysis as each regulator's immediate downstream neighborhood is analyzed individually, in contrast to more “global” approaches such as active subnetwork and related algorithms. Indeed, the optimal subnetworks determined self-consistently by the latter methods tend to be quite large in practice (hundreds of nodes) and may require further manual downsizing and cutting into smaller, more manageable, and more specific pathways.

Another promising global approach is provided by the Markov Random Field-based model [56]. Namely, it assumes that hidden true expression changes associated with genes in the network are observed, with some probability of an error, in the HT experiment. On the other hand, the true hidden values should be as consistent as possible with physical edges (interactions, regulation events, etc.) in the network. Finding the optimal assignment of true expression changes that provides simultaneously an explanation for the experimental observations and fits the network-imposed consistency constrains also represents a step toward mechanistic, pathway-level interpretation of the data.

5.7 CONCLUSIONS

Pathway analysis approach presented in this chapter is a broad framework that is still being actively developed. Overlap and group scoring methods are conceptually relatively simple, well understood, and already used commonly and successfully. Improvements and extensions in the existing collections of curated pathways should make these methods, especially more robust group scoring approaches, even more useful. We have also argued that “pathway” is a dynamic state of a system rather than a static annotation that is invariably true, exact, and always applicable. It should be thus stressed that overlap and group scoring methods are general statistical techniques for evaluating significance levels associated with some arbitrary class labels and it is only the interpretation and the conceptual framework that make these approaches so useful in pathway analysis. Pathway analysis is not reducible to these specific techniques, and neither are applications of these techniques limited to pathway analysis. Furthermore, while the methods outlined in this chapter were discussed in the context of analysis of microarray data, other HT datasets can be placed into a pathway context and analyzed similarly. For instance, a GSEA-like approach was recently suggested in the analysis of genome-wide association studies [57]. In this work, the authors argued that while traditional approaches seek significant associations of each individual SNP with the disease (very much like “traditional” microarray analysis seeks individual significantly differentially expressed genes), additional insights can be offered by identifying sets of SNPs occurring in the same pathway, so that even if they do not exhibit strong enough association individually, there is consistent association at the pathway level.

We have also considered pathway reconstruction methods that do not rely on the existing pathway annotations but directly use the underlying biomolecular network. These methods are in general more complex and less validated. However, they are also the most powerful and unbiased, and the methods of this class should be capable, in principle, to directly elucidate pathways as causal sequences of molecular events in every specific situation. Similar to group scoring methods, different datasets can be analyzed and various questions answered by pathway reconstruction algorithms using different regulatory and interaction networks. For example, in a recent study, an updated modern version of Active Subnetwork algorithm was used to identify markers correlated with breast cancer metastasis as differentially expressed (i.e. affected) pathways rather than individual differentially expressed genes [58]. A modified subnetwork scoring function that takes into account metastatic/non-metastatic sample class annotation label was designed, and it was found that subnetwork markers associated with metastasis are more reproducible than individual marker genes.

It is important that with the presently available amount of HT datasets and extensive biomolecular networks also available from HT experiments and from mining of peer-reviewed biomedical literature, the pathway reconstruction methods can now be broadly applied and should become a part of everyday arsenal of methods available for a researcher. While there is no "perfect" or "established" pathway reconstruction method yet, the algorithms for network-based pathway analysis are being actively developed and tested. Given the availability of the data and computational power, several different pathway reconstruction methods can be easily applied for every specific problem or project, and meta-analysis of the results should provide additional insights into the underlying mechanisms.

REFERENCES

1. Galperin MY. The Molecular Biology Database Collection: 2008 update. Nucleic Acids Res 2008;36(Database issue):D2–4.
2. Walsh G. Biopharmaceutical benchmarks 2006. Nat Biotechnol 2006;24(7): 769–776.
3. Food and Drug Administration. *Innovation or Stagnation? Challenge and Opportunity on the Critical Path to New Medical Products*. US Department of Health and Human Services, March 2004.
4. Reichert JM. Trends in US approvals: new biopharmaceuticals and vaccines. Trends Biotechnol 2006;24(7):293–298.
5. Clarke R, et al. The properties of high-dimensional data spaces: implications for exploring gene and protein expression data. Nat Rev Cancer 2008;8(1): 37–49.
6. Huang S. Rational drug discovery: what can we learn from regulatory networks? Drug Discov Today 2002;7(Suppl 20):S163–169.
7. Davidov E, et al. *Advancing drug discovery through systems biology*. Drug Discov Today 2003;8(4):175–183.

8. Apic G, et al. Illuminating drug discovery with biological pathways. FEBS Lett 2005;579(8):1872–1877.
9. Sivachenko AY, Yuryev A. Pathway analysis software as a tool for drug target selection, prioritization and validation of drug mechanism. Expert Opin Ther Targets 2007;11(3):411–421.
10. Araujo RP, Liotta LA, Petricoin EF. Proteins, drug targets and the mechanisms they control: the simple truth about complex networks. Nat Rev Drug Discov 2007;6(11):871–880.
11. Barabasi AL, Albert R. Emergence of scaling in random networks. Science 1999;286(5439):509–512.
12. Jeong H, et al. The large-scale organization of metabolic networks. Nature 2000;407(6804):651–654.
13. Wagner A, Fell DA. The small world inside large metabolic networks. Proc Biol Sci 2001;268(1478):1803–1810.
14. Goh KI, et al. Classification of scale-free networks. Proc Natl Acad Sci U S A 2002;99(20):12583–12588.
15. Rzhetsky A, Gomez SM. Birth of scale-free molecular networks and the number of distinct DNA and protein domains per genome. Bioinformatics 2001;17(10): 988–996.
16. Albert R, Barabási AL. Statistical mechanics of complex networks. Rev Mod Phys 2002;74(1):47.
17. Ravasz E, et al. Hierarchical organization of modularity in metabolic networks. Science 2002;297(5586):1551–1555.
18. Girvan M, Newman ME. Community structure in social and biological networks. Proc Natl Acad Sci U S A 2002;99(12):7821–7826.
19. Spirin V, Mirny LA. Protein complexes and functional modules in molecular networks. Proc Natl Acad Sci U S A 2003;100(21):12123–12128.
20. Newman ME, Girvan M. Finding and evaluating community structure in networks. Phys Rev E Stat Nonlin Soft Matter Phys 2004;69(2 Pt 2):026113.
21. Brohee S, van Helden J. Evaluation of clustering algorithms for protein-protein interaction networks. BMC Bioinformatics 2006;7:488.
22. Bader GD, Hogue CW. An automated method for finding molecular complexes in large protein interaction networks. BMC Bioinformatics 2003;4:2.
23. Fortunato S, Barthelemy M. Resolution limit in community detection. Proc Natl Acad Sci U S A 2007;104(1):36–41.
24. Yuryev A, et al. Automatic pathway building in biological association networks. BMC Bioinformatics 2006;7:171.
25. Pimienta G, Pascual J. Canonical and alternative MAPK signaling. Cell Cycle 2007;6(21):2628–2632.
26. Kitano H. Towards a theory of biological robustness. Mol Syst Biol 2007;3: 137.
27. Causton HC, et al. Remodeling of yeast genome expression in response to environmental changes. Mol Biol Cell 2001;12(2):323–337.
28. Breitkreutz A, et al. Phenotypic and transcriptional plasticity directed by a yeast mitogen-activated protein kinase network. Genetics 2003;165(3):997–1015.

29. Kafri R, Bar-Even A, Pilpel Y. Transcription control reprogramming in genetic backup circuits. Nat Genet 2005;37(3):295–299.
30. Stern S, et al. Genome-wide transcriptional plasticity underlies cellular adaptation to novel challenge. Mol Syst Biol 2007;3:106.
31. Fay JC, Wittkopp PJ. Evaluating the role of natural selection in the evolution of gene regulation. Heredity 2008;100(2):191–199.
32. Dahlquist KD, et al. GenMAPP, a new tool for viewing and analyzing microarray data on biological pathways. Nat Genet 2002;31(1):19–20.
33. Zeeberg BR, et al. GoMiner: a resource for biological interpretation of genomic and proteomic data. Genome Biol 2003;4(4):R28.
34. Doniger SW, et al. MAPPFinder: using Gene Ontology and GenMAPP to create a global gene-expression profile from microarray data. Genome Biol 2003; 4(1):R7.
35. Zhong S, et al. GoSurfer: a graphical interactive tool for comparative analysis of large gene sets in Gene Ontology space. Appl Bioinform 2004;3(4):261–264.
36. Pandey R, Guru RK, Mount DW. Pathway Miner: extracting gene association networks from molecular pathways for predicting the biological significance of gene expression microarray data. Bioinformatics 2004;20(13):2156–2158.
37. Pavlidis P, Lewis DP, Noble WS. Exploring gene expression data with class scores. Pac Symp Biocomput 2002:474–485.
38. Pavlidis P, et al. Using the gene ontology for microarray data mining: a comparison of methods and application to age effects in human prefrontal cortex. Neurochem Res 2004;29(6):1213–1222.
39. Mirnics K, et al. Molecular characterization of schizophrenia viewed by microarray analysis of gene expression in prefrontal cortex. Neuron 2000;28(1):53–67.
40. Mootha VK, et al. PGC-1alpha-responsive genes involved in oxidative phosphorylation are coordinately downregulated in human diabetes. Nat Genet 2003; 34(3):267–273.
41. Subramanian A, et al. Gene set enrichment analysis: a knowledge-based approach for interpreting genome-wide expression profiles. Proc Natl Acad Sci U S A 2005;102(43):15545–15550.
42. Kim SY, Volsky DJ. PAGE: parametric analysis of gene set enrichment. BMC Bioinformatics 2005;6:144.
43. Dinu I, et al. Improving gene set analysis of microarray data by SAM-GS. BMC Bioinformatics 2007;8:242.
44. Jiang Z, Gentleman R. Extensions to gene set enrichment. Bioinformatics 2007;23(3):306–313.
45. Gunsalus KC, et al. Predictive models of molecular machines involved in *Caenorhabditis elegans* early embryogenesis. Nature 2005;436(7052):861–865.
46. Hanisch D, et al. Co-clustering of biological networks and gene expression data. Bioinformatics 2002;18(Suppl 1):S145–154.
47. Segal E, Wang H, Koller D. Discovering molecular pathways from protein interaction and gene expression data. Bioinformatics 2003;19(Suppl 1):i264–271.
48. Tornow S, Mewes HW. Functional modules by relating protein interaction networks and gene expression. Nucleic Acids Res 2003;31(21):6283–6289.

49. Blatt M, Wiseman S, Domany E. Superparamagnetic clustering of data. Phys Rev Lett 1996;76(18):3251–3254.
50. Rapaport F, et al. Classification of microarray data using gene networks. BMC Bioinformatics 2007;8:35.
51. Ideker T, et al. Discovering regulatory and signalling circuits in molecular interaction networks. Bioinformatics 2002;18(Suppl 1):S233–240.
52. Ideker T. A systems approach to discovering signaling and regulatory pathways—or, how to digest large interaction networks into relevant pieces. Adv Exp Med Biol 2004;547:21–30.
53. Sohler F, Hanisch D, Zimmer R. New methods for joint analysis of biological networks and expression data. Bioinformatics 2004;20(10):1517–1521.
54. Rajagopalan D, Agarwal P. Inferring pathways from gene lists using a literature-derived network of biological relationships. Bioinformatics 2005;21(6):788–793.
55. Sivachenko A, et al. Identifying local gene expression patterns in biomolecular networks. In: *Computational Systems Bioinformatics (CSB)*, Proceedings, pp. 180–184. Palo Alto, CA: Stanford University, 2005.
56. Wei Z, Li H. A Markov random field model for network-based analysis of genomic data. Bioinformatics 2007;23(12):1537–1544.
57. Wang K, Li M, Bucan M. Pathway-based approaches for analysis of genomewide association studies. Am J Hum Genet 2007;81(6):1278–1283.
58. Chuang HY, et al. Network-based classification of breast cancer metastasis. Mol Syst Biol 2007;3:140.

6

INTEGRATIVE PATHWAY ANALYSIS OF DISEASE MOLECULAR DATA

ANDREJ BUGRIM, ZOLTAN DEZSO, YURI NIKOLSKY, AND TATIANA NIKOLSKAYA

Table of Contents

6.1 HIGH-THROUGHPUT DATA IN DISEASE BIOLOGY

Since the introduction of microarray technology in the nineties, high-throughput molecular data and its analysis have become an integral part of disease biology research. Even though the field is almost 20 years old, several new developments have occurred in the last 2 or 3 years which promise to revolutionize its applications. First, many technical quirks in microarrays and other gene expression assays had been worked out, the data becoming more reliable and reproducible. At the same time, the cost of running microarray assays has been dramatically reduced. These two factors alone made

Pathway Analysis for Drug Discovery, Edited by Anton Yuryev

microarray-based diagnostics an economically and clinically viable proposition. Meanwhile, other types of "OMICs" data have become more widely available in both academic and industrial research laboratories. The recently embraced technologies include proteomics, metabolomics, genotyping, and other platforms for generating high-throughput molecular data. As a result, today as never before, it is within technological and financial reach of many companies and academic centers to concurrently interrogate mechanisms of complex diseases on multiple molecular levels: from DNA, to gene expression, to protein behavior and metabolite imbalances. Tellingly, there is also an upsurge of offerings in the area of "consumer genomics"—generation and analysis of genomic data for individuals. In recent months at least three vendors started to offer such services: Decode, 23andMe, and Navigenics.

As a result of these developments, the problem of functional analysis of "OMICs" data has taken a center stage and become one of the main subjects of "systems biology," a novel integrative discipline aimed at understanding functioning of a biological system as a whole. Research in this area often combines mining of knowledge bases, statistical analysis, graph theory, and mathematical modeling. The field of systems biology progresses rapidly, with the number of publications growing exponentially over the last 2 years.

In this chapter, we describe how pathway analysis and pathway analysis tools are applied to different types of disease-related high-throughput molecular dataset in order to gain better understanding of functional processes underlying these diseases. Our focus will be on how different types of data are analyzed in the context of pathways and how they can be combined in the analysis. Our first study explores the functional landscape of the so-called "cancer stem cells" based on comprehensive set of Serial Analysis of Gene Expression (SAGE) data from different types of cancer cells. The second study describes how concurrent analysis of metabolomic and microarray gene expression data helps to understand functional pathways involved in atherosclerosis. Finally in our third example, we explore how cancer genotyping data could be investigated in the context of pathways and introduce a notion of a "mutated pathway."

6.2 FUNCTIONAL LANDSCAPE OF CANCER "STEM CELLS" ELUCIDATED FROM SAGE DATA

Cancer is the most complex and the most comprehensively studied disease area in molecular medicine. Thousands of "small experiment" articles have been published, focusing on genetics, epidemiology, biochemistry, and the molecular biology of multiple types of cancers. Since the dawn of "high-throughput" biology a decade ago, cancer biology has become the favorite field for "genome-wide" experimentation, such as microarray and SAGE gene expression, proteomics, gene copy number, and methylation assays, and, lately, multi-parallel sequencing. Hundreds of oncogenomics studies were accumulated in the public domain and at drug companies, with hundreds of thousands of individual datasets and billions of data points potentially available for analysis.

In our recent work [1], we have investigated how pathway analysis can be applied to gene expression data in order to understand important functional characteristics of the so-called "cancer stem cells." The concept of cancer stem cells is based on the observation that regular stem cells share many characteristics with cancer cells, such as self-renewing capacity, ability to migrate and invade into surrounding tissue, and giving rise to heterogeneous progeny. Correlating with this, several pathways and genes required for normal stem cell function have demonstrated to be activated in cancer cells and to play an essential role in tumorigenesis. Cancer stem cells are defined as a subset of tumor cells with stem cell-like properties that are thought to be responsible for the growth, progression, and recurrence of the tumor [2–4]. This hypothesis was proposed many years ago and revived in recent years with new experimental approaches and purification protocols [2–4]. Putative cancer stem cells have been purified from various human tumor types generally using cell surface markers specific for the normal stem cells of the same organ. The tumorigenicity and the "stemness" of the isolated cells are usually demonstrated by performing *in vitro* clonogenicity and *in vivo* tumorigenicity studies. In breast cancer, Al-Hajj et al. demonstrated that lin-/CD44+/CD24–/low (subsequently referred as CD44+) cell population isolated from malignant pleural effusion samples of breast cancer patients was more tumorigenic than CD44–/CD24+ (subsequently referred as CD24+) cells and the xenografts reproduced the heterogeneity observed in the original tumors, leading to the hypothesis that CD44+ cells are breast cancer stem cells. To determine the identity of these cells at the molecular level, we have purified these putative stem cells as well as CD24+ (differentiated) cells from breast carcinomas and determined their global gene expression and genetic profiles. Cells with the same properties were also isolated from normal human mammary epithelium and characterized using the same methods for comparison. Gene signatures specific for stem or differentiated cells were correlated with clinical outcome and activity of signaling pathways.

To determine the comprehensive gene expression profiles of the purified cells, we generated SAGE libraries from CD24+ and CD44+ cells purified from normal mammary epithelium and pleural effusion and ascites samples collected from breast cancer patients. The SAGE data further strengthened the hypothesis that CD24+ and CD44+ cells represent differentiated luminal epithelial and stem cells, respectively, since known markers of these cells were mutually exclusively found in the respective SAGE libraries.

To identify signaling pathways that are specifically activated in putative breast stem cells, we utilized the MetaCore™ data mining technology [5,6]. First, we selected statistically significantly differentially expressed genes demonstrating at least twofold difference in abundance between CD44+ and CD24+ cells from the same sample and having at least five SAGE tag counts/200,000. We mapped these genes onto the folders for Gene Ontology (GO) functional processes and canonical pathways and ranked these according to their statistical significance for fitting the data. The three differentially expressed gene sets were remarkably consistent in terms of the best-fitting processes and pathways. Two out of 10 top-ranked cell processes were shared among all three sets (cell motility and cell adhesion), and three additional processes were common between one tumor and the normal set (protein biosynthesis, protein folding, and cell proliferation). At the same time, cell motility and cell adhesion were among the 15 processes demonstrating the highest dispersion between the datasets, i.e. processes with the largest differences in genes differentially expressed between differentiated and stem cells in normal and tumor samples. This means that within the same cellular process, different pathways are implicated in cancer and normal stem cells. Similar findings were observed for individual pathways and genes within them. Five out of 12 top-ranked pathway maps were common in all three sets of differentially expressed genes: TGF-β and WNT signaling, cytoskeleton remodeling, integrin-mediated processes, reverse signaling by ephrin B, and chemokines and cell adhesion. One of the most complete maps the "TGF-β and WNT signaling and cytoskeleton remodeling" and a more detailed view of the TGF-β pathway are depicted in Figure 6.1A,B. Several extracellular matrix (ECM) proteins, including collagen IV, PAI1, and plasminogen, were overexpressed both in cancer and normal stem cells, while others such as TGFB1, vitronectin, and fibronectin were overexpressed in cancer but not in normal stem cells. Both cancer and normal stem cells overexpressed various integrins. ITGA3 and ITGA6 were specifically up-regulated in cancer but not in normal stem cells while ITGAV was down-regulated in normal but not in cancer stem cells. Furthermore, PLAT is overexpressed in both cancer stem cells but is down-

Figure 6.1 Maps and networks of signaling pathways specifically or commonly up- or down-regulated in normal and cancer stem cells. (A) Mapping expression data of cancer and normal mammary epithelial stem cells on pathway map of "TGF-β and WNT signaling and cytoskeletal remodeling." Rectangles indicate up-regulation of the indicated genes in breast cancer (red rectangle) or normal (blue rectangle) CD44+ cells or in both (yellow rectangle) compared to their corresponding CD24+ counterparts. (B) More detailed view of the TGF-β signaling pathway with up- or down-regulation of the genes indicated as described above. (C) Direct interactions network centered around TGF-β for the genes differentially expressed between breast cancer CD44+ and CD24+ cells. Genes downstream and upstream of TGF-β are marked with cyan and yellow, respectively. (D–F) The "direct interactions" network for the genes differentially expressed in the normal stem cells dataset but not in either of cancer stem cells datasets. The input list for the network was calculated as the NCD44/NCD24 gene lists with subtracted AscCD44/LPECD24 and LPECD44/LPECD24 gene lists. Red circles mark the genes that are up-regulated in NCD44/NCD24. Blue circles mark the genes that are down-regulated in NCD44/NCD24. Normalized SAGE tag count >20; Fold change >2 for all three figures.

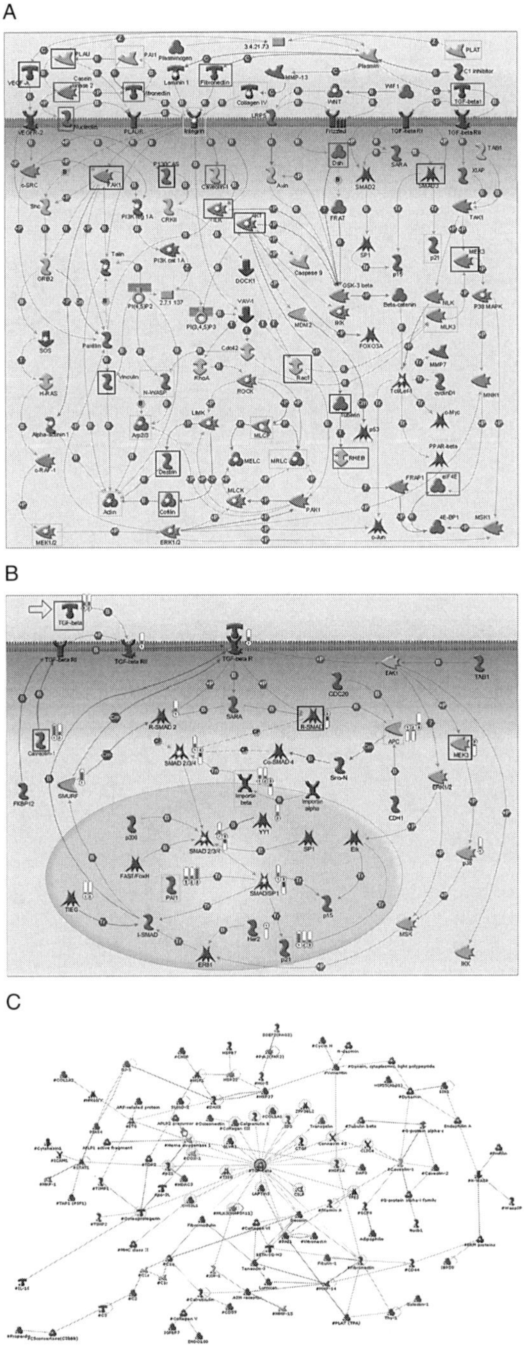
A
B
C

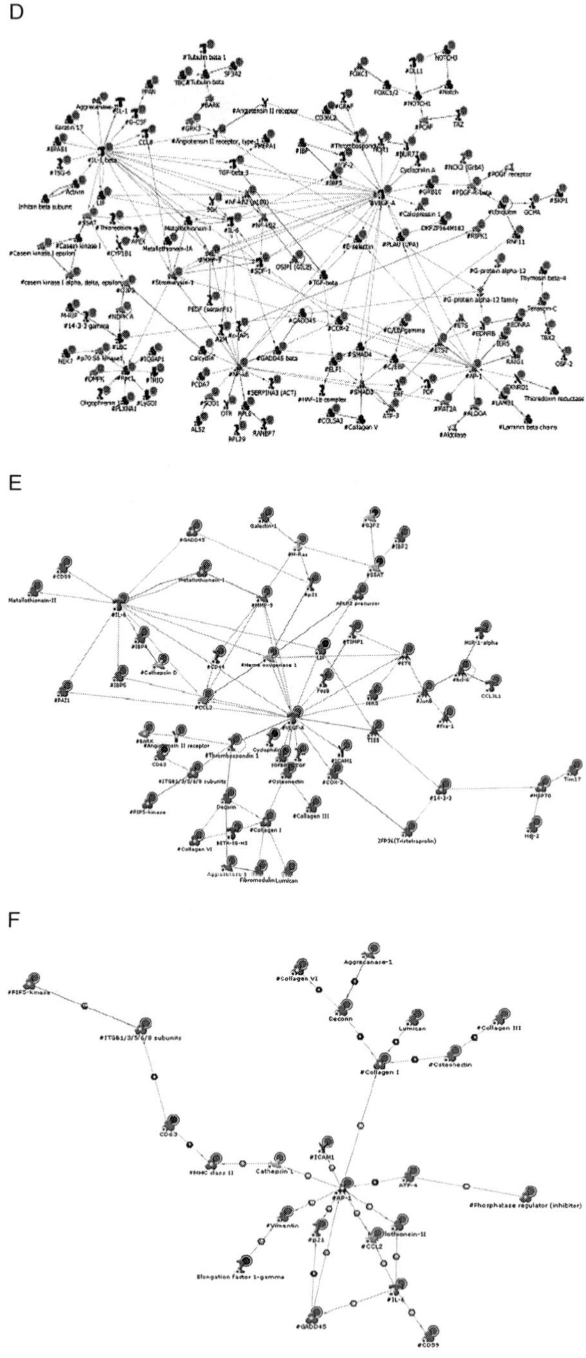

Figure 6.1 (*Continued*)

regulated in normal stem cells, while PLAU is overexpressed in normal but not in cancer stem cells.

Next we used the MetaCore data mining technology to perform a more detailed analysis of which pathways are differentially regulated in breast stem and differentiated cells, by building the direct interactions (DI) networks around sets of differentially expressed genes identified as described above. First, we generated the DI network for the set of genes differentially expressed between CD24+ and CD44+ cells in both cancers (Figure 6.1C). This network is centered around TGF-β1 (up-regulated in stem cells) as the central hub with 22 interactions. Other up-regulated hubs in this network include heme oxygenase and fibronectin, 7 decorin, and caveolin 1. Next, we identified the modules in this network that are differentially expressed in cancer stem cells only or in both cancer and normal stem cells. The modules around TGF-β1, dynamin, fibronectin, and caveolin, and casein kinase II are up-regulated specifically in cancer stem cells, while modules around collagen 1 and transcription factors HIF1A and ETS are up-regulated in cancer and normal stem cells. We also built DI networks with different abundance ratios for the genes differentially expressed between normal CD44+ and CD24+ cells but not in cancer cells (Figure 6.1D–F). The network topology for normal stem cells is different from that of cancer stem cells, with up-regulated VEGF-A, IL-1, NF-kβ, and AP-1 and down-regulated Rac1 and SMAD3 as the major hubs. This network also features activation of the Notch pathway and TGF-β3.

Due to the known importance of TGF-β signaling in regulating the pluripotency of human embryonic stem cells [7], as well as its role in tumorigenesis and metastasis [8], we investigated the role of this pathway in breast stem cells in further detail. We analyzed the expression of selected genes involved in TGF-β signaling by semiquantitative RT-PCR in the cell fractions. Our results suggest that the cellular response to TGF-β is determined by the cellular context and cells with different phenotype even within the same tumor and tissue type respond differently. Intriguingly we found that, at least in the tumors analyzed, the specific activation of TGF-β signaling in putative breast cancer stem cells is defined by the restricted expression of the TGFBR2 signaling receptor in these cells associated with its epigenetic silencing in the more differentiated CD24+ cells. Correlating with this treatment with a TGFBR kinase inhibitor specifically affected the phenotype of the CD44+ cells leading to their cellular differentiation. These findings have immediate therapeutic implications due to the current testing of TGF-β pathway and DNMT (DNA methyl-transferase) inhibitors in clinical trials [9]. Based on our results, we have proposed that treatment of tumors with inhibitors of TGF-β signaling may lead to the differentiation of the putative cancer stem cells, and thus, inhibit metastatic spread and recurrence. On the other hand, treatment with nonselective demethylating agents such as DNMT inhibitors may lead to the acquisition of cancer stem cell phenotype by the more differentiated tumor cells resulting in a more aggressive tumor and worse clinical outcome.

6.3 ELUCIDATING ATHEROSCLEROSIS PATHWAYS FROM INTEGRATED METABOLOMIC AND TRANSCRIPTOMIC DATASETS

In our next example, we show how pathway analysis can leverage data coming from two different platforms in reconstructing disease pathways of atherosclerosis. Atherosclerosis is a multifactorial disease of the large arteries and the leading cause of morbidity and mortality in industrialized countries [10]. There is ample evidence that *hypercholesterolemia* (i.e. elevated plasma levels of VLDL and LDL) induced by genetic modification or enhanced intake of dietary lipids is a major causative factor in atherogenesis [11,12]. It is equally clear that from the very beginning of lesion formation, atherogenesis requires an inflammatory component which is thought to drive the progression of the disease [13,14]. Indeed, some of the variation in the rate of lesion progression in different individuals may relate to variations in their basal inflammatory state [15,16]. However, while the inflammatory processes in the complex evolution of the lesion from the early fatty streak to a fibrous plaque are considered self-perpetuating phenomena, the initial trigger and origin of the inflammatory component in hypercholesterolemia remain enigmatic [15,17]. Recent observations suggest that the liver plays a key role in the inflammatory response evoked by dietary constituents [17,18]. For example, liver-derived inflammation markers such as C-reactive protein (CRP) and serum amyloid A (SAA) increase rapidly (within days) after consumption of an excess amount of dietary lipids [17,19], and thus by far precede the onset of early aortic lesion formation [17]. These findings suggest that nutritional cholesterol itself may contribute to the evolution of the inflammatory component of atherogenesis. In our recent work [20], we investigated a hypothesis that proatherogenic inflammatory factors originate at least partly from the liver. We also hypothesized that these factors come into play at high dietary cholesterol doses because of the exponential rather than linear nature of the relationship between cholesterol intake and atherosclerotic lesion size [21,22]. In our recent study [20], we sought evidence for the hypothesis that inflammation and hypercholesterolemia are not separate factors, but closely related features of the same trigger, dietary cholesterol. In particular, we addressed the question of how the liver responds to increasing dietary cholesterol loads at the gene transcription level and analyzed how hepatic cholesterol metabolism is linked to the hepatic inflammatory response, including underlying regulatory mechanisms. Notably, all analyses were performed at a very early stage of the atherogenic process (i.e. after 10 weeks of cholesterol feeding) to limit potential feedback reactions from the vessel wall. An established model for cholesterol-induced atherosclerosis, ApoE3Leiden transgenic (E3L) mice, allowed the application of experimental conditions that mimic the human situation: E3L mice display a lipoprotein profile similar to that of humans suffering from *dysbetalipoproteinemia* and develop atherosclerotic lesions that resemble human lesions with regard to morphology and cellular composition

[23,24]. E3L mice were exposed to increasing doses of dietary cholesterol (as the only dietary variable modulated), and liver genome and metabolome datasets were analyzed in a unique context, i.e. at the time point of first lesion development. Advanced functional analysis allowed us to integrate metabolome and transcriptome datasets and to analyze pathways and biochemical processes comprehensively.

To get insight into the complex traits underlying the (patho) physiological response of the liver to dietary cholesterol, whole-genome and metabolome measurements were executed. Compared to Control (cholesterol free), a relatively small number of genes (551) significantly changed with LC (low-cholesterol) treatment. HC (high-cholesterol) treatment modulated most (440 out of 551) of these genes and, additionally, affected 1,896 other genes. The individual gene expression profiles within a treatment group were very similar and formed clusters as confirmed by hierarchical clustering analysis Differences in gene expression between the treatment groups were validated and confirmed for a selected group of genes by RT-PCR. Standard Gene Ontology (GO) Biological Process annotation allowed categorization of 52% of the differentially expressed genes based on their biological function. LC treatment predominantly affected genes belonging to lipid and lipoprotein metabolism, protein metabolism, carbohydrate metabolism, energy metabolism, and transport. HC affected the same GO groups but, additionally, also genes relevant to immune and inflammatory responses, cell proliferation, apoptosis, cell adhesion, and cytoskeleton integrity. To refine the liver transcriptome data analysis and to define which biological processes are switched on/off with increasing dietary cholesterol loads, we performed gene enrichment analysis in four different functional ontologies: biological processes, canonical pathway maps, cellular processes, and disease categories using MetaCore™ pathway analysis platform. This allowed us to analyze functionally related genes (e.g. genes belonging to a specific biochemical process) on the level of detailed biochemical mechanisms. Table 6.1 summarizes the significantly changed biological processes for LC and HC. Four key ("master") process categories were affected by cholesterol feeding: lipid metabolism, carbohydrate and amino acid metabolism, transport, and immune and inflammatory responses. In the LC group, most significant effects occurred within the master process of lipid metabolism. Important subprocesses (i.e. processes in which more than 10% of process-related genes changed significantly) were lipid biosynthesis, lipoprotein metabolism, cholesterol metabolism, and cholesterol biosynthesis (Table 6.1). The overall functional effect for LC can be summarized as a substantial down-regulation of cholesterol and lipid metabolism. This adaptive response of the liver indicates metabolic liver resilience up to doses of 0.25% (w/w) cholesterol. A further increase of dietary cholesterol (1% w/w; HC) intensified the changes in gene expression seen with LC, indicating further metabolic adaptation. For example, all individual genes of the cholesterol biosynthesis pathway were down-regulated to a greater extent by HC than by LC (see pathway map in Figure 6.2A): the gene of the rate-limiting

TABLE 6.1 Master Processes and Their Subprocesses (Child Terms) Are Listed Together with the Number of Genes Measured (Third Column)

Analysis of Processes that Are Changed Significantly upon Treatment with Dietary Cholesterol

Master Process	Subprocess (Child Terms)	Number of Genes Measured	Differentially Expressed (%) LC	Differentially Expressed (%) HC
Lipid metabolism		264	8.7*	24.2*
	Fatty acid metabolism, fatty acid beta-oxidation	8	0.0	50.0*
	Triacylglycerol metabolism	7	0.0	57.1*
	Cholesterol metabolism	27	33.3*	33.3*
	Cholesterol biosynthesis	7	71.4*	57.1*
	Lipoprotein metabolism	18	16.7*	44.4*
	Lipid biosynthesis	105	11.4*	23.8*
Immune response		297	3.0	12.1*
	Antigen presentation, exogenous antigen	10	10.0	70.0*
	Antigen processing	17	5.9	35.3*
	Acute-phase response	11	9.1	36.4*
General metabolism		3,600	3.3	13.1*
	Cellular polysaccharide metabolism	19	5.3	26.3*
	Polysaccharide biosynthesis	9	0.0	33.3*
	Cofactor metabolism	116	5.2	21.6*
	Regulation of translational initiation	9	0.0	44.4*
	Amino acid metabolism	103	2.9	20.4*
Transport		1,119	2.9	14.3*
	Intracellular protein transport	161	3.7	19.9*
	Golgi vesicle transport	16	6.3	37.5*
	Mitochondrial transport	11	18.2*	54.5*

Percentages reflect the fraction of genes differentially expressed (within a specific process or pathway) in the LC and HC groups compared to the Con group. Relevant biological processes were identified in GenMAPP by comparison of the set of differentially expressed genes (ANOVA; $p < 0.01$ and FDR < 0.05) with all genes present on the array. *Biological processes with a z-score > 2 and a PermuteP < 0.05.

enzyme of this pathway, *HMG CoA reductase* (HDMH), was down-regulated 2.8-fold and 10.6-fold by LC and HC, respectively. Similarly, genes relevant for lipid and lipoprotein metabolism, *LDL receptor* (LC 1.3-fold down, HC 1.9-fold down) and *lipoprotein lipase* (LC 1.8-fold up, HC 5.5-fold up), were dose-dependently modulated.

Besides marked effects on "lipid metabolism," HC treatment induced significant changes in the master processes: "general metabolism," "transport,"

A

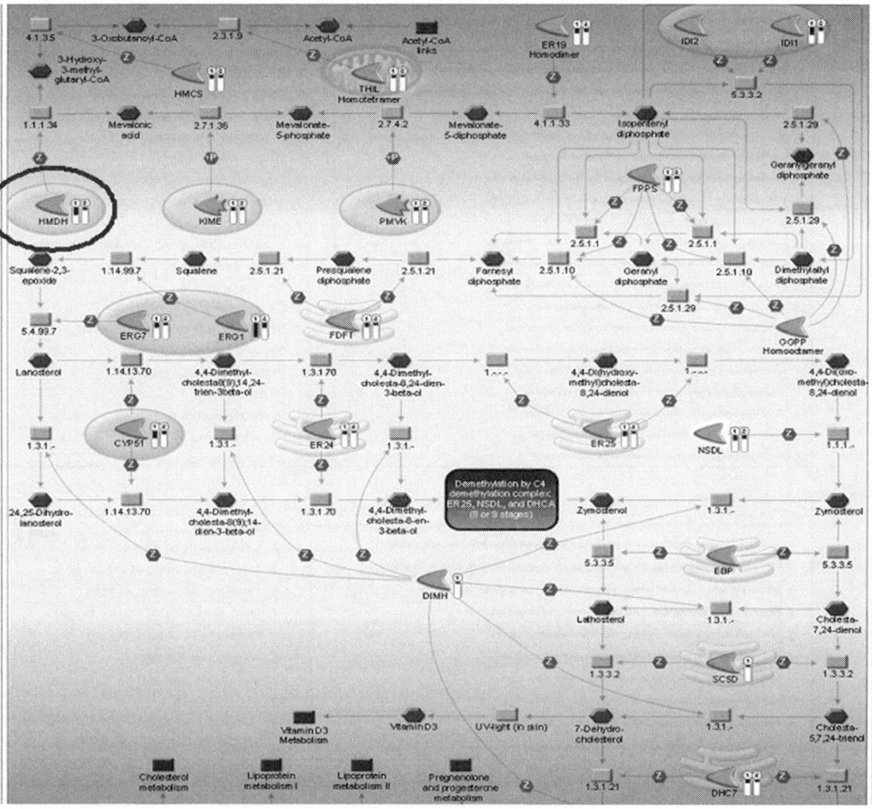

Figure 6.2 (A) Canonical pathway analysis of cholesterol metabolism, which is the top scored map of the enrichment analysis using canonical pathways. Gene expression changes are visualized on the map as thermometer-like figures. Down-regulation of the genes (dark color) by HC and LC is indicated with (1) and (2), respectively. The rate-limiting step in the pathway to cholesterol synthesis (and a major site of regulation) is the conversion of hydroxymethylglutaryl-CoA to mevalonate, a reaction catalyzed by HMG-CoA reductase (HMDH; in circle). Metabolites are depicted as hexagons. (B) Global analysis of changes in gene expression in LC and HC using disease categories based on over 500 human diseases with gene content annotated in MetaCore software (GeneGO). Top 20 diseases are shown for LC and HC including their *p*-value.

and "immune and inflammatory response" (Table 6.1). In particular, HC enhanced the subprocesses involved in translational initiation, Golgi vesicle transport, mitochondrial transport, antigen presentation, antigen processing, and acute phase response by affecting the expression of more than 35% of the genes in these subprocesses. Significantly, HC but not LC dietary stress activated specific inflammatory pathways (i.e. the platelet-derived growth factor [PDGF], interferon-γ [IFNγ], interleukin-1 [IL-1]) and tumor necrosis factor-α

B

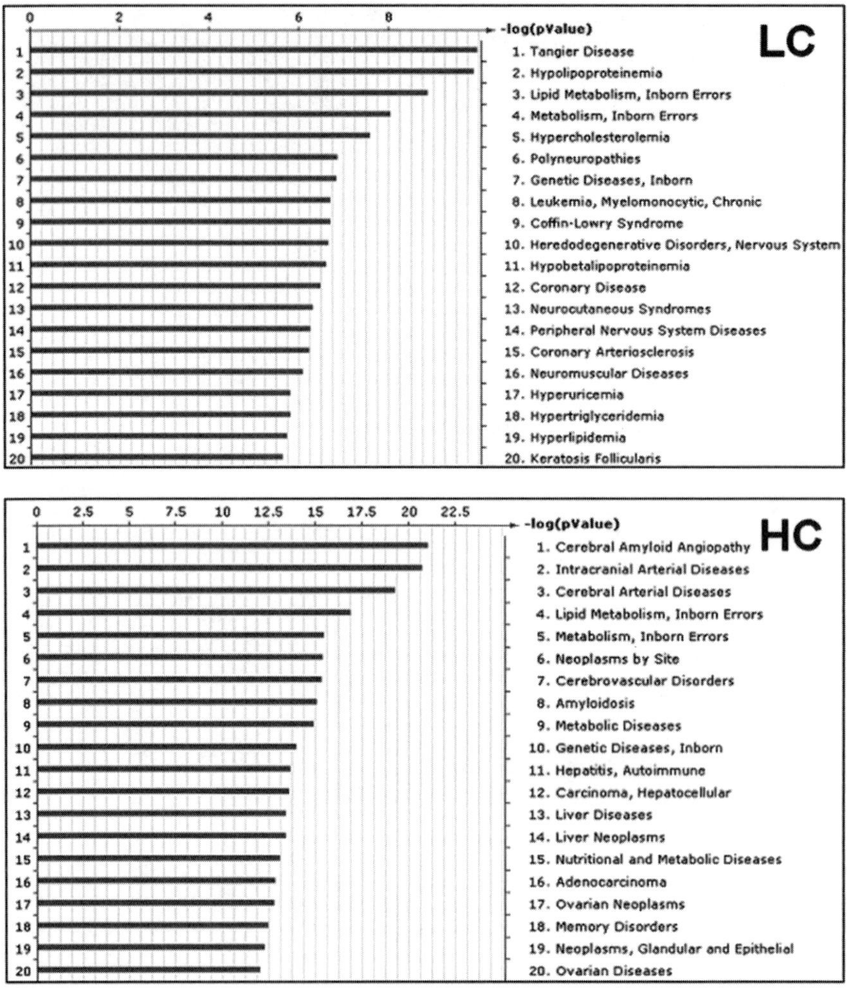

Figure 6.2 (*Continued*)

(TNFα) signaling pathways (Figure 6.3). More generally, HC treatment induced many genes, the gene products of which reportedly or putatively initiate or mediate inflammatory events, including genes encoding for proteases, complement components, chemokines and their receptors, heat shock proteins, adhesion molecules and integrins, acute phase proteins, inflammatory transcription factors, altogether indicating a profound reprogramming of the liver toward an inflammatory state not observed for LC.

We have also investigated enrichment of different disease-related categories by genes affected by either high- or low-cholesterol diets. In this test, differentially expressed genes are compared to the sets of genes annotated in

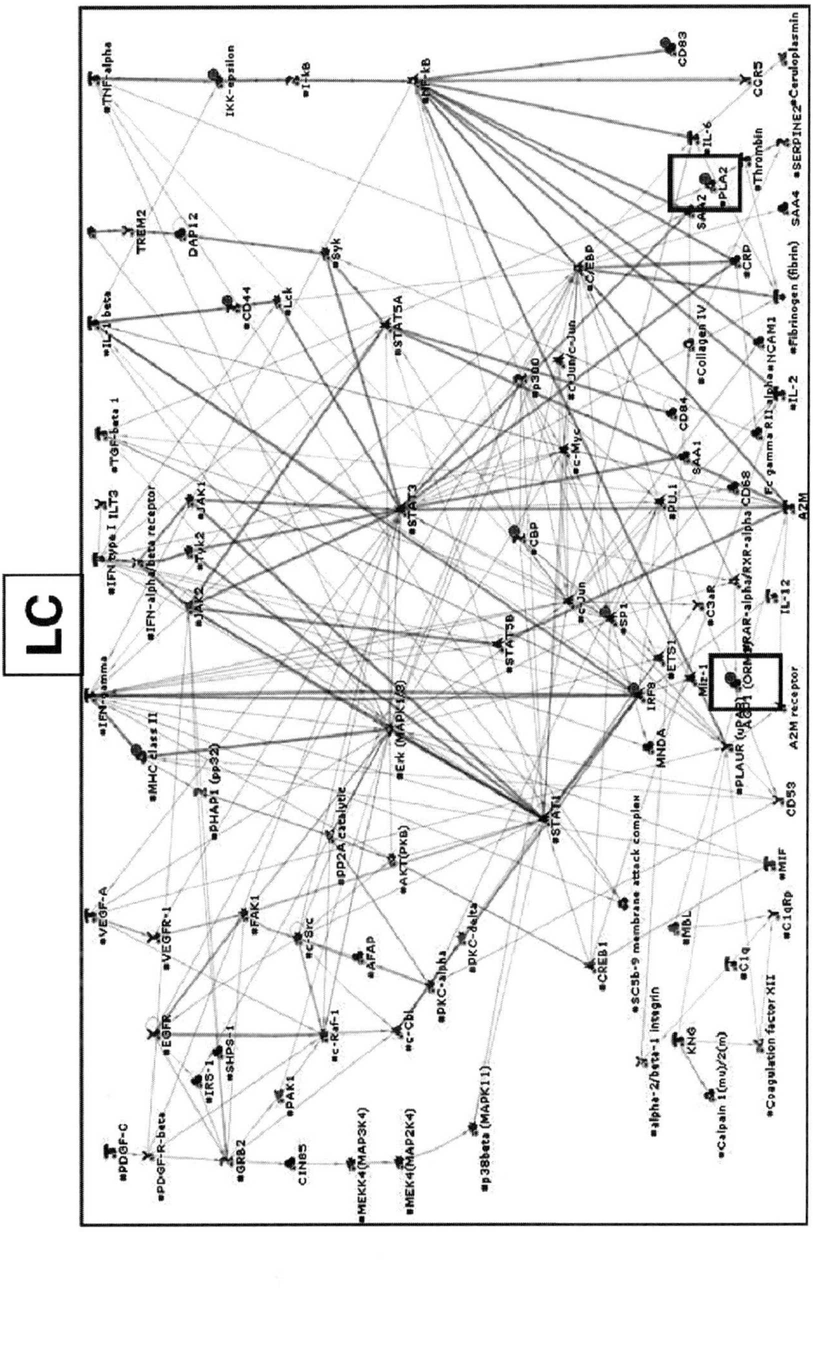

Figure 6.3 Analysis of the inflammatory pathways activated by LC and HC diet. A master inflammatory network was generated in MetaCore™ by combining relevant inflammatory pathways. Differentially expressed genes in response to LC and HC were mapped into this master network. The activation of the network by LC treatment was minimal, whereas HC treatment resulted in a profound activation of specific proinflammatory pathways (marked with blue arrows). Circles indicate transcriptional node points and rectangles highlight representative downstream target genes that were up-regulated.

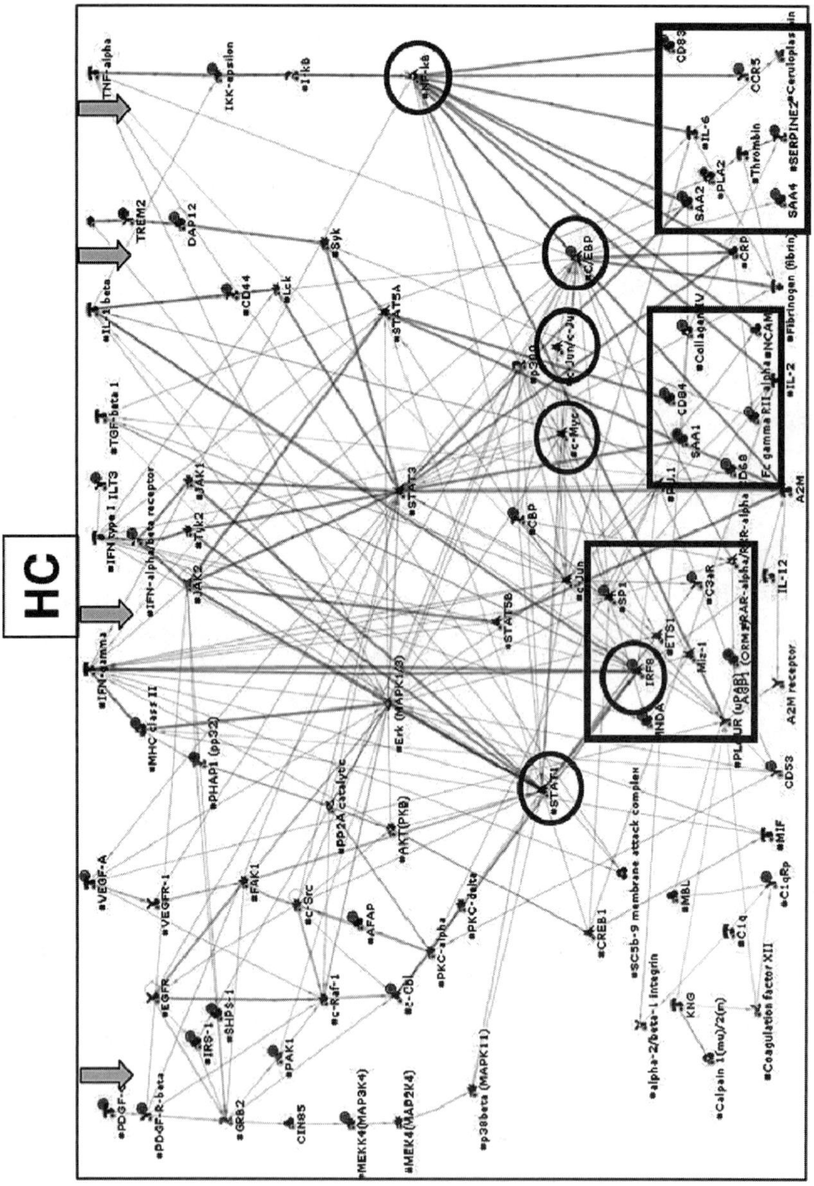

Figure 6.3 *(Continued)*

MetaCore™ as markers for various diseases. Enrichment analysis with these disease-related categories confirmed activation of many signaling and effector pathways relevant for inflammation and immunity by HC, but not by LC, treatment. The most affected (i.e. activated at the gene expression level) disease categories with HC treatment were interrelated cardiovascular disorders and (auto) immune diseases, including cerebral and intracranial arterial diseases, cerebral amyloid angiopathy, hepatocellular carcinoma, and hepatitis (Figure 6.2B).

To verify whether the switch from metabolic adaptation (with LC treatment) to hepatic inflammatory stress (with HC treatment) is also reflected at the metabolite level, we performed a comprehensive HPLC/MS-based lipidome analysis (measurement in total of about 300 identified di- and triglycerides, phosphatidylcholines, lysophosphatidylcholines, cholesterol esters) on liver homogenates of Con, LC and HC groups, and corresponding plasma samples.

We found that individual metabolite fingerprints within a treatment group were similar and formed clusters as assessed by principal component analysis. The clusters of the Control and LC groups overlapped partly demonstrating that the Control and LC groups have a similar intrahepatic lipid pattern. This indicates that the metabolic adjustments on the gene level in LC were efficacious and enabled the liver to cope with moderate dietary stress. The HC cluster did not overlap with the clusters of Control group, demonstrating that the switch to a proinflammatory liver gene expression profile is accompanied by development of a new metabolic hepatic state, which differs significantly from the metabolic state at baseline.

To identify the transcription factors and underlying regulatory mechanisms that govern the hepatic response to LC and HC, we performed a combined analysis of the liver transcriptome and metabolome dataset. Both datasets were simultaneously loaded into MetaCore™ software and functional networks were built from integrated data using "Analyze Network" algorithm. In this algorithm, both differentially expressed genes and metabolites are mapped onto nodes of the same global network of biological processes and tightly connected modules are identified. The global network itself contains all kinds of biomolecular events: protein–protein interactions, binding of small molecules to proteins, metabolic reactions, etc., allowing for concurrent analysis of data from different sources. Functional networks reconstructed from integrated dataset allowed identification of transcriptional key ("master") regulators relevant for liver resilience and liver inflammation. The adaptation of hepatic lipid metabolism to LC stress was mainly controlled by retinoid X receptor (RXR), SP-1, peroxisome proliferator activated receptor-α (PPARα), sterol regulatory element binding protein-1 (SREBP1), and SREBP2, which are established positive regulators of genes involved in cholesterol biosynthesis [25]. Combined analysis of genome and metabolite datasets revealed that the intrahepatic level of eicosapentaenoic acid, a suppressor of SREBP1 [26], was increased, providing a molecular explanation for the observed down-regulation of genes involved in cholesterol biosynthesis (Figure 6.4D).

A subsequent network analysis of HC-modulated genes allowed the identification of transcription factors that mediate the evolution of hepatic inflammation and that are ultimately responsible for the effects on the process level. HC-evoked changes require specific transcriptional master regulators, some of which are established in this context (nuclear factor kappa B, NF-κB; activator protein, AP-1; CAAT/enhancer-binding protein, C/EBPβ; p53), and others new (CREB-binding protein, CBP; hepatocyte nuclear factor-4α, HNF4α; SP-1; signal transducer and activator of transcription-3/-5, STAT-3/-5); Yin Yang-1, YY1) (Figure 6.4A–C).

Consistent with this, the identified transcription factors control the expression of genes encoding for acute phase response proteins, complement factors, growth factors, proteases, chemokine receptors, and factors stimulating cell adhesion, as confirmed by data mining. Most importantly, HC induced genes, the gene products of which can act extracellularly and possess reportedly proatherogenic properties. Examples include complement components (C1qb, C1qR, C3aR1, C9), chemoattractant factors (ccl6, ccl12, ccl19), chemoattractant receptors (CCR2, CCR5), cytokines inducing impaired endothelial barrier function (IFN-γ), adhesion regulators (integrin β2, integrin β5, CD164 antigen/sialomucin, junction adhesion molecule-2), growth factors (PDGF, VEGF-C, TGF-β), proteases involved in matrix remodeling during atherogenesis (cathepsin B, L, S and Z; matrix metalloprotease-12), and cardiovascular risk factors/inflammation markers (haptoglobin, orosomucoid 2, fibrinogen-like protein 2, α1-microglobulin). This up-regulation of proatherogenic candidate genes in the HC group is consistent with the observed enhanced early atherosclerosis found in this group. Expansion of the lipid and inflammatory networks revealed that hepatic lipid metabolism is linked to the hepatic inflammatory response via specific transcriptional regulators that control both processes. Among these dual regulators were CBP, C/EBPs, PPARα, and SP-1. The presence of molecular links between lipid metabolism and inflammation raises the possibility that specific intervention with an anti-inflammatory compound may in turn affect plasma cholesterol levels.

Thus, merging metabolome and transcriptome datasets within the context of functional pathways and global networks of biomolecular processes allowed identification of transcriptional master regulators, which control gene alteration and which are ultimately responsible for effects at the process level. It helped us to understand mechanisms by which increased doses of dietary cholesterol affect liver homeostasis and evoke hepatic inflammation. The following important findings were made: The liver absorbs escalating doses of dietary cholesterol primarily by adjusting the expression level of genes involved in lipid metabolism, as revealed by advanced gene expression analysis. This metabolic resilience is confirmed by analysis of metabolites in liver. At high doses of dietary cholesterol, the liver also develops an inflammatory stress response, which is characterized by up-regulation of proatherogenic candidate

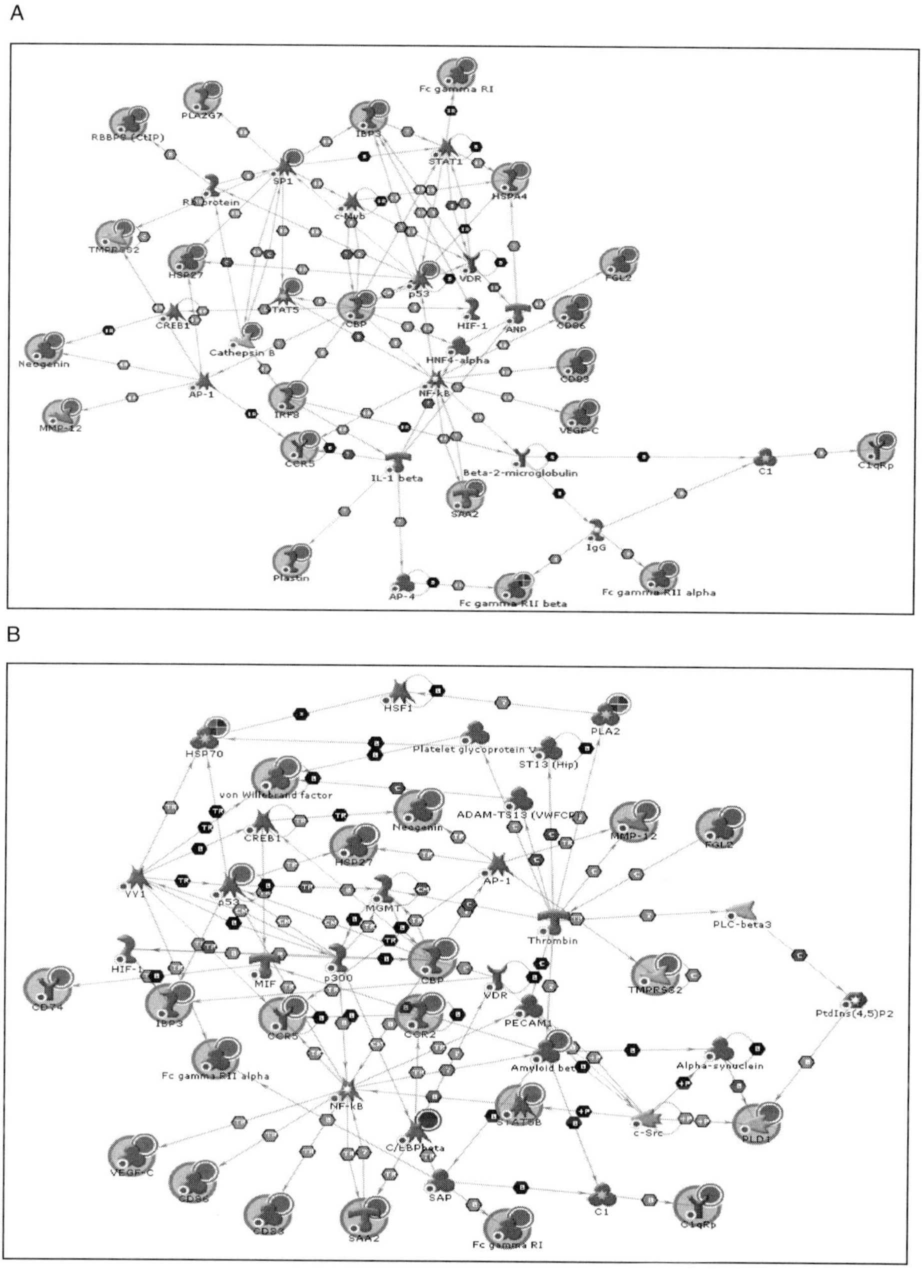

Figure 6.4 Biological networks of differentially expressed genes for HC allowing the identification of transcriptional master regulators. (A–C) Red (blue) dots in right corner of a gene indicate up-regulation (down-regulation) of a particular gene. Networks shown are representative networks used to identify transcriptional (master) regulators that control the gene expression changes under HC conditions (threshold of significance for networks $p < 0.01$). Metabolites are indicated as hexagons. (D) Concurrent network analysis of metabolomic and expression data reveals consistency between elevated level of eicosapentaenoic acid (red circle), an inhibitor of SREBP1 and underexpression of genes whose transcription is activated by SREBP1 (blue circles). See color insert.

C

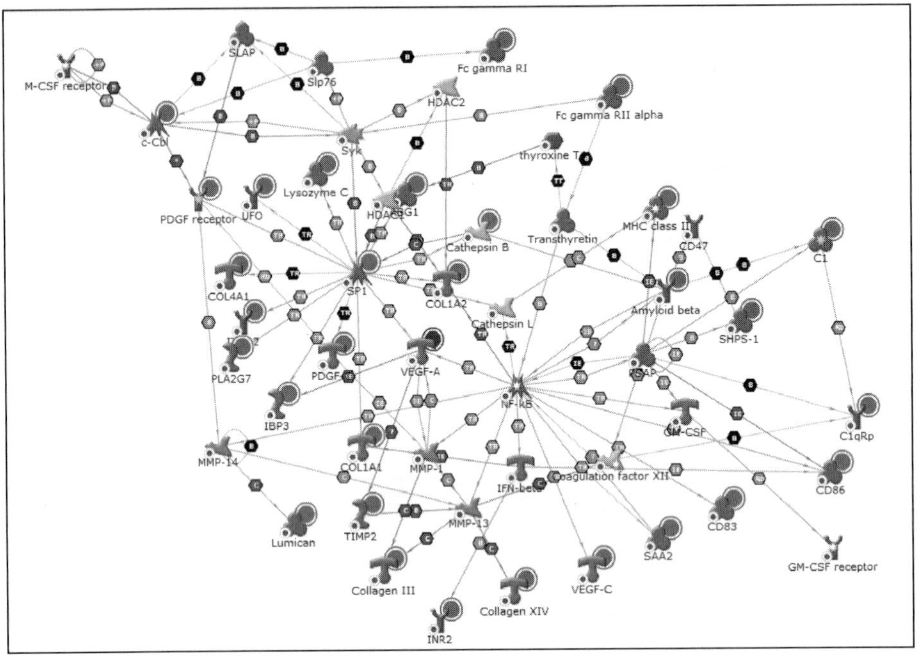

D

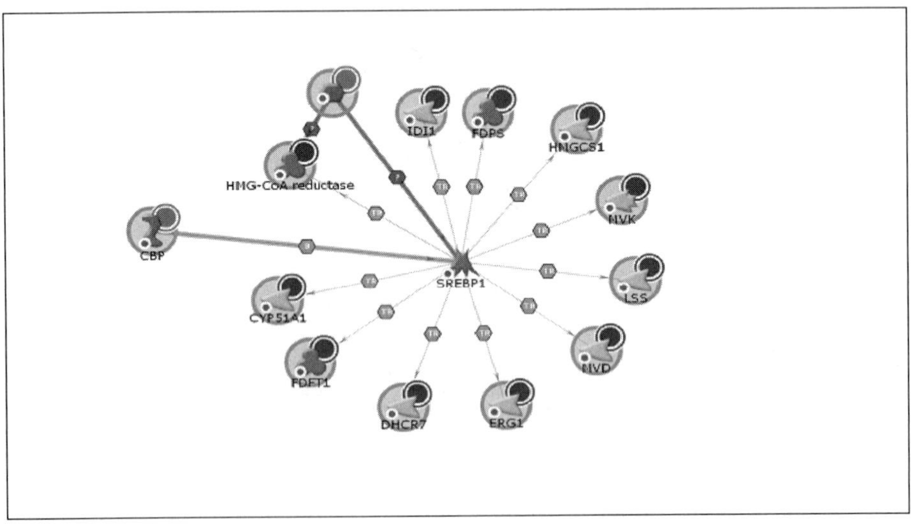

Figure 6.4 (*Continued*)

genes and activation of (at least four distinct) inflammatory pathways. The evolution of hepatic inflammation involves specific transcriptional regulators, several of which have been newly identified. Interestingly, some of these transcription factors have a dual role and control both hepatic lipid metabolism and hepatic inflammation, indicating that the same regulatory mechanisms underlie these processes and thereby link the two processes.

6.4 ANALYSIS OF MUTATED PATHWAYS IN CANCER

Discovery of the genes mutated in human cancer has provided key insights into the mechanisms underlying tumorigenesis and has been proven useful for the design of a new generation of targeted approaches for clinical intervention [27]. With the determination of the human genome sequence and improvements in sequencing and bioinformatic technologies, systematic analyses of genetic alterations in human cancers have become possible [28–30].

Using such large-scale approaches, we recently studied the genomes of breast and colorectal cancers by examining of all of the Reference Sequence (RefSeq) genes. The RefSeq database is a comprehensive, nonredundant collection of annotated gene sequences that represents a consolidation of gene information from all major gene databases [31]. The RefSeq database is believed to include the great majority of human gene sequences and represents the gold standard in the field.

Combining the data from two of the recent studies [31,32], it was found that 1,718 genes (9.4% of the 18,191 genes analyzed) had at least one non-silent mutation in either a breast or colorectal cancer. We identified candidate cancer genes (*CAN*-genes) that are most likely to be drivers and are therefore most worthy of further investigation. A gene was considered to be a *CAN*-gene if it harbored at least one non-synonymous mutation in both the Discovery and Validation Screens and if the total number of mutations per nucleotide sequenced exceeded a minimum threshold. Using these criteria, we identified a total of 280 *CAN*-genes, equally distributed between colorectal and breast cancers. In the discussed study, we also used a metric, called the cancer mutation prevalence (CaMP) score, to rank genes by the number and nature of the mutations observed. To assess the likelihood that each of these genes is mutated at a frequency higher than the passenger mutation rate, we devised a method based on Empirical Bayes' simulations. Though the likelihoods depend on the passenger rates, the rankings of the genes by CaMP scores are similar regardless of the assumed passenger mutation rates (rank correlations >0.9). CaMP scores thereby provide priorities for future studies that are independent of many of the assumptions required to calculate passenger probabilities.

It is becoming increasingly clear that pathways rather than individual genes govern the course of tumorigenesis [27]. Mutations in any of several genes of a single pathway can thereby cause equivalent increases in net cell proliferation. Accordingly, we devised a method to determine whether the genes within specific pathways were mutated more often than predicted by chance. The resultant "pathway CaMP" score incorporated the total number of mutations from all genes within each group, the number of different genes mutated, the combined sizes of the genes in each group, and the total number of tumors examined.

Using this metric, we analyzed a highly curated database (Metacore, GeneGo, Inc.) that includes human protein–protein interactions, signal transduction and metabolic pathways, and a variety of cellular functions and processes. By including the number of mutated genes in addition to the total number of mutations as parameters, we excluded pathways that simply contained one gene that was mutated at high frequency (e.g. pathways containing only *TP53* mutations). There were 108 pathways that were found to be preferentially mutated in breast tumors. Many of the pathways involved PI3K signaling (Figure 6.5).

Mutations in *PIK3CA* are frequent in multiple tumor types, including breast cancers [33–36]. In the current study, we identified mutations not only in *PIK3CA* but also previously unreported mutations in *GAB1, IKBKB, IRS4, NFKB1, NFKBIA, NFKBIE, PIK3R1, PIK3R4*, and *RPS6KA3*, implicating both the PI3K pathway in general and NF-κB signaling in particular in breast tumorigenesis. Within the 38 significant colorectal cancer pathways that appeared to be mutated in a statistically significant manner, there were also many that centered on PI3K. The pathway components mutated in colorectal cancers differed from those in breast, with mutations found in *IRS2, IRS4, PIK3R5, PRKCZ, PTEN, RHEB*, and *RPS6KB1* in addition to *PIK3CA*. Additional pathways altered in colorectal cancer were related to cell adhesion, the cytoskeleton, and the extracellular matrix, supporting the idea that interactions between the cancer cell and the extracellular environment are important steps in the neoplastic process.

Finally, there were nine examples of mutated genes whose protein products were predicted to interact with other mutated genes more often than predicted by chance. We extracted the protein interaction data of Metacore and calculated the interactions of proteins within a set of interest (such as colon *CAN*-genes for example) and compared that with the number of connection in the global protein "interactome." The goal of the analysis was to identify proteins with a significantly large number of interactions within the set of interest. We assigned statistical significance by using the cumulative hypergeometric distribution. The average number of mutant gene products with which these nine mutant genes interacted was 25. These results illustrate the potential utility of pathway-based analyses and highlight a variety of different gene groups and pathways that can help focus further investigations on these tumor types.

To understand the value of pathway analysis of genotyping data, consider "genomic landscape" of a typical cancer genome depicted in Figure 6.6. In these landscapes, every RefSeq gene is given a location on a 2-dimensional map corresponding to its chromosomal position, and all mutated genes in that

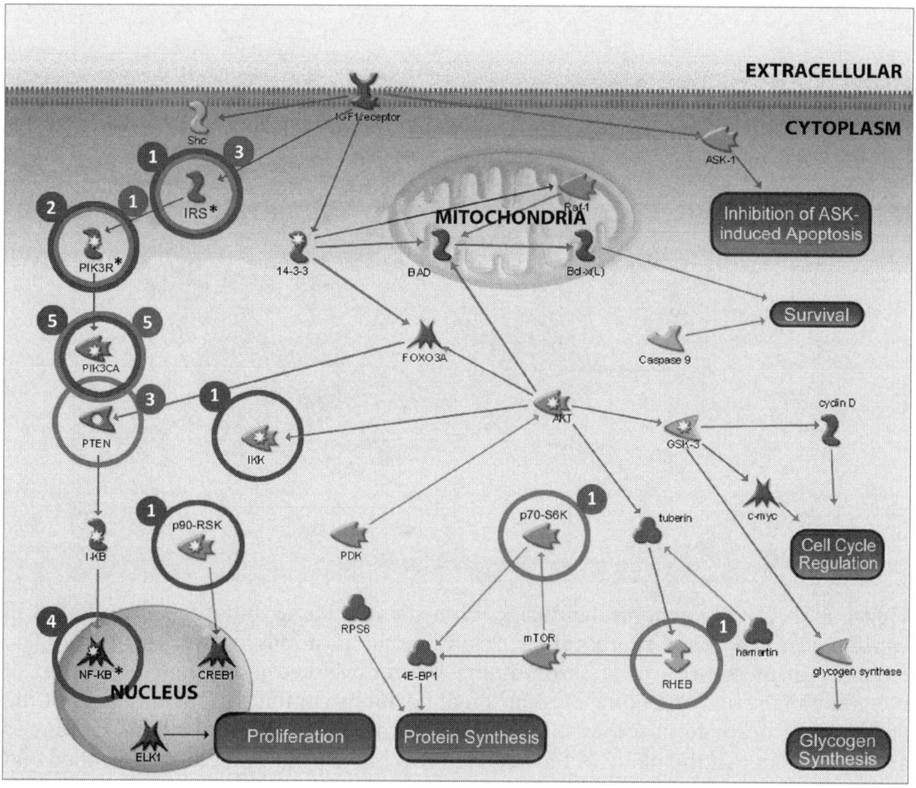

Figure 6.5 PI3K pathway mutations in breast and colorectal cancers. The identities and relationships of genes that function in PI3K signaling are indicated. Circled genes have somatic mutations in colorectal (red) and breast (blue) cancers. The number of tumors with somatic mutations in each mutated protein is indicated by the number adjacent to the circle. Asterisks indicate proteins with mutated isoforms that may play similar roles in the cell. These include insulin receptor substrates IRS2 and IRS4; phosphatidylinositol 3-kinase regulatory subunits PIK3R1, PIK3R4, and PIK3R5; and nuclear factor kappa-B regulators NFKB1, NFKBIA, and NFKBIE. See color insert.

tumor are indicated by a bright dot. The relief feature of the map is provided by the *CAN*-genes with the 60 highest CaMP scores. Just as topographical maps contain geological features of varying elevations, the cancer genome landscape consists of relief features (mutated genes) with heterogeneous heights (determined by CaMP scores). There are a few "mountains" representing individual *CAN*-genes mutated at high frequency. However, the landscapes contain a much larger number of "hills" representing the *CAN*-genes that are mutated at relatively low frequency. It is notable that this general genomic landscape (few gene mountains and many gene hills) is a common feature of both breast and colorectal tumors.

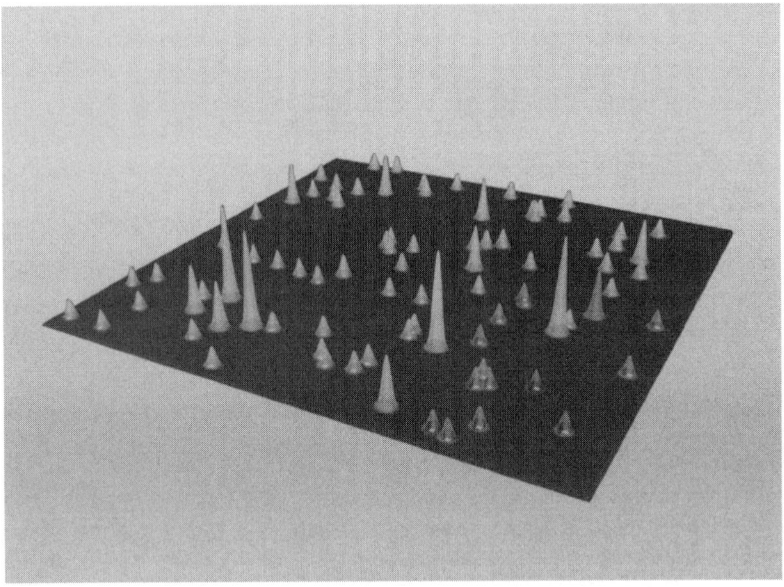

Figure 6.6 Cancer genome landscape. Non-silent somatic mutations are plotted in two-dimensional space representing chromosomal positions of RefSeq genes. The telomere of the short arm of chromosome 1 is represented in the rear left corner of the green plane and ascending chromosomal positions continue in the direction of the arrow. Chromosomal positions that follow the front edge of the plane are continued at the back edge of the plane of the adjacent row and chromosomes are appended end to end. Peaks indicate the 60 highest-ranking *CAN*-genes for, with peak heights reflecting CaMP scores. The largest "mountains" represent *TP53*, *PIK3CA*, and the other highly mutated genes. See color insert.

Historically, the focus of cancer research has been on the gene mountains, in part because they were the only alterations identifiable with available technologies. The ability to analyze the sequence of virtually all protein-encoding genes in cancers has shown that the vast majority of mutations in cancers, including those that are most likely to be drivers, do not occur in such mountains and emphasize the heterogeneity and complexity of human neoplasia. This new view of cancer is consistent with the idea that a large number of mutations, each associated with a small fitness advantage, drive tumor progression [37]. But is it possible to make sense out of this complexity? When all the mutations that occur in different tumors are summed, the number of potential driver genes is large. But this is likely to actually reflect changes in a much more limited number of pathways, numbering no more than 20 [27]. This interpretation is consistent with virtually all screens in model organisms, which have generally shown that the same phenotype can arise from alterations in any of several genes. Other recent studies lend support to this interpretation. For example, sequencing studies of the kinome in large numbers of tumors have shown that specific kinases are sometimes mutated in a small fraction of tumors of a given type [29,30,38–40]. We cannot be certain that the bulk of the low-frequency mutations observed in our study are not passengers. However, in the kinome studies, the position of mutations within the activation loop and the demonstrated effects of the target residues on kinase function unambiguously implicate many of these rare mutations as drivers. Similarly, recent analyses of myelomas suggest that there are multiple genes, each mutated in a small proportion of tumors, which can alter the same signal transduction pathway.

Regardless of whether this pathway-centric interpretation is correct, it is clear that the "easy" part of future cancer genome research will be the identification of genetic alterations. The vast majority of subtle mutations in individual patients' tumors can now be identified with existing technology (Figure 6.6), making personal cancer genomics a reality. Though understanding the precise role of these genetic alterations in tumorigenesis will be challenging, opportunities for exploiting such personal genomic data on cancers are already apparent. For example, many of the genes altered in breast cancers appear to affect the NF-κB pathway, suggesting that drugs targeting this pathway could be efficacious in breast cancers with such mutations [41,42]. Furthermore, our data indicate that individual cancers accumulate multiple unique epitopes predicted to interact with common HLA-alleles, thereby providing a large number of potential targets for immunotherapy [43]. Finally, any mutation identified in an individual cancer, whether driver or passenger, can be used as an exquisitely specific biomarker to guide patient management [44].

6.5 THE FUTURE OF PATHWAY ANALYSIS IN MOLECULAR DIAGNOSTICS AND PERSONALIZED MEDICINE

Applications of pathway analysis to disease "OMICs" data described above represent only initial efforts in this area. As computational methods mature and data acquisition technologies become even more robust and widely available, prediction of drug response in individual patients may become a reality. At that time, pathway analysis is likely to become an important part of molecular diagnostics pipelines. One of the greatest opportunities for diagnostic markers is development of personalized therapies for complex disease such as cancers. In these areas, new drugs are becoming increasingly more expensive, usually work only in the fraction of all patients and carry significant side effects and morbidity risks [45]. The typically high cost of molecular diagnostics tests would be more than offset by savings associated with elimination of unnecessary therapies and treating side effects. Substantial progress has already been made in using gene expression data in distinguishing different subsets of cancers, predicting outcomes and response to therapy. Diagnostics products based on gene expression profiling are already offered by such companies as Andendia BV and Genomics Health. We believe however, that more significant advances could be achieved on the route of developing complex biomarkers, incorporating molecular data of different types into their signatures. Pathways and molecular networks are the major analytical means of correlating different types of data and building these complex signatures. However, further development of algorithms will be needed. The major step forward should be incorporation of information on pathways and molecular interactions earlier in the analysis pipeline. As one might have noticed, in all the examples given in this chapter individual datasets are mapped onto the database of molecular pathways. Even though this allowed revealing important functional features and relations between the data of different nature, each set in itself was produced without taking into account any of the pathway information. When for example, we determine the set of genes differentially expressed between two conditions, we apply statistical analysis such as *t*-test or ANOVA on expression profiles of individual genes. No information about molecular interactions among gene products is taken into account. Yet, it is easy to imagine that small but consistent change of expression of all key genes in a given pathway is more telling than substantial change in the expression of disconnected genes. Nevertheless, standard statistical procedures would favor selection of the latter set of genes.

Thus we believe that the next generation of pathways analysis tools should incorporate pathway information early on the stage of selecting molecular descriptors, rather than applying already selected descriptors onto pathway data. The statistical procedures should take into account not only molecular profiles but also the connectivity information (Figure 6.7). This approach is

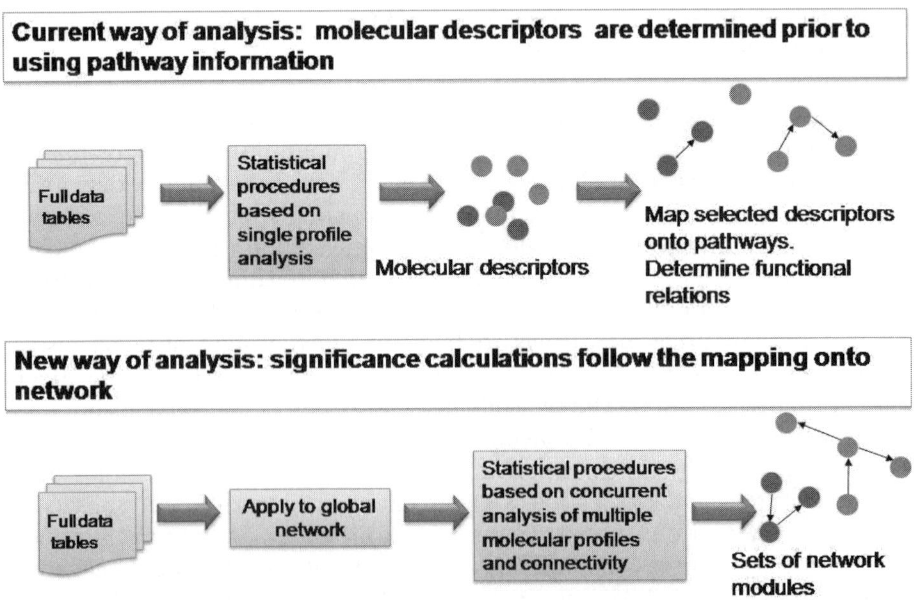

Figure 6.7 Current and future ways of analyzing “OMICs” data.

especially important when different types of data are analyzed concurrently. For example, it is well-established fact that sets of differentially expressed genes have only modest intersection with the sets of proteins whose levels change under the same conditions. If pathway information is incorporated early on, biomarker selection procedures could take into account regulatory cascades with consistent change in protein activity and expression of corresponding target genes.

REFERENCES

1. Shipitsin M, Campbell LL, Argani P, Weremowicz S, Bloushtain-Qimron N, Yao J, Nikolskaya T, Serebryiskaya T, Beroukhim R, Hu M, Halushka MK, Sukumar S, Parker LM, Anderson KS, Harris LN, Garber JE, Richardson AL, Schnitt SJ, Nikolsky Y, Gelman RS, Polyak K. Molecular definition of breast tumor heterogeneity. Cancer Cell 2007;11(3):259–273.
2. Wicha MS, Liu S, Dontu G. Cancer stem cells: an old idea—a paradigm shift. Cancer Res 2006;66:1883–1890; discussion 1895–1896.
3. Clarke MF, Fuller M. Stem cells and cancer: two faces of eve. Cell 2006; 124:1111–1115.
4. Polyak K, Hahn WC. Roots and stems: stem cells in cancer. Nat Med 2006; 12:296–300.
5. Nikolsky Y, Nikolskaya T, Bugrim A. Biological networks and analysis of experimental data in drug discovery. Drug Discov Today 2005;10:653–662.
6. Nikolsky Y, Ekins S, Nikolskaya T, Bugrim A. A novel method for generation of signature networks as biomarkers from complex high throughput data. Toxicol Lett 2005;158:20–29.
7. James D, Levine AJ, Besser D, Hemmati-Brivanlou A. TGFbeta/activIn/nodal signaling is necessary for the maintenance of pluripotency in human embryonic stem cells. Development 2005;132:1273–1282.
8. Siegel PM, Massague J. Cytostatic and apoptotic actions of TGF-beta in homeostasis and cancer. Nat Rev Cancer 2003;3:807–821.
9. Arteaga CL. Inhibition of TGFbeta signaling in cancer therapy. Curr Opin Genet Dev 2006;16:30–37.
10. Braunwald E. Shattuck lecture—cardiovascular medicine at the turn of the millennium: triumphs, concerns, and opportunities. N Engl J Med 1997;337:1360–1369.
11. Blum CB, Levy RI. Role of dietary intervention in the primary prevention of coronary heart disease. Individuals with high-normal or elevated serum cholesterol levels should be placed on cholesterol-lowering diets. Cardiology 1987;74:20–21.
12. Steinberg D. Hypercholesterolemia and inflammation in atherogenesis: two sides of the same coin. Mol Nutr Food Res 2005;49:995–998.
13. Steinberg D. Atherogenesis in perspective: hypercholesterolemia and inflammation as partners in crime. Nat Med 2002;8:1211–1217.
14. Willerson JT, Ridker PM. Inflammation as a cardiovascular risk factor. Circulation 2004;109:II2–10.

15. Libby P, Ridker PM, Maseri A. Inflammation and atherosclerosis. Circulation 2002;105:1135–1143.
16. Verschuren L, Kleemann R, Offerman EH, Szalai AJ, Emeis SJ, Princen HM, Kooistra T. Effect of low dose atorvastatin versus diet-induced cholesterol lowering on atherosclerotic lesion progression and inflammation in apolipoprotein E*3-Leiden transgenic mice. Arterioscler Thromb Vasc Biol 2005;25:161–167.
17. Kleemann R, Kooistra T. HMG-CoA reductase inhibitors: effects on chronic subacute inflammation and onset of atherosclerosis induced by dietary cholesterol. Curr Drug Targets Cardiovasc Haematol Disord 2005;5:441–453.
18. Rein D, Schijlen E, Kooistra T, Herbers K, Verschuren L, Hall R, Sonnewald U, Bovy A, Kleemann R. Transgenic flavonoid tomato intake reduces C-reactive protein in human C-reactive protein transgenic mice more than wild-type tomato. J Nutr 2006;136:2331–2337.
19. Tannock LR, O'Brien KD, Knopp RH, Retzlaff B, Fish B, Wener MH, Kahn SE, Chait A. Cholesterol feeding increases C-reactive protein and serum amyloid A levels in lean insulin-sensitive subjects. Circulation 2005;111:3058–3062.
20. Kleemann R, Verschuren L, van Erk M, Nikolsky Y, Cnubben N, Verheij E, Smilde A, Hendriks H, Zadelaar S, Smith G, Kaznacheev V, Nikolskaya T, Melnikov A, Camejo EH, Van der Greef J, Van Ommen B, Kooistra T. Atherosclerosis and liver inflammation induced by increased dietary cholesterol intake: a combined transcriptomics and metabolomics analysis. Genome Biol 2007;8(9): R200 17892536.
21. Groot PH, van Vlijmen BJ, Benson GM, Hofker MH, Schiffelers R, Vidgeon-Hart M, Havekes LM. Quantitative assessment of aortic atherosclerosis in APOE*3Leiden transgenic mice and its relationship to serum cholesterol exposure. Arterioscler Thromb Vasc Biol 1996;16:926–933.
22. Zadelaar S, Kleemann R, Verschuren L, de Vries-van der Weij, van der HJ, Princen HM, Kooistra T. Mouse models for atherosclerosis and pharmaceutical modifiers. Arterioscler Thromb Vasc Biol 2007; May 31 [epub ahead of print].
23. van Vlijmen BJ, Van den Maagdenberg AM, Gijbels MJ, Van Der BH, HogenEsch H, Frants RR, Hofker MH, Havekes LM. Diet-induced hyperlipoproteinemia and atherosclerosis in apolipoprotein E3-Leiden transgenic mice. J Clin Invest 1994; 93:1403–1410.
24. Kleemann R, Princen HM, Emeis JJ, Jukema JW, Fontijn RD, Horrevoets AJ, Kooistra T, Havekes LM. Rosuvastatin reduces atherosclerosis development beyond and independent of its plasma cholesterol-lowering effect in APOE*3-Leiden transgenic mice: evidence for antiinflammatory effects of rosuvastatin. Circulation 2003;108:1368–1374.
25. Shimano H. Sterol regulatory element-binding protein-1 as a dominant transcription factor for gene regulation of lipogenic enzymes in the liver. Trends Cardiovasc Med 2000;10:275–278.
26. Zaima N, Sugawara T, Goto D, Hirata T. Trans geometric isomers of EPA decrease LXRalpha-induced cellular triacylglycerol via suppression of SREBP-1c and PGC-1beta. J Lipid Res 2006;47:2712–2717.
27. Vogelstein B, Kinzler KW. Cancer genes and the pathways they control. Nat Med 2004;10:789–799.

28. Futreal PA, Coin L, Marshall M, Down T, Hubbard T, Wooster R, Rahman N, Stratton MR. A census of human cancer genes. Nat Rev Cancer 2004;4:177–183.

29. Bardelli A, Velculescu VE. Mutational analysis of gene families in human cancer. Curr Opin Genet Dev 2005;15(1):5–12.

30. Greenman C, et al. Patterns of somatic mutation in human cancer genomes. Nature 2007;446:153–158.

31. Pruitt KD, Tatusova T, Maglott DR. NCBI reference sequences (RefSeq): a curated non-redundant sequence database of genomes, transcripts and proteins. Nucleic Acids Res 2007;35:D61–D65.

32. Wood LD, et al. The genomic landscapes of human breast and colorectal cancers. Science 2007;318:1108–1113.

33. Samuels Y, Wang ZH, Bardelli A, Silliman N, Ptak J, Szabo S, Yan H, Gazdar A, Powell DM, Riggins GJ, Willson JKV, Markowitz S, Kinzler KW, Vogelstein B, Velculescu VE. High frequency of mutations of the PIK3CA gene in human cancers. Science 2004;304:554–554.

34. Bachman KE, Blair BG, Brenner K, Bardelli A, Arena S, Zhou SB, Hicks J, De Marzo AM, Argani P, Park BH. p21(WAF1/cIP1) mediates the growth response to TGF-beta in human epithelial cells. Cancer Biol Ther 2004;3:221–225.

35. Broderick DK, Di C, Parrett TJ, Samuels YR, Cummins JM, McLendon RE, Fults DW, Velculescu VE, Bigner DD, Yan H. Mutations of PIK3CA in anaplastic oligodendrogliomas, high-grade astrocytomas, and medulloblastomas. Cancer Res 2004;64:5048–5050.

36. Lee JW, Soung YH, Kim SY, Lee HW, Park WS, Nam SW, Kim SH, Lee JY, Yoo NJ, Lee SH. PIK3CA gene is frequently mutated in breast carcinomas and hepatocellular carcinomas. Oncogene 2005;24:1477–1480.

37. Beerenwinke N, Antal T, Dingli D, Traulsen A, Kinzler KW, Velculescu VE, Vogelstein B, Nowak MA. Genetic progression and the waiting time to cancer. Plos Comput Biol 2007;3(11):e225.

38. Stephens P, et al. A screen of the complete protein kinase gene family identifies diverse patterns of somatic mutations in human breast cancer. Nat Genet 2005;37:590–592.

39. Parsons DW, Wang TL, Samuels Y, Bardelli A, Cummins JM, DeLong L, Silliman N, Ptak J, Szabo S, Willson JKV, Markowitz S, Kinzler K, Vogelstein B, Lengauer C, Velculescu VE. Colorectal cancer: mutations in a signalling pathway. Nature 2005;436:792–792.

40. Thomas RK, et al. High-throughput oncogene mutation profiling in human cancer. Nat Genet 2007;39:347–351.

41. Annunziata CM, Davis RE, Demchenko Y, Bellamy W, Gabrea A, Zhan F, Lenz G, Hanamura I, Wright G, Xiao W, Dave S, Hurt EM, Tan B, Zhao H, Stephens O, Santra M, Williams DR, Dang L, Barlogie B, Shaughnessy JD, Kuehl WM, Staudt LM. Cancer Cell 2007;12:115–130.

42. Keats JJ, Fonseca R, Chesi M, Schop R, Baker A, Ching WJ, Van Wier S, Tiedemann R, Shi CX, Sebag M, Braggio E, Henry T, Zhu YX, Fogle H, Price-Troska T, Ahmann G, Mancini C, Brents LA, Kumar S, Greipp P, Dispenzieri A, Bryant B, Mulligan G, Bruhn L, Barrett M, Valdez R, Trent J, Stewart AK,

Carpten J, Bergsagel PL. Promiscuous mutations activate the noncanonical NF-kappa B pathway in multiple myeloma. Cancer Cell 2007;12:131–144.

43. Diehl F, Diaz LA. Digital quantification of mutant DNA in cancer patients. Curr Opin Oncol 2007;19:36–42.
44. Sanchez R, Sali A. Large-scale protein structure modeling of the *Saccharomyces cerevisiae* genome. Proc Natl Acad Sci U S A 1998;95:13597–13602.
45. Hartwell L, Mankoff D, Paulovich A, Ramsey S, Swisher E. Cancer biomarkers: a systems approach. Nat Biotechnol 2006;24(8):905–908.

7

WHOLE-GENOME EXPRESSION PROFILING OF PAPILLARY SEROUS OVARIAN CANCER: ACTIVATED PATHWAYS, POTENTIAL TARGETS, AND NOISE

JOHN FARLEY, LAURENT L. OZBUN, AND MICHAEL J. BIRRER

Table of Contents

7.1 INTRODUCTION

Ovarian cancer is the most lethal gynecologic malignancy [1]. In 2008 it is expected that 21,650 cases of ovarian cancer will occur resulting in 15,520

Pathway Analysis for Drug Discovery, Edited by Anton Yuryev

deaths [1]. The high case fatality rate for this tumor results in part from the advanced stage at the time of diagnosis. The overall 5-year survival for ovarian cancer is approximately 45%; however, if this cancer is diagnosed at an early stage the survival approaches 93%. Unfortunately, only 19% of patients will present with early stage disease. For those unfortunate women who present with advanced stage disease, the 5-year survival is approximately 29.6% [1]. Although ovarian cancer is responsive to multiple chemotherapeutic agents, with objective response rates of up to 80%, over three quarters of patients who have a response to initial chemotherapy treatment relapse within 2 years of primary therapy. These patients then become candidates for further therapy of their recurrent disease [2]. Although there are many treatment options for the ovarian cancer patient in the recurrent setting, practically, second line therapy for ovarian cancer is not very effective [3,4]. The consequence of ineffectual second line therapeutic options is that there has been no substantial increase in ovarian cancer survival over the past 30 years [1].

The pathogenesis of ovarian cancer remains poorly understood. There are four main histologic subtypes of ovarian cancer, including serous, clear cell, endometrioid, and mucinous [5]. Of these, serous adenocarcinomas account for approximately 60% of the ovarian cancer cases diagnosed. Ovarian surface epithelium (OSE) is thought to be the source of most epithelial ovarian cancer cases (papillary serous histology), while evidence suggests that some endometrioid and clear cell cancers arise from endometriosis [6,7]. Unfortunately, there are no generally accepted precursor lesions for ovarian cancer and intermediate lesions such as those seen in colon cancer have not been identified. Thus, a stepwise progression for ovarian carcinogenesis, such as the paradigm that has been established for cervical and colorectal cancer, has yet to be formulated [6,8]. The etiology of ovarian cancer likely involves multiple different pathways and molecular etiologies [5,6].

7.1.1 Unique Features of Ovarian Cancer

Epithelial ovarian cancer tumors demonstrate a wide spectrum of pathologic grade. Ten percent to 15% of tumors diagnosed as having a serous histology are categorized as low malignant potential (LMP) malignancies (grade 0) [5]. LMP tumors represent a clinical enigma as they display atypical nuclear architecture on a histologic level, have the clinical ability for metastatic behavior, and yet this subset of malignancies is considerably less aggressive than high-grade serous tumors and is largely refractory to adjuvant chemotherapy (Figure 7.1) [5,9–11]. The 5-year survival for patients with LMP tumors is 95% in contrast to a less than 45% survival for advanced high-grade disease over the same period [10,11]. The origin and role of LMP tumors in the development of invasive epithelial cancer of the ovary remains to be defined [10,11]. The key question for ovarian cancer carcinogenesis is whether LMP tumors represent intermediate lesions between normal epithelium and invasive high-

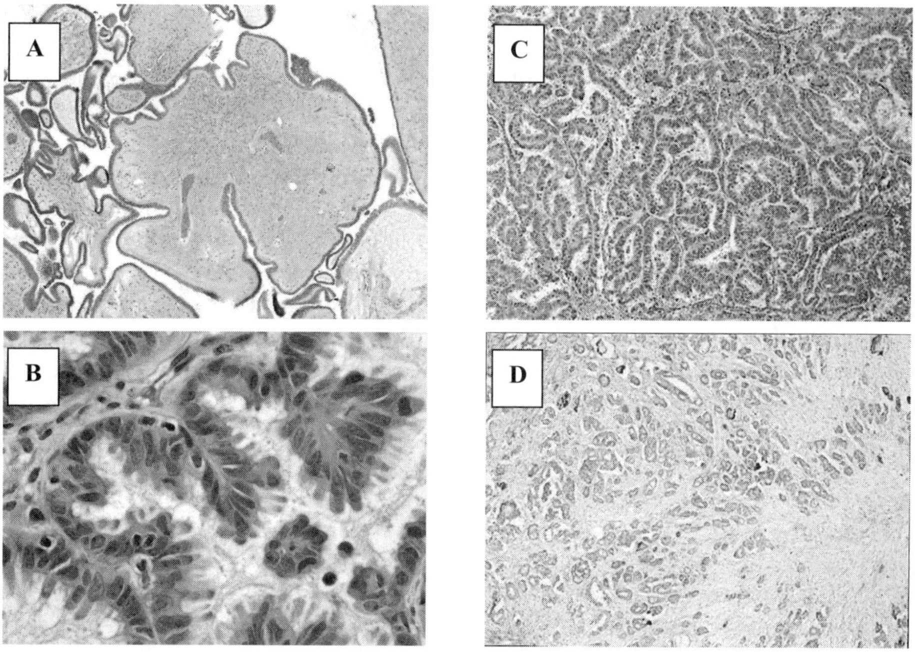

Figure 7.1 Papillary serous ovarian cancer photomicrographs of low malignant potential tumors (A,B) and high-grade invasive carcinoma (C,D) at low (A,C) and high magnification (B,D).

grade tumors. Can LMP tumors progress to invasive cancer, or are they an entirely different histologic subtype?

In contrast to LMP ovarian cancers, the vast majority of high-grade invasive ovarian cancers present as advanced stage disease. The 5-year survival for advanced epithelial ovarian cancer is considered much worse than LMP tumors with most of these patients succumbing from their disease. The survival is extremely hereterogenous with overall survival ranging from 29% to 69% depending on the stage of presentation and residual disease present after surgery [1]. There currently is no prognostic tool available to stratify these patients and the molecular pathways that distinguish these tumors remain completely unknown.

7.2 EXPRESSION PROFILING OF OVARIAN CANCER

The development of advanced genomic technologies, such as oligonucleotide microarray analysis, has provided a means to capture global gene expression patterns for a large number of tissue samples. Oligonucleotide microarrays have the capability to determine the expression of all the genes expressed within a cell simultaneously [5,8]. This gene expression pattern can be correlated with many clinically relevant characteristics of an individual tumor. These approaches have been used to characterize the biological relationships among histologic subtypes of ovarian cancer and identify genes whose altered expression is important in the development of ovarian cancer [8,12]. One of the most common array platforms is the Affymetrix® expression platform. With this technology, total RNA is extracted and purified. Biotin-labeled cRNA is then prepared for each sample. Labeled cRNA is fragmented, combined with a hybridization cocktail containing biotinylated hybridization controls, and incubated on the oligonucleotide array. Laser excitation then stimulates fluorescence emission of labeled probes bound to target sequences. These emissions generate a specific image for the sample analyzed. Array images are then acquired and analyzed with Genechips Operating Software (GCOS).

7.2.1 What's "Normal?"

As with many cancers of epithelial origin, it is important to establish an appropriate control for evaluating differential gene expression. Expression profiling studies of ovarian cancer have relied on a variety of sources of "normal" cells for comparison with tumors, including whole ovary samples (WO), ovarian surface epithelium (OSE) exposed to short-term culture, and immortalized OSE cell lines (IOSE). Comparison of the gene expression profiles generated from OSE brushings, whole ovary (WO) samples, short-term cultures of normal OSE (NOSE), immortalized OSE cell lines (IOSE), and telomerase-immortalized OSE (TIOSE) cell lines revealed that all of these "normal" groups formed robust, and distinct bands in hierarchical clustering (Figure 7.2) [12]. The correlation coefficient for all combinations of any two of the groups only ranged from 0.04 to 0.54, emphasizing the disparity of the groups. These findings emphasize concerns regarding the source of the "normal" controls. Exposing cells to tissue culture conditions significantly alters gene expression, either by directly affecting transcriptional regulation or by selecting for a subset of cells that are not representative of the original culture [12]. The WO samples form a distinct cluster likely due to the large stromal component in the WO profile. The brushing technique allows the collection of OSE without stroma, and provides a relatively pure sample of OSE that is not exposed to culture conditions. As a result, the OSE samples represent the most straightforward collection technique of the tissue felt to be the most biologically relevant to the development of epithelial ovarian cancer. These results suggest that the selection of a normal control to compare to epithelial ovarian cancer samples in microarray studies strongly influences the genes that are identified as differentially expressed. OSE collected by the brushing technique provides the most accurate sample [12].

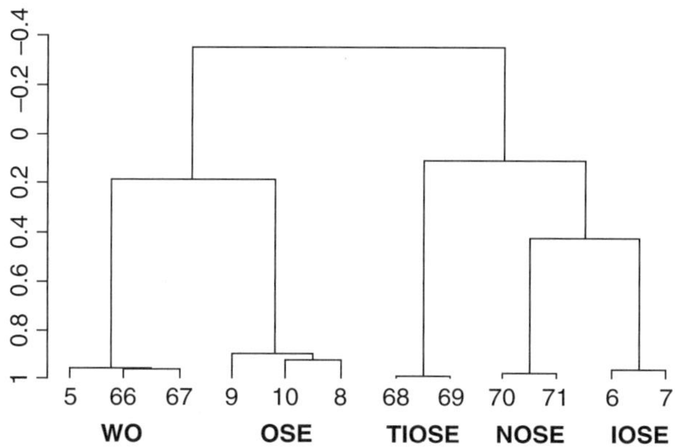

Figure 7.2 Agglomerative hierarchical clustering of the samples based on the 446 genes that discriminate between the five normal groups at $\alpha = 0.0001$ using centered correlation and average linkage. Samples that merge into clusters low on the dendrogram are more similar than those that merge at a higher level.

7.3 DEFINING THE BIOLOGIC RELATIONSHIP AMONG OVARIAN TUMORS OF VARYING GRADE

To establish the biological relationship among LMP tumors, low-grade, and high-grade invasive serous ovarian carcinomas and identify genes whose expression accounts for their phenotypes, 90 microdissected serous ovarian tumors that spanned the pathologic spectrum and normal ovarian surface epithelium (OSE) brushings were interrogated using the 47,000 transcript Affymetrix U133 Plus 2.0 oligonucleotide array [5]. Included were invasive low-grade, early-stage high-grade and late-stage high-grade serous ovarian tumors, and LMP tumors. Unsupervised analysis showed a distinct separation between LMP tumors and high-grade cancer (Figure 7.3A). Furthermore, when low-grade invasive tumors were included in the analysis, they closely aligned with LMP lesions rather than their high-grade invasive counterparts (Figure 7.3B). The dissimilarity between LMP and high-grade tumors was substantiated using binary tree prediction and expression data from an independent set of microarrays (Figure 7.3C). The identification of two unique branches containing LMP tumors and high-grade carcinomas is consistent with the distinct clinicopathologic aspects of the two diseases and prior molecular studies [5,13,14]. The alignment of invasive low-grade serous tumors with serous LMP tumors instead of high-grade invasive serous cancers strongly argues that low-grade invasive serous tumors are more similar to serous LMP tumors than high-grade tumors (Figure 7.3B,C).

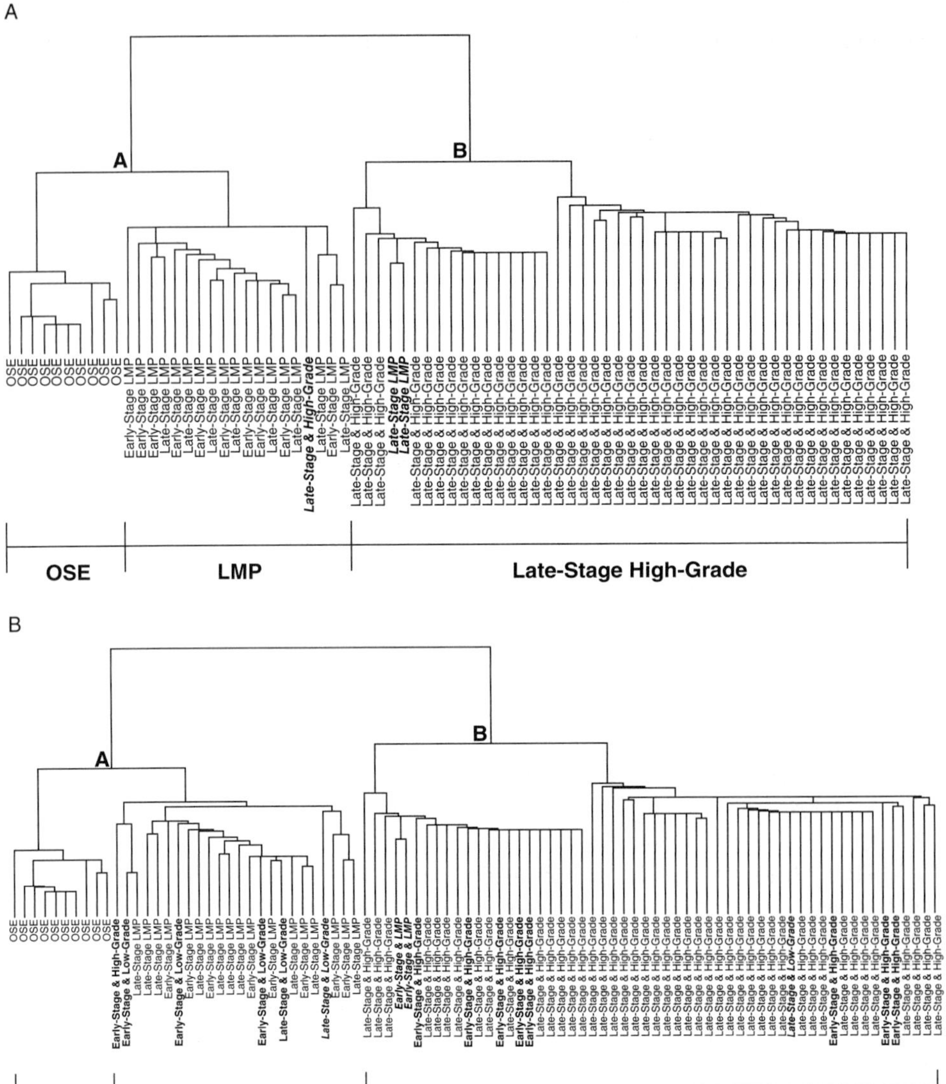

Figure 7.3 (A) Hierarchical clustering analysis of the 16,178 probe sets passing the filtering criteria for LMP tumors, late-stage high-grade cancers, and OSE. OSE specimens grouped independently from LMP specimens (node A), whereas late-stage high-grade tumors clustered in two distinct groups (node B). Misclassified specimens are bold italicized. (B) Hierarchical clustering analysis of the 14,119 probe sets passing the filtering criteria for LMP, low-grade, high-grade, and OSE specimens and binary tree validation. Overall tree structure was retained despite the association low-grade tumors with LMP tumors and the grouping of early-stage and late-stage high-grade lesions. Low-grade and early-stage high-grade samples are indicated in bold. Misclassified specimens are bold italicized. (C) Binary tree analysis confirmed the hierarchical clustering results. The diagram was generated using binary tree prediction followed by leave-one-out cross-validation to estimate the error associated with the tree building process. OSE samples were classified as basal to the ovarian cancer specimens. LMP tumors and low-grade cancers were more closely aligned to each other, as were early-stage and late-stage high-grade tumors. Percentages indicate the misclassification error associated with each node.

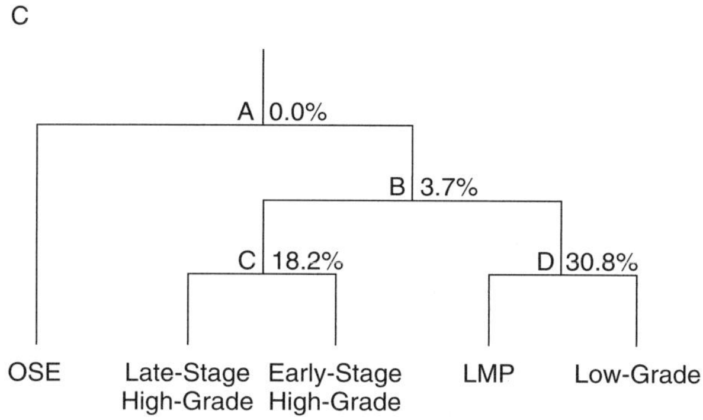

Figure 7.3 (*Continued*)

7.4 DEFINING THE BIOLOGIC RELATIONSHIP AMONG OVARIAN CANCERS OF DIFFERENT HISTOLOGIES

Unsupervised hierarchical clustering clearly distinguishes serous LMP and low-grade disease from high-grade advanced cancer [5]. Within high-grade disease there are subsets of patients displaying distinct clinical phenotypes (e.g. survival or chemoresponse) that are driven by unique genes/pathways. Some of these pathways could also distinguish high-grade papillary serous disease from other histotypes (i.e. endometrioid, clear cell, and mucinous). Analyzing all advanced tumors in aggregate could identify genes contributing to the disease in a large proportion of the patients.

In an effort to further identify the molecular signatures of specific ovarian cancer histologies, the gene expression profiles of serous, endometrioid, and clear cell cancers were examined [6]. A total of 24 papillary serous, 11 endometrioid, and nine clear cell ovarian tumors were analyzed. Comparing the histosubtypes of ovarian cancer directly to one another, 166 genes differentiated the samples into the three subtypes. When clear cell ovarian cancer was compared with non-clear cell ovarian cancer (serous and endometrioid ovarian cancer grouped together), 171 differentially expressed genes were identified. Serous and endometrioid cancer were distinguished from the other histologic subtypes by 62 and 66 differentially expressed genes, respectively [6].

To identify specific genes involved in the development of individual histologic types of ovarian cancers, separate comparisons of each histologic subtype to normal OSE brushings were completed. These comparisons yielded lists of 94 genes for clear cell cancer, 422 genes for endometrioid cancer, and 467 genes for serous cancer [6]. Forty-three genes were common to all three lists and therefore displayed consistent differential expression between normal OSE and ovarian cancer regardless of histologic subtype. Twenty-nine genes have increased expression in ovarian cancer compared with normal OSE, whereas 14 have decreased expression in cancer. Among the genes with increased expression in cancer are homogentisate oxidase (HGD), peroxisome proliferative activated receptor gamma (PPARG), v-rel reticuloendotheliosis viral oncogene homologue B (RELB), and p21-activated kinase 1 (PAK1) [6]. Decreased expression was documented for tenascin XB (TNXB), galectin 8 (LGALS8), post-meiotic segregation increased 2-like 8 and 2-like 9 (PMS2L8 and PMS2L9), deafness autosomal dominant 5/inversely correlated with estrogen receptor expression 1 (DFNA5/ICERE1), disabled homologue 2/differentially expressed in ovarian cancer 2 (DAB2/DOC2), and retinoic acid receptor responder 1 (RARRES1/TIG1) [6].

This group of 43 genes comprised the common genes appearing on each ovarian cancer subtype's comparison with normal OSE. This suggests that at least part of the transformation process might be shared among endometrioid, serous, and clear cell ovarian cancer, as evidenced by the common genes distinguishing them from normal OSE. However, the question of whether the OSE serves as a common precursor is not necessarily clarified. It is conceivable that tumors of different histologies may arise from different precursor cells but undergo similar transformation events.

7.5 COMPLEXITIES OF WHOLE-GENOME PROFILING

As seen earlier in our identification of multiple differentially expressed genes involving the distinct histologic subtypes of ovarian cancer, genomic profiling experiments frequently provide bewildering numbers of differentially expressed genes. This could include thousands to tens of thousands of differentially expressed genes. How does a scientist sort out which genes are important and what pathways might be critical in ovarian cancer development or clinical-pathologic characteristics (Figure 7.4A)? One way to accomplish this is through Gene Ontology (GO) classification [5]. Various tools have been developed by groups external to the GO Consortium to perform this sort of analysis, such as GoMiner, Onto-Express16, and GO Term Finder [15,16]. To determine whether particular functional categories of genes are highly enriched in a specific gene list, we identify the GO categories that are statistically significant among the lists of differentially regulated genes. The GO analysis tools allow you to upload your full gene set. A list of all "interesting" genes within that set, usually those that have been up- or down-regulated in an expression experiment, is elucidated. The tool then allows you to view which GO categories have been enriched for your genes of interest, and usually provides some sort of statistical measure to guard against GO categories that appear by chance alone. Subsequently, the ontological system describes gene products in terms of their biological process, cellular component, and molecular function. This permits categorization of differentially regulated genes into distinct functional groupings (Figure 7.4B).

A

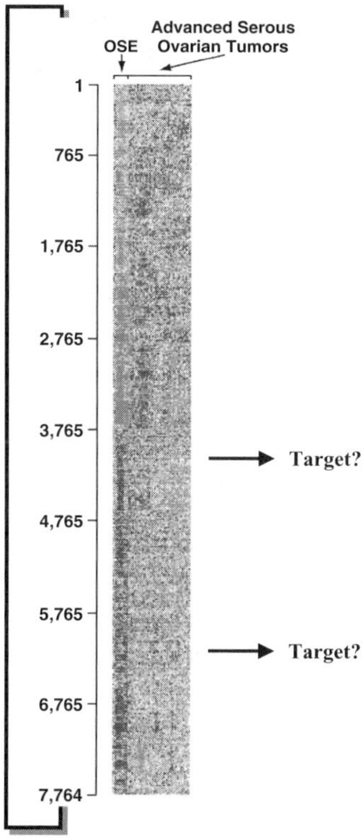

B

	Adv Serous Ovarian Cancer	
Gene Ontology Category	**Present**	**Number of Genes**
Mitotic Cell Cycle	Yes	70
M Phase	Yes	66
Mitosis	Yes	51
G2/M Transition	Yes	14
Cytokinesis	Yes	28

Figure 7.4 (A) Identification of critical signaling pathways. (B) Calculation of significance of probe set with GO category.

7.6 *IN SILICO* PATHWAY IDENTIFICATION

One of the more challenging aspects of expression profiling is determining the biologic interactions and relevance from the large number of differentially expressed genes discovered by whole genomic profiling. Software packages such as PathwayAssist version 3.0 software (Iobion Informatics LLC, La Jolla, CA) aid in identifying co-regulated pathways. This software package contains >500,000 documented protein interactions acquired from PubMed using the natural language processing algorithm MEDSCAN. PathwayAssist allows the enhancement of identification algorithms used to derive protein–protein interactions, a continual increase in the amount of scientific knowledge, and an ability to develop more detailed user/system specific databases. This proprietary database can be used to develop a biological association network (BAN) to identify putative signaling pathways. By overlaying expression data over the BAN, co-regulated genes defining specific signaling pathways can be identified. In this way, one can select genes and pathways that might be activated within individual patients, and perhaps distinguish between genes that are causal versus markers.

Our initial characterization of papillary serous ovarian cancer sought to differentiate gene expression profiles between advanced-stage serous ovarian cancers and normal OSE [8]. A comparison of 37 papillary serous advanced ovarian cancers to six normal surface epithelium brushings revealed 1,191 differentially expressed genes that differed by 1.5-fold or greater [8]. Of the 1,191 differentially regulated genes, slightly more were underexpressed (54%) in ovarian cancer compared to normal ovary brushings than overexpressed (46%). Over half (56%) of the differentially expressed genes code for proteins whose functions have not been characterized and the remaining 44% of the genes encode proteins involved in numerous biologic functions including cell adhesion, apoptosis, growth, and differentiation [8]. Utilizing PathwayAssist, we were able to identify a series of interacting pathways which clearly contribute to the development of this disease (Figure 7.5).

Whole-genome expression data from advanced-stage papillary serous ovarian cancer identified PAR1, PAR2, MT-SP1, SNX1, GPRK5, MAGP2, HEF1, FAK, VAV3, YES, CDC42, RECK, ET-1, IAP, and MT1-MMP genes as coordinately differentially regulated between cancer and normal OSE [8]. PAR1, PAR2, HEF1, VAV3, CDC42, MAGP2, RECK, SNX1, and GPRK5 had not been previously identified as being dysregulated in serous ovarian cancer. Pathways were identified by incorporating these microarray results (genes that are differentially expressed between normal and malignant ovarian epithelial cells) into PathwayAssist (Figure 7.5). Green filled symbols represent genes that are down-regulated in cancer compared to normal ovarian epithelium, red solid symbols are genes that are up-regulated in cancer specimens compared to normal, and gray shaded symbols represent genes that did not show a significant difference between cancer and normal specimens. These data suggest that positive signaling through FAK coupled to the down-

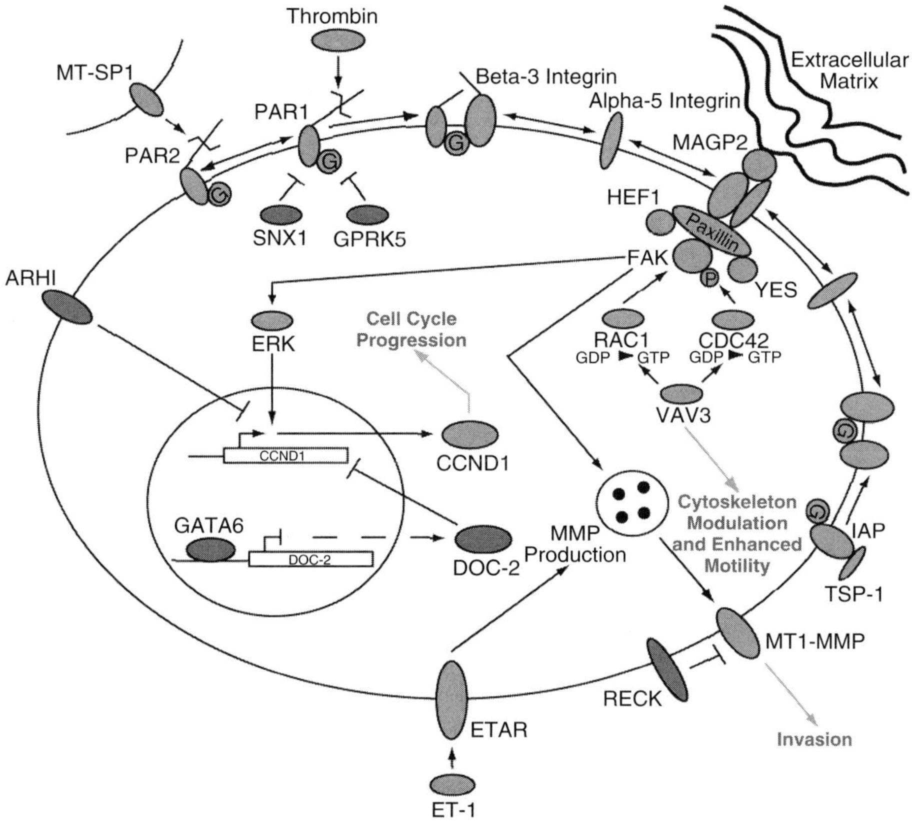

Figure 7.5 Schematic representation of potential signaling pathways involved in advanced serous ovarian cancer. See color insert.

regulation of RECK enhances MT1-MMP activity, resulting in increased invasion [8]. It is likely the deregulation of this pathway is important for the development of advanced-stage ovarian epithelial cancer.

PathwayAssist can identify specific activated pathways that expose potential mechanisms underlying clinicopathologic characteristics of tumors. For instance, the comparison of differentially expressed genes between serous low malignant potential (LMP) tumors and serous high-grade tumors identified molecular pathways that are unique to each group of tumors. Serous high-grade tumors revealed pathways involving cellular proliferation, metastasis, and chromosomal instability [5] (Figure 7.6A). In contrast, growth control pathways, such as the p53 pathway, characterize LMP tumors (Figure 7.6B). These pathway diagrams were generated with the assistance of PathwayAssist software using gene expression data. Genes included in either pathway were required to have a fold change value of 1.5 or more. Multiple Affymetrix probe sets for a gene were averaged. Differentially expressed genes identified

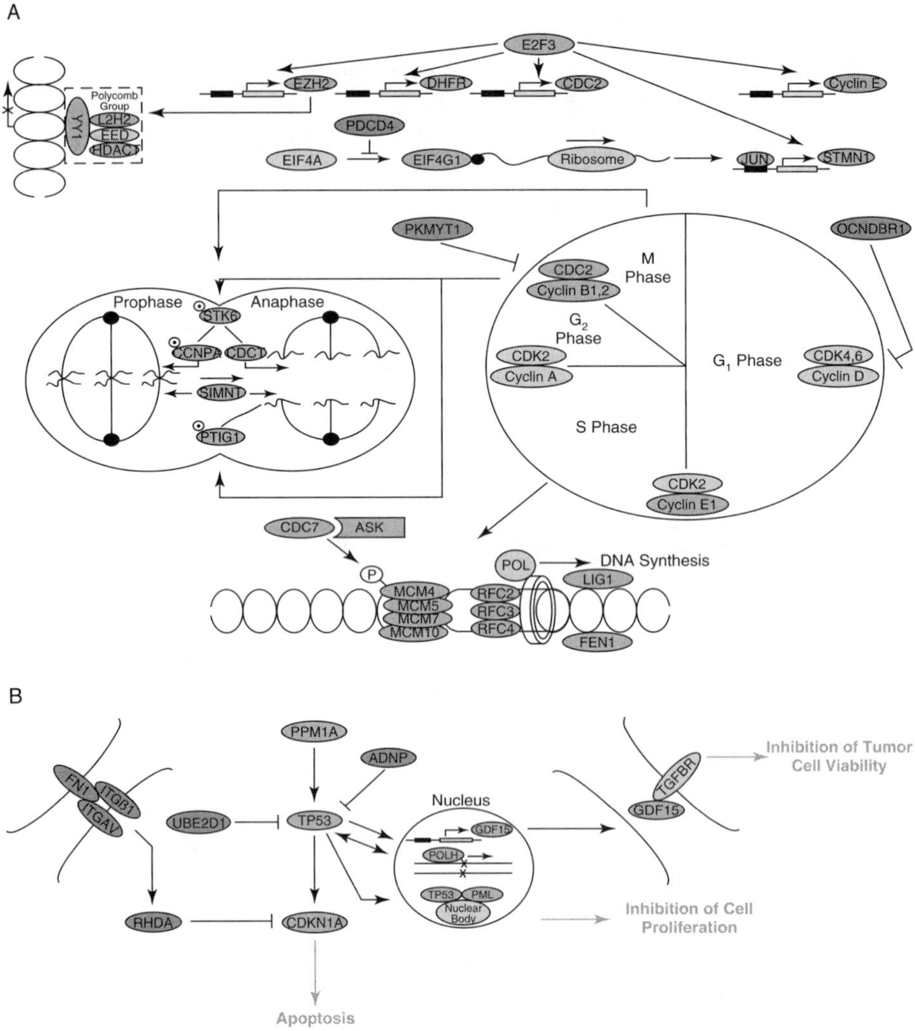

Figure 7.6 (A) Pathway analysis for proliferation and chromosomal instability in high-grade serous ovarian cancers versus OSE. (B) Pathway analysis of differentially regulated genes unique to LMP tumors versus OSE. See color insert.

for these tumors and their associated interactions are described as follows: green filled oval, the gene is down-regulated versus OSE; red filled oval, up-regulated; gray oval, genes displaying no change in expression (Figure 7.6A,B). For example, two negative regulators of p53, UBE2D1, and ADNP were down-regulated in LMP tumors [5]. This has important implications concerning the origin of these tumors and pathways that ultimately may be targeted for novel therapeutics. These results suggest that serous LMP tumors and low-grade ovarian carcinomas may represent a distinct classification of tumor rather than an early precursor in the development of the advanced high-grade malignancy.

7.7 NOISE

Once biologically relevant pathways have been identified, there still is a possibility that spurious gene expression data are present. "Noise" in microarray experiments can result from multiple sources: poor hybridizations, distortion during amplification, tumor heterogeneity, contaminating nonmalignant cells. To eliminate the "noise," independent validation is critical for expression profiling experiments. Real-time PCR quantification of gene expression can provide direct validation of the array result (Figure 7.7A). This technique is more quantitative than the array and therefore provides a much better estimate of true expression levels. However, since there is frequently a difference between RNA and protein expression levels, direct evaluation of protein levels is ideal. Immunohistochemical staining validation of differentially expressed genes from array results is commonly used to determine the expression levels of the protein products if appropriate antibodies are available (Figure 7.7B).

Another independent validation of differentially expressed genes contributing to the transformation process is the determination of DNA copy numbers. Comparative genome hybridization (CGH) is a powerful whole-genome assay available for detecting gene copy number at a given genomic complement [17]. Multiple CGH platforms have been developed, and have been used to identify prognostic markers for ovarian cancer [18,19]. cDNA array analysis is an alternative platform that can be used to assess DNA copy number variation on a gene-by-gene basis. In the general method of CGH, tumor cells from serous adenocarcinomas are procured by laser-based microdissection (Figure 7.8) [17,19]. DNA is extracted, amplified, labeled, and hybridized onto a 60-mer 22K oligonucleotide array platform. Scanning and signal quantification is then performed. Normalized tumor-to-control ratios are used for analysis [17,19]. Finally, to identify chromosome segments associated with cancer survival, copy number abnormalities (CNA), as well as the boundaries within the genome where these changes occur, are determined using a changing point algorithm.

In conclusion, Pathway analysis is a vital tool for identifying mutually reinforcing signaling networks in large gene expression datasets. In advanced serous ovarian cancer, this approach has identified pathways implicated in cell proliferation, invasion, motility, chromosomal instability, and gene silencing that may contain therapeutic targets. Serous LMP tumors have been characterized by activated p53 signaling leading to a low proliferative potential. Combined with validation data and additional genomic analyses, pathway analysis may aid in the identification of reliable causative genes contributing to the disease.

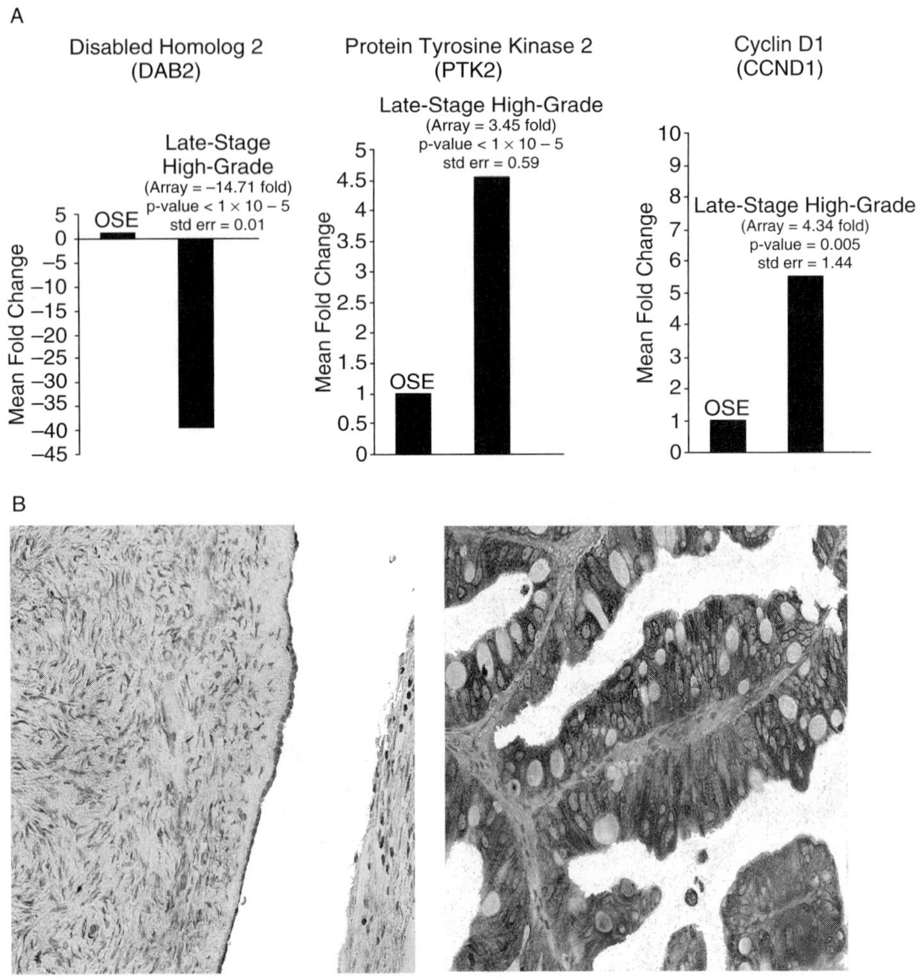

Figure 7.7 (A) Quantitative RT-PCR validation of microarray gene expression data. (B) Immunohistochemistry validation of differential protein expression.

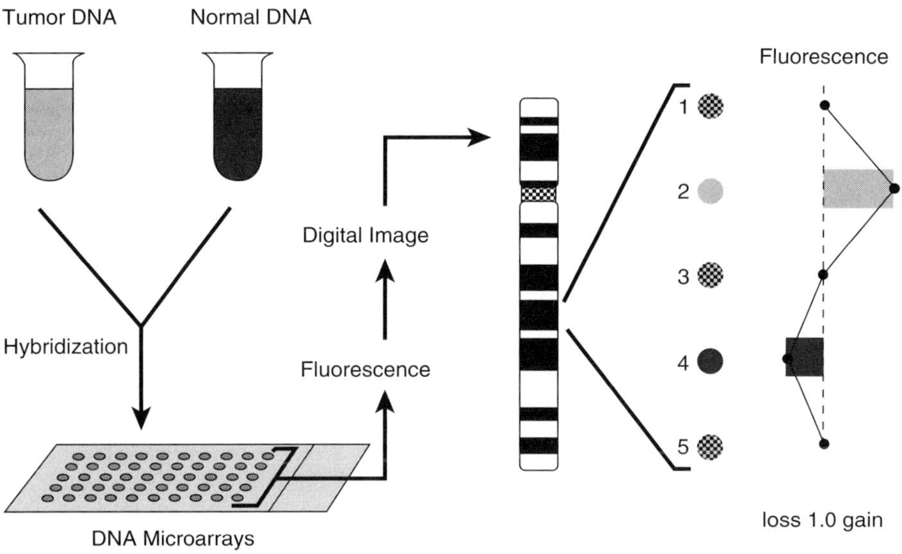

Figure 7.8 Comparative Genome Hybridization (CGH).

REFERENCES

1. Jemal A, Siegel R, Ward E, Murray T, Xu J, Thun MJ. Cancer statistics, 2008. CA Cancer J Clin 2008;58:71–96.
2. Tummala MK, McGuire WP. Recurrent ovarian cancer. Clin Adv Hematol Oncol 2005;3:723–736.
3. Roland PY, Barnes MN, Niwas S, Robertson MW, Alvarez R, Austin JM, et al. Response to salvage treatment in recurrent ovarian cancer treated initially with paclitaxel and platinum-based combination regimens. Gynecol Oncol 1998;68: 178–182.
4. Alberts DS. Treatment of refractory and recurrent ovarian cancer. Semin Oncol 1999;26:8–14.
5. Bonome T, Lee JY, Park DC, Radonovich M, Pise-Masison C, Brady J, et al. Expression profiling of serous low malignant potential, low-grade, and high-grade tumors of the ovary. Cancer Res 2005;65:10602–10612.
6. Zorn KK, Bonome T, Gangi L, Chandramouli GV, Awtrey CS, Gardner GJ, et al. Gene expression profiles of serous, endometrioid, and clear cell subtypes of ovarian and endometrial cancer. Clin Cancer Res 2005;11:6422–6430.
7. Feeley KM, Wells M. Precursor lesions of ovarian epithelial malignancy. Histopathology 2001;38:87–95.
8. Donninger H, Bonome T, Radonovich M, Pise-Masison CA, Brady J, Shih JH, et al. Whole genome expression profiling of advance stage papillary serous ovarian cancer reveals activated pathways. Oncogene 2004;23:8065–8077.

9. Berman JJ. Borderline Ovarian Tumor Workshop, Bethesda, Maryland, August 27–28, 2003. Hum Pathol 2004;35:907–909.
10. Bell DA, Longacre TA, Prat J, Kohn EC, Soslow RA, Ellenson LH, et al. Serous borderline (low malignant potential, atypical proliferative) ovarian tumors: workshop perspectives. Hum Pathol 2004;35:934–948.
11. Silverberg SG, Bell DA, Kurman RJ, Seidman JD, Prat J, Ronnett BM, et al. Borderline ovarian tumors: key points and workshop summary. Hum Pathol 2004;35:910–917.
12. Zorn KK, Jazaeri AA, Awtrey CS, Gardner GJ, Mok SC, Boyd J, et al. Choice of normal ovarian control influences determination of differentially expressed genes in ovarian cancer expression profiling studies. Clin Cancer Res 2003;9:4811–4818.
13. Singer G, Oldt R, III, Cohen Y, Wang BG, Sidransky D, Kurman RJ, et al. Mutations in BRAF and KRAS characterize the development of low-grade ovarian serous carcinoma. J Natl Cancer Inst 2003;95:484–486.
14. Chan WY, Cheung KK, Schorge JO, Huang LW, Welch WR, Bell DA, et al. Bcl-2 and p53 protein expression, apoptosis, and p53 mutation in human epithelial ovarian cancers. Am J Pathol 2000;156:409–417.
15. Lomax J. Get ready to GO! A biologist's guide to the Gene Ontology. Brief Bioinform 2005;6:298–304.
16. Zeeberg BR, Feng W, Wang G, Wang MD, Fojo AT, Sunshine M, et al. GoMiner: a resource for biological interpretation of genomic and proteomic data. Genome Biol 2003;4:R28.
17. Birrer MJ, Johnson ME, Hao K, Wong KK, Park DC, Bell A, et al. Whole genome oligonucleotide-based array comparative genomic hybridization analysis identified fibroblast growth factor 1 as a prognostic marker for advanced-stage serous ovarian adenocarcinomas. J Clin Oncol 2007;25:2281–2287.
18. Cheng KW, Lahad JP, Kuo WL, Lapuk A, Yamada K, Auersperg N, et al. The RAB25 small GTPase determines aggressiveness of ovarian and breast cancers. Nat Med 2004;10:1251–1256.
19. Tsuda H, Ito YM, Ohashi Y, Wong KK, Hashiguchi Y, Welch WR, et al. Identification of overexpression and amplification of ABCF2 in clear cell ovarian adenocarcinomas by cDNA microarray analyses. Clin Cancer Res 2005;11:6880–6888.

8

MAMMALIAN PROTEOME AND TOXICANT NETWORK ANALYSIS

SEAN EKINS AND CRAIG N. GIROUX

Table of Contents

8.1 INTRODUCTION

We are beginning to understand organisms in the context of computationally generated networks in which individual interacting proteins are tightly connected together as functional modules, whose higher-level associations form the "small world" topology of the intact cellular system [1]. In this emerging network paradigm, traditional units of cellular structure and function are being revisualized as modular components of the global genetic network. A

Pathway Analysis for Drug Discovery, Edited by Anton Yuryev

disease state is then visualized as a perturbation in the topology of the normal or "healthy" cellular network. For example, the centrality of the mitochondrion as a key cellular site for the interaction of metabolic growth control and death signaling pathways has made it a focus for investigating how protein networks are perturbed in chemical or drug-induced toxicity, and consequently for elucidating the role of toxicant-induced oxidative stress in the etiology of complex diseases [2].

The human heart mitochondrial proteome has recently been characterized [3,4] and reconstructed by high-throughput techniques [5]. Comprehensive proteome datasets for the mouse liver mitochondrial inner membrane [6] and rat liver [7] have similarly been described. Although all of these descriptions of mammalian proteomes should be considered incomplete works in progress, they are presently sufficient to demonstrate the power of network analysis to visualize the functional organization of a set of structurally co-localized protein components and to predict novel, previously undetected protein interactions in this module. In addition, public databases have been established specifically for toxicogenomics data to facilitate comparison of the growing number of mammalian transcription profiling studies of exposure to chemicals or drugs [8,9]. Unsupervised methods of data mining, such as statistical clustering analyses, are commonly performed to compare different toxicant exposures [10]. Recently however, both academic and commercially available software to generate network-based or statistical models from these toxicogenomic profiles have become available [11–13]. Toxicant response networks can be derived with Bayesian statistical methods, as recently demonstrated using lung airway epithelial cells exposed to TCDD which resulted in two networks associated with the Ah-receptor and the retinoic acid receptor beta [14]. As an alternative approach, gene expression data can be mapped to a global database of literature-curated gene interactions and graph theoretic methods that can be used to construct toxicant-specific networks from these mapped data [12,15,16].

The first attempts at consolidation of co-expression clustering and the deduction of functional networks from literature databases were performed with metabolic pathway data from model genomic organisms, *E. coli*, and yeast [17]. Despite this proof of concept, we are still at the early stages of the functional modeling of expression data, and relatively few studies have applied these new computational tools to the analysis of mammalian gene networks. A developing approach to facilitate mammalian network modeling exploits the observation that gene–gene interactions and patterns of gene regulation are highly conserved between species. The initial studies examined to date suggest that the topology of network connections between genes in a functional module may be even more highly conserved than are the sequences of the orthologous components [18]. This emerging principle is consistent with our biological understanding of the regulation of evolutionarily conserved processes such as the core metabolism, cell cycle, and cellular defense mechanisms including the oxidative stress response [19–21].

Biological functions are performed by modules of closely interacting proteins [22] that can be identified by applying similarity grouping algorithms to microarray expression data, such as has been demonstrated for yeast [23,24]. Regulatory modules can be identified by algorithms that link expressed genes with known transcriptional factors via the genome-wide identification of DNA binding sites [25,26] or enable building of condition-specific probabilistic models [27]. Many different algorithms have been used to computationally parse large networks into modules including those based on network connectivity and clustering methods such as Monte Carlo optimization [28], shortest path length distribution [29], and other graph topology-based algorithms [30]. Clusters identified in this way generally correspond to known protein complexes or metabolic pathways [28]. An additional graphical approach to discover smaller, functionally interacting sets of genes relies on identification of gene network motifs [31]. Gene network motifs are elemental subgraphs with topological features that confer specific functionality, such as feedback or feed-forward loops. Certain motif patterns are highly enriched in cellular networks over random node connection topologies, and local gene network motifs appear to be conserved between species, suggesting that they perform critical informational processing functions within the context of a larger cellular process-specific module [32].

To date, motif-structured modules have been identified in regulatory networks of *E. coli* [33,34] and in yeast–protein interaction networks [35]. A common feature of these regulatory networks are hub nodes which have a high degree of connectivity and which are linked to many low-degree nodes [1]. This topological organization gives networks the property of robustness such that removal of even a substantial fraction of non-hub nodes still allows the network to retain comprehensive connectivity [36]. Hub nodes in the yeast protein–protein interaction network have been classified further into either party hubs that constitutively interact with multiple partners or data hubs that conditionally bind different partners dependent upon the specific conditions of place or time [37]. Compared to non-hub nodes, party and date hubs are more likely to be essential, suggesting that there is a hierarchy of hub function, at least in yeast. Hub nodes therefore provide candidate "druggable" target genes for therapeutic intervention and functional pathway modulation.

The well-established network analysis of functional modules in yeast and the apparent conservation of network architecture between species suggests that it should be possible to apply the analytical methods described above in order to map toxicant-specific networks in both yeast and in mammals. One study has assessed the effect of four carcinogens on nearly 5,000 yeast gene deletion strains in order to define functionally validated protein subnetworks that are important for toxicity [38]. In this yeast study, profiling data from tert-butyl hydroperoxide exposure, which causes oxidative stress, was used to construct a toxicant-specific network that identified key toxicity-modulating gene nodes in the endosome, ER, actin, and vacuolar membrane modules [38].

It is important to note that although there is a highly conserved cellular response to oxidative stress [21,39], differences in stress-induced gene expression between yeast and cultured human cells have been reported [40,41]. Principally, the strong general toxicant stress response that is observed in yeast has not yet been reported in cultured human cells, and secondly both cell type and toxicant-specific stress responses are observed in human cells [40,41]. There have been numerous studies to assess oxidative stress by transcription or protein expression profiling in mammalian cells, including studies in epithelial lens cells [42,43], human hepatocyte cell line Hep-G2 [44,45], rat hepatocyte cell line H4IIE [45], porcine kidney cell line LLC-PK1 [45], lymphocyte cell line K562 [45], prostate cell line DU-145 [46], breast carcinoma cell line MCF-7 [46], and additional tumor cell lines [41,47]. These studies have typically used microarray profiling of gene expression data and clustering algorithms to identify gene sets that constitute the mammalian antioxidant defense system. In addition, many recent toxicogenomic profiling studies of known hepatotoxicants and nephrotoxicants demonstrate that a broad spectrum of these agents induce components of the oxidative stress response [48–54]; this work has been reviewed recently [55]. While similar to the studies of known oxidative stressors, these drug and chemical toxicogenomic studies also relied on microarray expression profiling and clustering algorithms to identify a subset of toxicant-responsive genes.

Notably to date, there are few applications of gene network analysis to mammalian expression profiling datasets. Although a recent analysis utilizing a Bayesian network algorithm further implicates oxidative stress as a mechanism of acetaminophen toxicity in a rat exposure model [56], there have been no specific analyses of mammalian oxidative stress response networks. We address this deficiency in this present chapter, in which we demonstrate that our gene network approach is able to map a significant number of components from the human heart mitochondrial proteome onto the human global gene interaction database. We further demonstrate that our approach supports comparative network analysis and enables predictive toxicology by our mapping of yeast orthologs involved in the oxidative stress response onto a human gene interactome database [57]. Finally, we demonstrate that toxicant-specific gene subsets can be visualized by network mapping of gene expression profiling data from exposure studies of known agents that cause oxidative stress, including tert-butyl hydroperoxide, acetaminophen, furan, benzene, carbon tetrachloride, and cisplatin. For these examples, a subset of toxicant-specific genes was first selected using a traditional hierarchical clustering algorithm, then mapped onto the human interactome database, and finally visualized in the context of the local network of interacting genes. Using our network-based approach, we are able to identify biologically important hub genes whose significance may be missed when using conventional methods of statistical analysis. This report continues our discussion of the application of gene network approaches to the analysis of global profiling data in toxicology [12,13,55,58–63].

8.2 METHODS

8.2.1 Database, Network Algorithms, and Filters

The development of human interactome databases has been thoroughly described in detail previously [12,58,59] (e.g. for MetaCore™ and MetaDrug™, GeneGo, St. Joseph, MI). Within the MetaCore™ platform, networks are generated as a combination of binary single-step interactions (edges) that connect proteins and genes (nodes). The nodes and edges of the graph network are derived from corresponding interaction tables in the MetaCore™ database and are visualized as clusters of interconnected nodes with the Macromedia Flash Player Plug-in. The end nodes on the networks have only one edge; the internal nodes may have from two to several hundred edges. Condition-specific networks can be built from any input list of genes, proteins, and compounds corresponding to the components (network classes) in the database. The nodes in the input list are considered root nodes. The input list of toxicant-specific components can be selected by any classification algorithm, which thus enables testing of multiple data preprocessing schemes. Identifiers (gene and protein names) of the input component list are recognized in the software platform by a built-in synonyms dictionary. Input gene lists can be imported as text or Excel files prepared by the investigator or they can be directly parsed from the output files of Affymetrix, Agilent, GeneSpring, and other microarray analysis software. MetaCore™ recognizes most of the commonly used gene and protein identifier formats such as LocusLink, SwissProt, RefSeq, and Unigene. For input components that are named with historically assigned Genbank accession numbers, the DAVID gene nomenclature database http://apps1.niaid.nih.gov/david/ [64] can be used to convert to contemporary LocusLink identifiers for data entry into the MetaCore™ platform.

First-pass data filtration: Prior to building networks, the interaction parameters are first preselected, based on the required level of trust, interaction direction, effects, mechanisms, and tissue specificity (i.e. only edges in which both nodes belong to a selected tissue are utilized in the network). Unlinked nodes from the input list, which have no connections to other nodes on the list, are removed. The edges of networks are then assigned weights that are dependent on the type of interactions between the components (e.g. enzyme and substrate, product and precursor, etc.). Once an investigator establishes an initial list of objects (genes or proteins translated into network "classes"), the relationships among these components are visualized as a graph network of nodes linked by edges that are specified by the connectivity relationships in the MetaCore™ database interactions table. To construct the final condition-specific network model, edges (connections determined by component interactions) are assembled into pathway-based clusters by using one of several available graph network algorithms, such as direct interactions, analyze networks, etc. [58,59].

8.2.2 Mapping Experimental Data and Network Statistical Analysis

The object identifier and interaction relationships of each node in the network (genes and proteins) are determined by reference tables in the general database schema. This database schema enables the computational mapping of HT experimental data from genomics or proteomics profiling experiments onto a human interactome-based network. Each experimental data point (for example, probe signal intensity from a microarray hybridization or sequence tag frequency from a SAGE distribution) represents an attribute of the unique gene or protein identifier. Thus, the HT data are linked with the corresponding identifier in the database table and their relative value is visualized on any network containing this corresponding object by a solid circle above the node (red and blue represent increased and decreased levels, respectively; for examples, see the figures in this chapter).

The "analyze networks" algorithm used in this study includes both network objects from the input gene list and closely connected network objects from the interactome database. This algorithm generates a large network that is then partitioned into subnetworks, each of which is ranked with a *z*-score and *p*-value according to its saturation with objects from the initial gene list. A second algorithm used in this study, the "Direct Interactions" method, produces a conservative condition-specific network by using only the network objects present in the input gene list [59], and has been described previously [12]. Gene ontology (GO) processes are also mapped to network objects (gene or protein) and the corresponding *p*-value rankings are calculated for the statistical representation of each GO process in a network. The equations for the *z*-score and *p*-value calculations have been described previously [59].

8.3 APPLICATIONS

8.3.1 Human Heart Mitochondrial Proteome Networks

The mitochondrion has a central role in the cellular response to oxidative stress, both as a critically sensitive cellular target for oxidative damage and as a mediator of programmed cell death responses [65,66]. However, genomics- and proteomics-based modeling of the human mitochondrion has lagged behind similar whole-cell analyses. Therefore, as a proof of concept for our network building approach, the available human heart mitochondrial proteome data (722 proteins) [3,4] were modeled using the MetaCore platform. In our analysis, 53.9% of the human mitochondrial proteome was mapped to described objects in the human MetaCore database. The "direct interaction" algorithm (previously described [12]) was then applied [3,4] to this dataset to identify interactions between the 388 mitochondrial components present in the interactome database and to enable visualization and further interrogation of the mitochondrial network. Many small connected subnetworks, nonconnected proteins, and one relatively large subnetwork were produced

(Figure 8.1A). The significantly enriched GO processes were then identified in this dataset, as a test of the ability of our network model to represent known mitochondrial pathways and processes. Reassuringly, the canonical mitochondrial functions including electron transport, the NADH to ubiquinone pathway

A

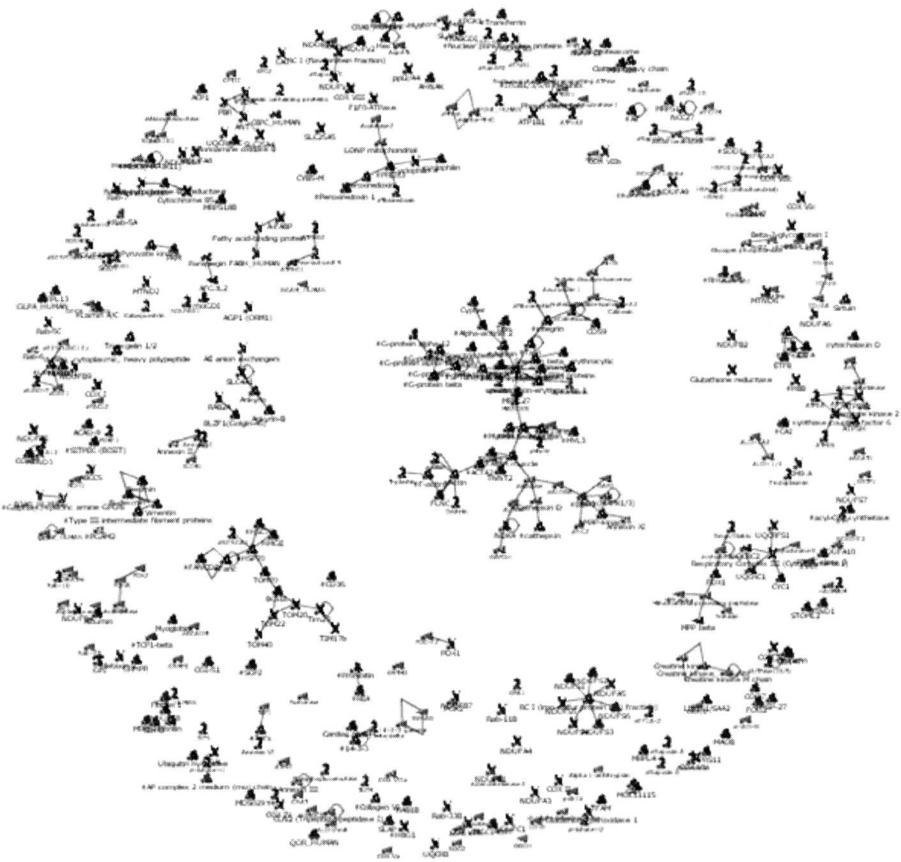

Figure 8.1 Human heart mitochondrial proteome networks. (A) A global "direct interaction" network generated with MetaCore™ (www.genego.com) [12] using 388 out of 630 uploaded protein identifiers (LocusLink identifiers obtained from DAVID http://apps1.niaid.nih.gov/david/) from an initial list of 722 proteins [3,4]. (B) Detail of the largest central network (in Figure 8.1A) observed with the interactions on edges (between nodes) hidden for clarity. Details of the node definitions have been previously published [12]. The solid blue dot highlighting ERM proteins represents a down-regulated gene identified in a published microarray profiling study of trovofloxacin-treated human hepatocytes [67]. (C) Subcellular localization of proteins for the most significant network using Ingenuity Pathways Analysis™ (https://analysis.ingenuity.com). Proteins highlighted in solid color are those that mapped from the list uploaded (solid lines = direct interaction; dashed lines = indirect interaction).

Figure 8.1 (*Continued*)

($p = 2.5 \times 10^{-28}$), and energy and metabolite production pathways were present with high statistical significance in our network (Table 8.1).

The largest subnetwork of protein interactions (Figure 8.1B) was then analyzed separately. In this large subnetwork, the integrin-mediated signaling ($p = 4.2 \times 10^{-71}$) and cell matrix adhesion ($p = 8.96 \times 10^{-68}$) pathways were the most statistically significant GO processes (Table 8.2), which suggests to us that the original mitochondrial preparation used for the proteomic analysis may have also contained fragments of the outer cell membrane. This can be seen more clearly in Figure 8.1C, which represents an analysis using Ingenuity Pathways Analysis (https://analysis.ingenuity.com) where proteins for the most significant network (network score = 56) are positioned in their subcellular locations including the outer cell membrane and extracellular space.

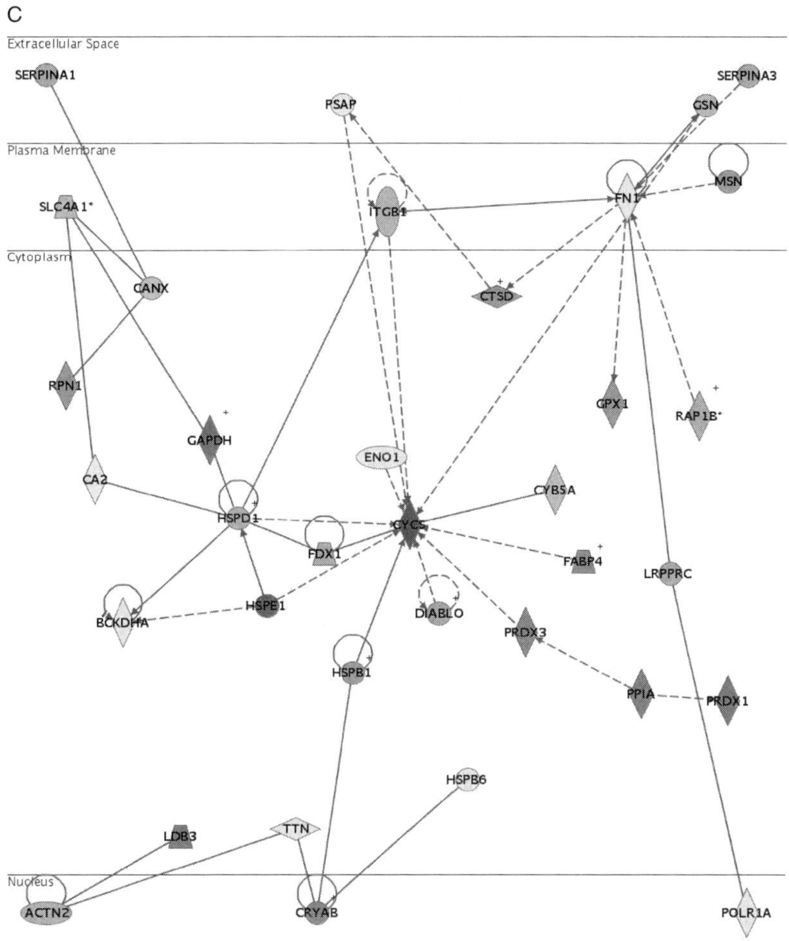

Figure 8.1 (*Continued*)

TABLE 8.1 Complete GO Process for All Proteins

Processes for Input List

#	Process	Ratio	*p*-value
1	Mitochondrial electron transport, NADH to ubiquinone	23/26	2.5131e-28
2	Electron transport	48/185	1.2775e-23
3	Glycolysis	21/37	3.6882e-19
4	Proton transport	17/34	1.6685e-14
5	Generation of precursor metabolites and energy	28/110	5.6823e-14
6	Muscle development	28/117	3.0304e-13
7	Tricarboxylic acid cycle	13/21	5.3712e-13
8	Metabolism	33/185	1.4309e-11
9	Muscle contraction	22/94	1.8001e-10
10	ATP synthesis coupled proton transport	11/21	4.2328e-10

TABLE 8.2 GO Process for the Largest Subnetwork

Processes for Input List

#	Process	Ratio	*p*-value
1	Integrin-mediated signaling pathway	48/73	4.2166e-71
2	Cell-matrix adhesion	48/81	8.9605e-68
3	Cellular extravasation	27/28	1.1258e-48
4	Cell-substrate junction assembly	21/21	6.2479e-39
5	Muscle development	37/117	7.2683e-39
6	Neutrophil chemotaxis	26/40	5.8860e-38
7	Homophilic cell adhesion	29/57	7.4972e-38
8	Antimicrobial humoral response (sensu Vertebrata)	32/110	3.1128e-32
9	Cell adhesion	47/337	7.3760e-32
10	G1/S transition of mitotic cell cycle	25/63	3.3484e-29

Indeed, this result demonstrates the utility of network analysis for the biological interpretation of molecular profiling data from experimental studies of cell structure and function. The fact that we observe multiple small subnetworks of the known mitochondrial functions with MetaCore™, rather than one large network is consistent with the limited extent of the literature for known interactions of the components of cellular organelles, as compared to the more extensively characterized set of known interactions for the soluble, cytoplasmic proteins. Connectivity gaps in the global mitochondrial network are expected since nearly 50% of this mitochondrial proteome dataset fails to map to the MetaCore™ interactome database. For comparative purposes, 78.4% of these objects are not mapped in the KEGG public database (http://apps1.niaid.nih.gov/david/) that is also curated from the literature. We have previously noted differences in object mapping between the KEGG and MetaCore databases [55]. Finally a comparison with Ingenuity Pathways Analysis v4.0 found that 60.6% of the objects failed to map to this database (as 285 out of 722 proteins mapped). These percentages will undoubtedly change as newer versions of the software are released.

Interactome databases and network building tools can also be used to interpret high-throughput toxicity profiling data [12,13,55,58,59,61]. Published transcription microarray data derived from human hepatocytes treated with trovofloxacin [67], a drug that is known to cause mitochondrial damage and result in idiosyncratic toxicity, were mapped onto our mitochondrial proteome network. Five toxicity-associated genes mapped onto the protein network in Figure 8.1A, including nuclear pore complex (up-regulated), cyclophilin, transcription initiation factors, PDC-E2, and ERM (all down-regulated). Only ERM was present on the largest network, which may be nonspecific for mitochondria (Figure 8.1B). PDC-E2 and the nuclear pore complex have also been implicated in biliary cirrhosis, which suggests that they may provide mitochondrial protein biomarkers for toxicity [68]. Thus, our study demonstrates that the use of a network model for targets of cellular toxicity, such as the mitochondrion, provides a novel tool to evaluate pathways and mechanisms of hepatotoxicity from drug toxicity profiling studies.

8.3.2 Yeast Ortholog Mapping and Comparative Network Modeling

Yeast cell-based assays are increasingly being used in high-throughput screening both for mechanistic toxicity testing of xenobiotics and for drug discovery of molecular therapeutics. In addition, the powerful yeast genetic system enables high-throughput determination of chemical genetic profiles of compounds, providing insights into their mechanisms of action [69,70]. The high degree of conservation between both individual cellular components and complex cellular processes in yeast and human cells, suggested to us that ortholog mapping could be used to model the human toxicant network from yeast high-throughput toxicity data. In particular, the yeast mitochondrion has served as the paradigm system for understanding molecular aspects of toxicity and human disease relevance of this organelle [71].

We have demonstrated previously that yeast cells recapitulate the highly conserved oxidative stress response [39] and that the yeast oxidative stress response network exhibits a modular organization that integrates cytoplasmic and mitochondrial components [72]. Therefore, we propose that an oxidative stress response network in yeast can be used to computationally model the corresponding mitochondrial stress interaction network in human cells, using orthologous genes as seeds to build the corresponding human network [57]. As an example of this comparative network approach, we have mapped a module of 77 yeast oxidative stress-response genes, which contains 29 human orthologs, onto the human interactome. In this specific example, 21 of 29 orthologs are highlighted on networks (Figure 8.2A). Furthermore, these closely connected, cross-network mapped orthologs were used with the "analyze network" algorithm to construct predictive subnetworks for the oxidative stress response in the human interactome. One such subnetwork of the human interactome consists of nine orthologs ($p = 8.791e^{-18}$) and predicts the key involvement of several redox-responsive transcription factors (p53, CREB1, MEK2, and PKC) that are known to be involved in the mammalian response to oxidative stress (Figure 8.2B).

8.3.3 Network Analysis of the Oxidative Stress Response in Hepatotoxicity

In order to illustrate the broad utility of our network-based approach to capture "signatures" for the oxidative stress response, we have analyzed toxicity profiling datasets from the literature that include diverse drug and chemical toxicants studied in several rodent or human exposure systems. For example, an expression profiling dataset from human fibroblasts treated with tert-butyl hydroperoxide [73] identified 80 toxicant-responsive genes, 68 of which mapped to the human interactome in networks. In this example, a "direct

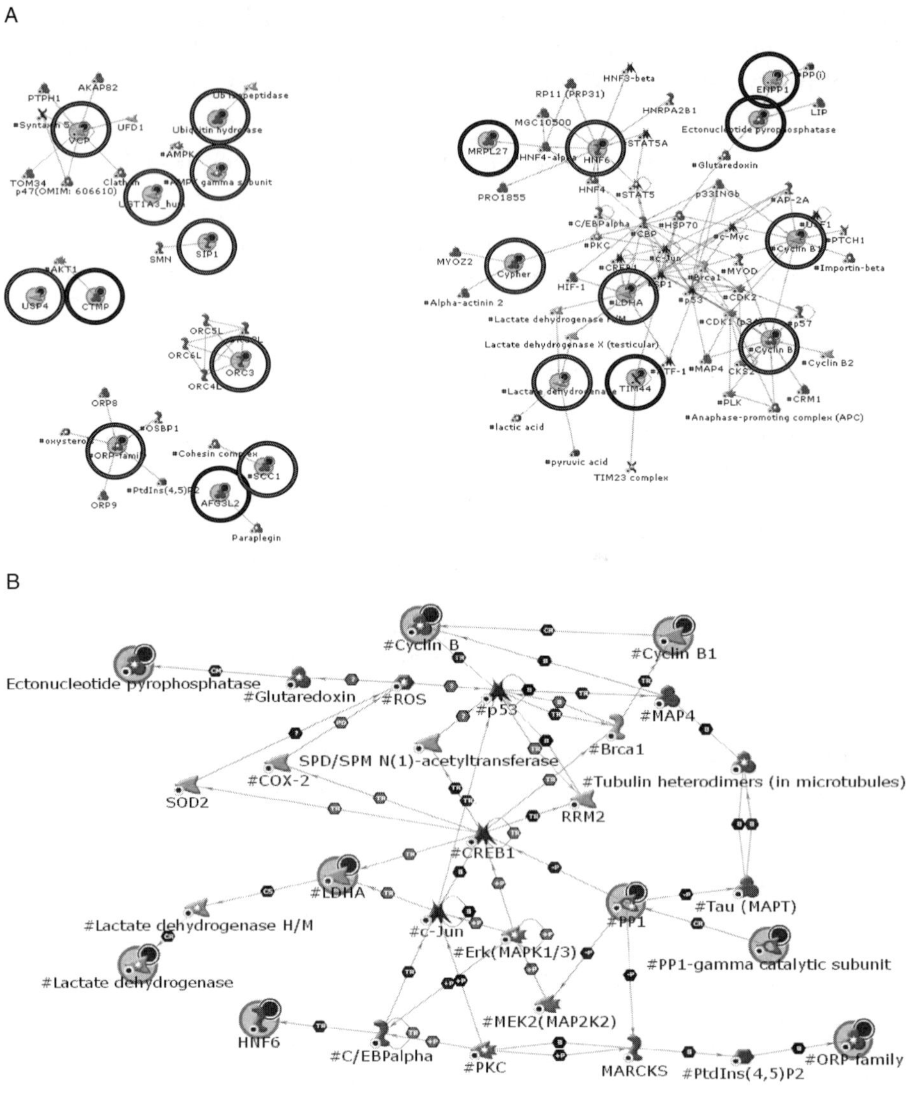

Figure 8.2 (A) Ortholog mapping of a yeast oxidative response module [57] to the human mitochondrial proteome network. For clarity, the PP1 gene is not shown. (B) Conserved human gene subnetwork for the oxidative stress response using the "analyze network" algorithm ($p = 8.791e^{-18}$).

interaction" algorithm-based network was generated, which highlights up-regulated subnetworks for the DNA repair and cell cycle as well as heat shock processes (Figure 8.3). This result is consistent with known key cellular stress responses and confirms our expectation that network-based analysis can reveal functional connections between co-responsive processes underlying cellular defense to toxicant insult.

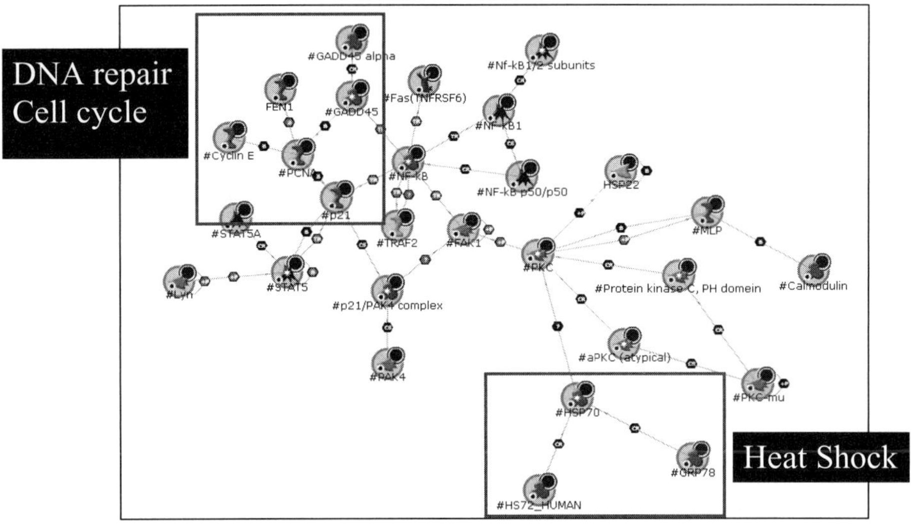

Figure 8.3 A toxicity subnetwork modeled in MetaCore™ and generated using microarray data from human fibroblasts treated with tert-butyl hydroperoxide [73]. This network demonstrates two known response pathways for oxidative stress.

To examine adverse drug effects resulting in hepatotoxicity, we constructed toxicant response networks from two separate studies using rats treated with increasing doses of acetaminophen, in which liver toxicity was profiled by microarray analysis [74]. The first study identified 30 toxicity-responsive genes, 23 of which we mapped to the rat interactome in MetaCore. Modeling with the "analyze network" algorithm for this small number of genes, yielded subnetworks related to glycolysis and mitochondrial function (z-score 41.58, Figure 8.4A). At a higher dose of acetaminophen, 89 differentially expressed genes were identified, 64 of which we mapped to the rat interactome. These data yielded a larger subnetwork containing key modules that are connected to p53 and SP1, key regulatory genes related to cell cycle control and oxidative stress (z-score 51.98, Figure 8.4B). Our network-based analysis is consistent with the postulated role of oxidative stress as an underlying mechanism of hepatotoxicity from acetaminophen treatment [75].

Microarray profiling of rats treated with the hepatotoxicant furan (45 mg/kg) [53] identified 181 differentially expressed genes, 139 of which were used to construct a "direct interaction" toxicity network (Figure 8.5A). This resulting large network is comprised of two separate modules or subnetworks, one of which contains up-regulated genes (on the right side of Figure 8.5A) and the other of which contains down-regulated genes (on the left side of Figure 8.5A). In particular, FXR was down-regulated along with many cytochrome P450 genes, while cell cycle and cell proliferation genes were up-regulated. When the "analyze network" algorithm is used (z-score 47.44, Figure 8.5B),

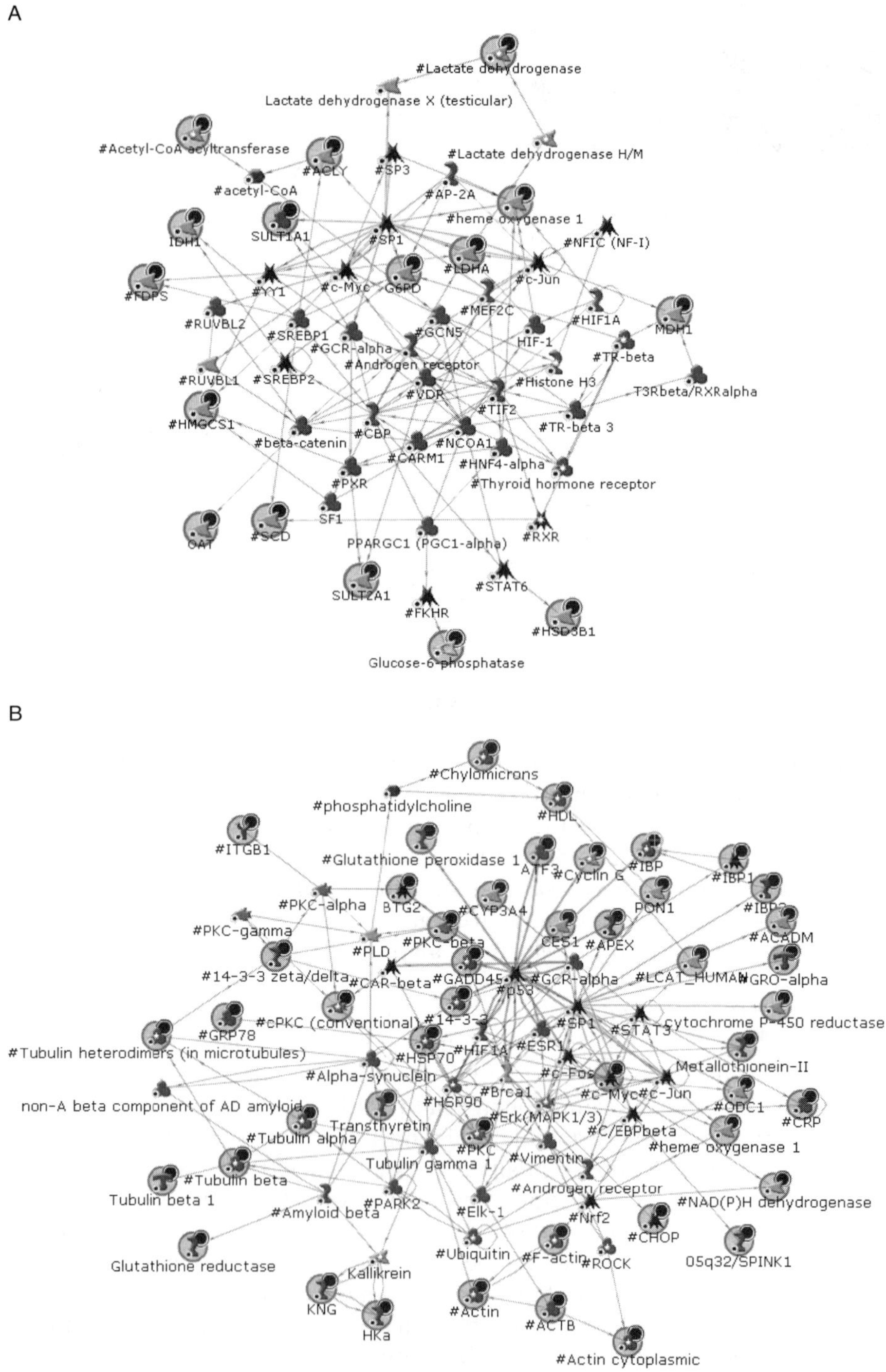

Figure 8.4 Networks for acetaminophen exposure in the rat liver. (A) A 23-gene network constructed from 30 genes mapped to MetaCore™ [74] (B) A 64-gene network constructed from 89 genes mapped to MetaCore™ [53].

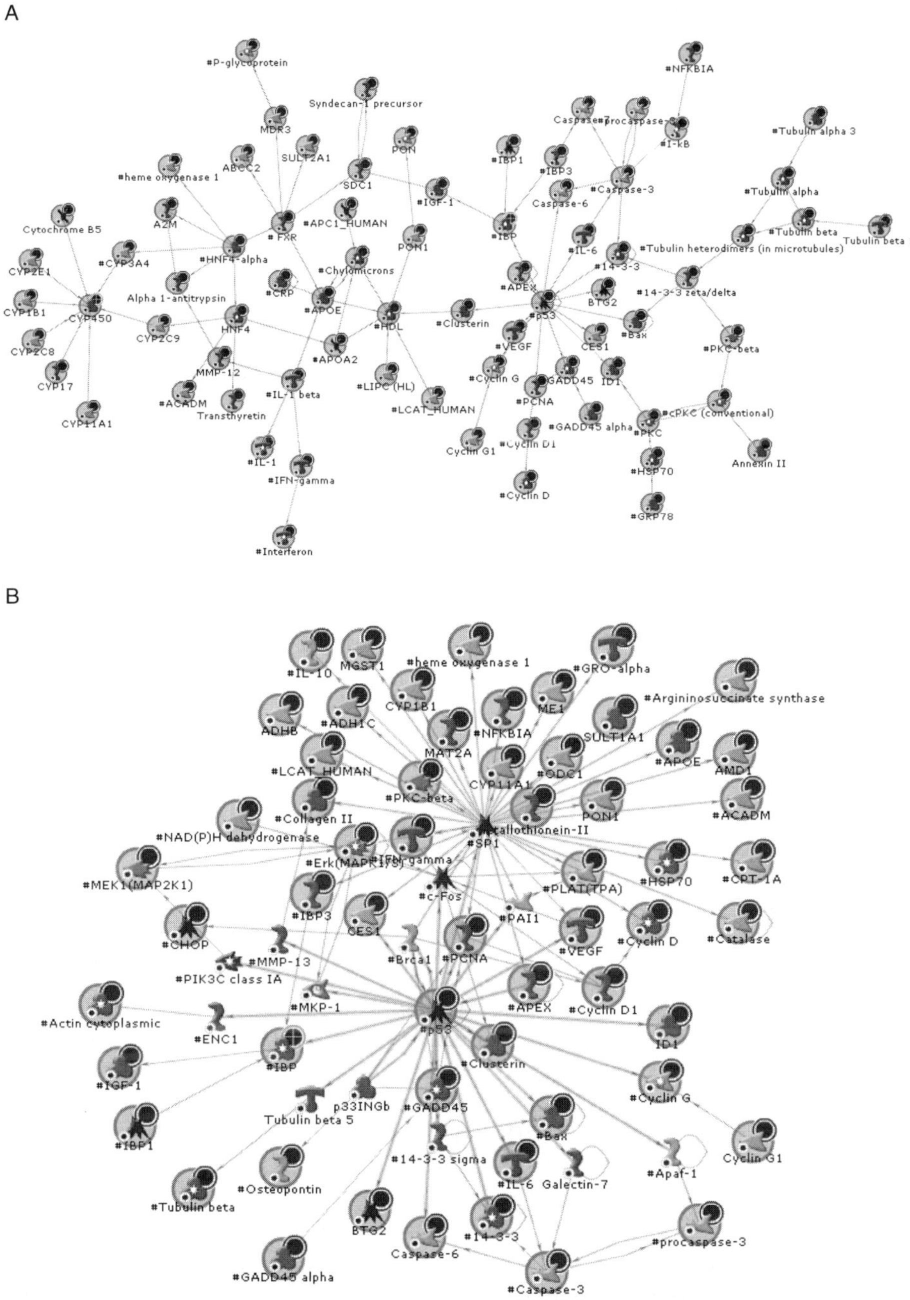

Figure 8.5 Networks for furan exposure in the rat liver. (A) A 139-gene network constructed from 181 genes mapped to MetaCore™ [53] using the "direct interaction" algorithm. (B) An expanded network constructed with the "analyze network" algorithm.

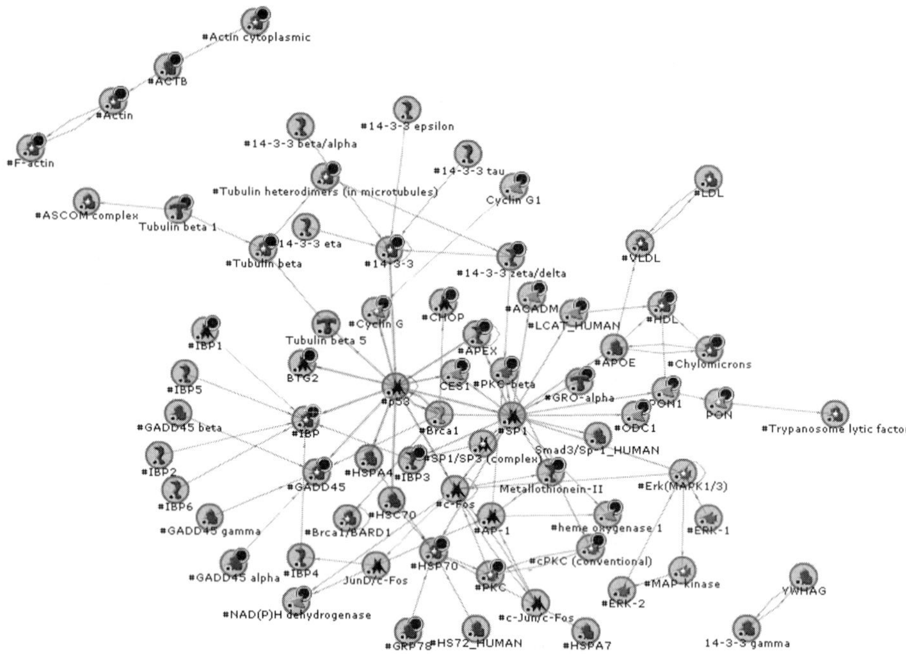

Figure 8.6 A shared hepatotoxicity network for acetaminophen and furan using MetaCore™.

the furan network genes are organized around two central hubs, p53 and SP1. Notably, SP1 was not identified on the direct interactions network and therefore exhibits different network connectivity than does p53 to the oxidative stress response genes. For example, heme oxygenase 1 is connected to SP1 on the "analyze network"-based network, whereas on the direct interactions network it is connected to HNF4α. Our analysis of the furan response network demonstrates the utility of network-based analysis to reveal key architectural features of the regulatory system underlying drug-induced hepatotoxicity.

Since one of the acetaminophen networks and both of the furan networks were generated with microarray data from the study of Huang et al. [53], it was possible to compare the overlap of these different toxicant networks for their shared genes. A "direct interaction" network was then generated from this shared gene list. In this common toxicity network, the p53 and SP1 genes have central roles at the core of a large network which has two much smaller, peripheral subnetworks (Figure 8.6). Thus, network-based visualization enables a comparative toxicogenomic analysis of the shared versus the specific cellular pathways that underlie drug-induced hepatotoxicity.

Two microarray profiling datasets from mice treated with benzene at either 300 ppm for 6 h/day or 5 days/wk [76] or 100 ppm for 6 h/day or 5 days/wk [77] were used to generate networks to examine differences between low-dose

A

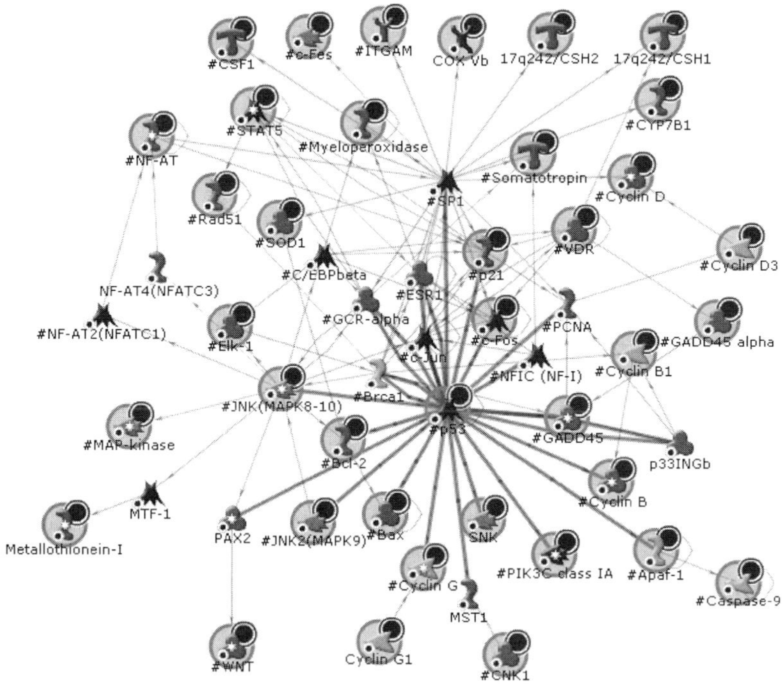

Figure 8.7 Benzene toxicity networks. (A) A 63-gene network constructed from 73 genes mapped to MetaCore [76]. (B) A 53-gene network constructed from 76 genes mapped to MetaCore™ using the "analyze networks" algorithm [77]. (C) Overlap of the low-dose versus high-dose gene lists on networks.

versus high-dose responses. The high-dose exposure datasets identified 73 toxicity-associated genes, of which 63 mapped in MetaCore™. The "analyze networks" algorithm generated a high-dose network that was also centered about SP1 and p53 hubs (z-score 53.44, Figure 8.7A). Similarly the low-dose exposure dataset identified 76 genes of which 53 mapped in MetaCore™. The "analyze networks" algorithm generated a low-dose network that was similarly centered about SP1 and p53 hubs (z-score 49.26, Figure 8.7B). The networks from both dose level exposures were compared to identify the shared GADD45alpha, GADD45, cyclinG, cyclinG1, Bax, and cFos genes which exhibit differential expression in both networks (Figure 8.7C). Thus, network-based analysis can address the long-standing problem of comparative dose evaluation in toxicological studies.

Gene network-based signatures can also be used to test candidate mechanisms of toxicity, such as the induction of oxidative stress, for compounds with complex mechanisms of action. For example, microarray profiling data from rats treated for 4 days with 1,000 mg/kg carbon tetrachloride [78] identified 37 toxicity-associated genes, 26 of which were then used to construct a toxicity

B

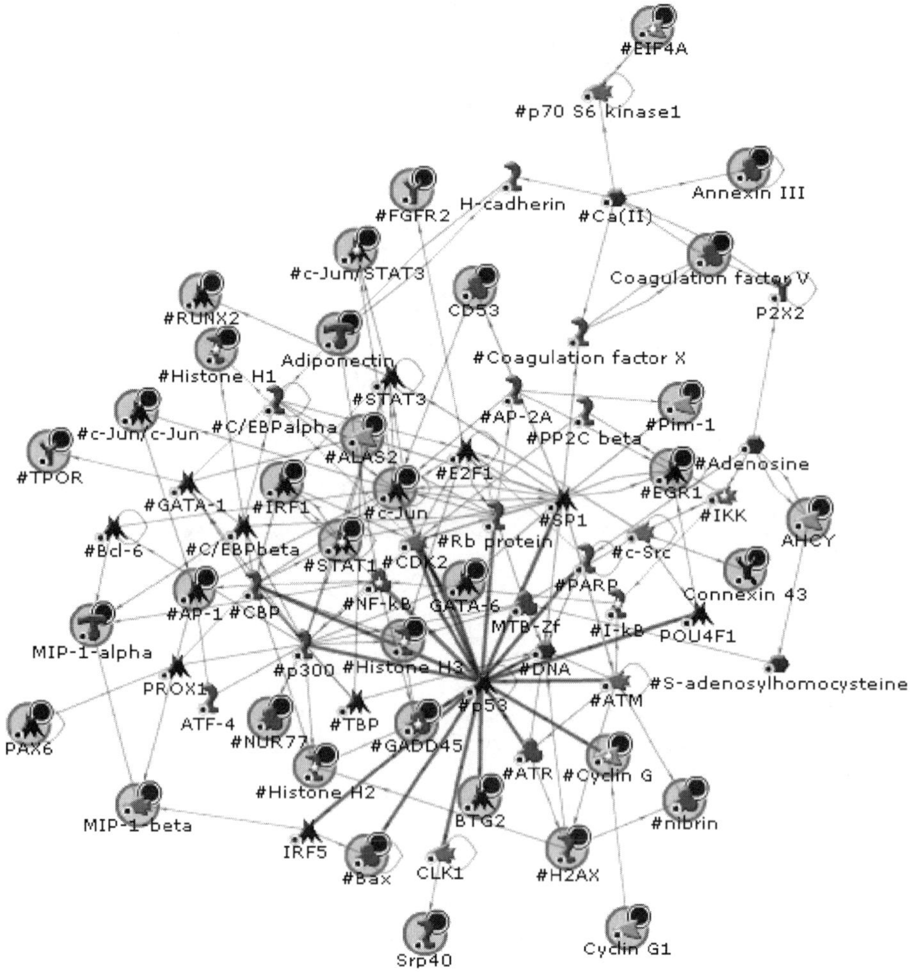

Figure 8.7 (*Continued*)

network with the "analyze network" algorithm. This response network visualizes the up-regulation of inflammatory genes CD44, SULT and down-regulation of Cytochrome P450 genes (z-score 45.35, Figure 8.8); both of these responses are commonly seen in oxidative stress. Note that there are genes predicted to be included in this toxicity network that were either not assayed on the microarray or that did not show a significant change in expression. As a second example of a toxicant with a complex mode of action, microarray profiling data from rat kidneys after treatment with the nephrotoxicant cisplatin [79] identified 75 toxicity-associated genes, 57 of which were mapped in MetaCore. These genes were used to construct a direct interaction network

C

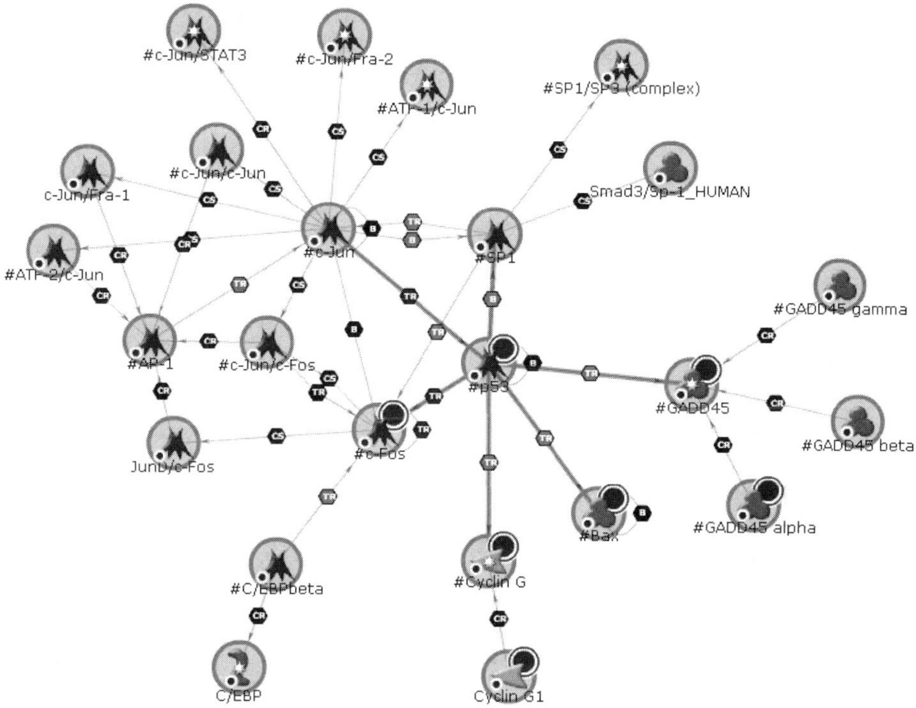

Figure 8.7 (*Continued*)

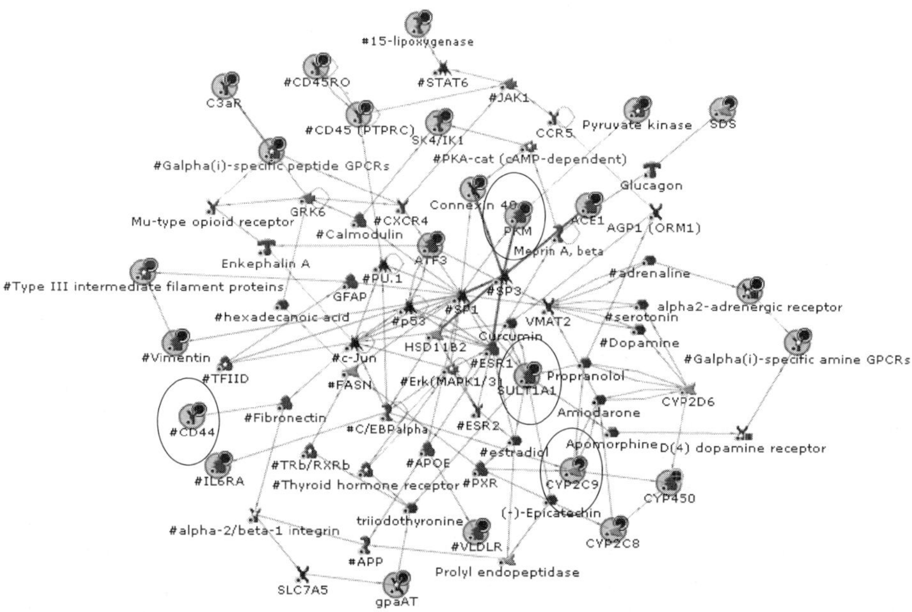

Figure 8.8 Rat toxicity response to carbon tetrachloride. A 26-gene network constructed from 37 genes mapped to MetaCore™ using the "analyze network" algorithm [78].

which contains two subnetworks centered about FXR and p53 hubs (Figure 8.9A). Expanding these subnetworks by application of the "analyze network" algorithm now links the FXR and p53 subnetworks via connections to SP1, cFos, STAT1, STAT3, etc. (Figure 8.9B, z-score 56.05). These two examples illustrate the predictive power of network-based analysis to infer functional relationships even when presented with relatively small datasets from exposure to complex toxicants. As an obvious caveat, network-based predictions from such limited datasets are less reliable and must be experimentally verified.

8.4 DISCUSSION

The body is exposed to numerous exogenous compounds, many of which are either direct oxidants or are metabolized to form reactive oxygen species [80]. Many drugs and chemicals are known to cause oxidative stress as a component of their mechanism of toxicity [54]. It is also widely appreciated that oxidative stress and mitochondrial damage are important in the etiology of many complex diseases [2]. Cells have adaptive gene expression mechanisms to respond to oxidative stress and activation of these antioxidant defenses is triggered by exposure to reactive oxygen species [81]. The unique physiological responses to toxicant challenge, both protective and pathological, reflect differential activity of the underlying cellular pathways. Although gene expression profiling has been used to classify distinct cellular states, it is still problematic to identify mechanisms of toxicity from these profiles. We have taken a network-based approach to understanding both changes in the human mitochondrial proteome (a key target and intracellular mediator of oxidative stress), and changes in the gene expression profiles of cells or animals exposed to compounds that induce oxidative stress. Modeling of gene expression microarray data to form toxicant-specific gene networks supports the identification of the genes and pathways mediating cellular toxicity and enables a functional integration of these data for predictive toxicology and mechanistic risk assessment.

First, we have demonstrated that the MetaCore™ database of manually curated gene and protein interactions maps more of the proteins in the human mitochondrial proteome than does the widely used KEGG public database (Figure 8.1). This result is an important practical consideration for network building and for integrated modeling of high-throughput toxicity exposure data, since a broad spectrum of xenobiotics and toxicants are known to impact critical mitochondrial functions. Second, in a conserved yeast model system, we have demonstrated that cellular responses to hydrogen peroxide exposure can be captured by microarray profiling and used to construct a phenotypic response network for oxidative stress (Figure 8.2A,B). Third, using data from published mammalian toxicity profiling studies, we have demonstrated that a computational platform for pathway and gene network analysis could be used

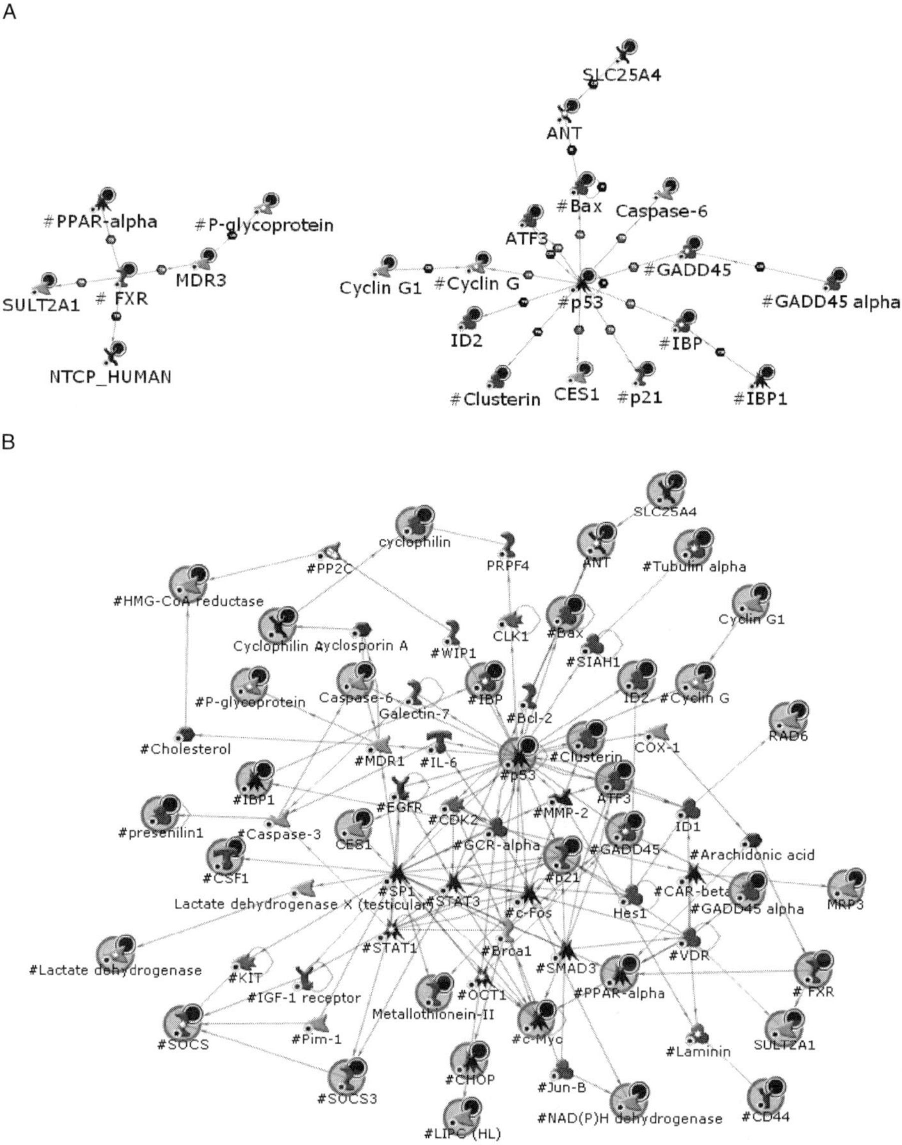

Figure 8.9 Rat kidney toxicity response to cisplatin. A 57-gene network constructed from 75 genes mapped to MetaCore™ from a profiling study of cisplatin exposure in rat kidney [79]: (A) "direct interaction" algorithm, (B) "analyze network" algorithm *z*-score 56.05 [79].

to construct a gene network that identifies the underlying cellular pathways and processes of chemical-induced toxicity.

We first demonstrated mammalian toxicity network analysis using microarray profiling data from human fibroblasts treated with tert-butyl hydroperoxide [73], a reference oxidant (Figure 8.3). We then extended this analysis to microarray profiling data of animal exposure models for chemically distinct hepatotoxicants (e.g. furan and acetaminophen) that have been proposed to exert their liver toxicity by induction of oxidative stress. When comparing the hepatic response to these two compounds, network analysis visualized subnetworks of oxidative stress response pathways revealing both commonality (Figure 8.6) and toxicant-specific differences. The furan-induced network links modules of up-regulated genes for cell proliferation and cell cycle progression with modules of down-regulated genes for nuclear receptors and Cytochrome P450s (Figure 8.5). In contrast, the toxicity network for acetaminophen links modules for the energy-producing pathways of cellular glycolysis and mitochondrial hydroxylation.

Comparison of the toxicity networks of mice treated with a low versus high dose of benzene (Figure 8.7A,B) identifies a common network aspect, cell cycle, and related genes are connected to SP1 and p53 hubs (Figure 8.7C). In addition, exposure profiling studies of carbon tetrachloride (Figure 8.8) and cisplatin (Figure 8.9) in rat liver and kidney, respectively, further support the central role of SP1 and p53 hubs in connecting differentially expressed genes in a toxicant network. These examples demonstrate the utility of gene networks to reveal the potential mechanisms of toxicity and present a novel method for visualizing complex tissue-level processes of the toxicant response.

In summary, we report our initial studies to develop network-based approaches to model the toxicant response of chemical and drug exposure in mammalian systems. As a developing technology, there are not yet standardized methods for applying network-based approaches to expression profile datasets. Methods of gene network analysis are limited by the currently incomplete content of the human interactome, as represented in either commercial databases (e.g. MetaCore™, MetaDrug™ and Ingenuity Pathways Analysis, Pathway Assist, etc.) or in publicly annotated databases (e.g. KEGG). Network analysis is similarly limited by the lack of fully developed algorithms for quantitative comparison of biological networks and by the paucity of suitably complete microarray profiling datasets for validation and benchmarking of proposed analytical methods. In this report, we have not provided an extensive comparison of other available network building tools and content databases as this represents work in progress. However, we feel that such an ongoing comparison is very much needed to assess the strengths and weaknesses of developing methods for network analysis and to ensure successful maturation of the network-based approach to understanding mechanisms of toxicity.

The development of computational technologies to enable the storage and comparison of toxicity networks, such as those described in this report, will add another dimension to the ease of utility of such methods. At present, rigorous comparison of relatively small functional network modules of 5–25 nodes is computationally tractable and several network comparison algorithms have recently been applied to the analysis of such intracellular pathways. Kanehisa and coworkers have described a heuristic graph comparison algorithm [82] that has been extended by others to determine the similarity between protein and gene expression networks [83]. A second useful algorithm for comparative network analysis identifies common interaction pathways by globally aligning protein interaction networks, for example, between metabolic processes in yeast *S. cerevisiae* and *H. pylori* [84]. These comparative methods may be applicable to larger toxicity networks in the future. In a recent study, we have presented a more detailed discussion of some of the challenges and potential solutions to comparative network analysis in the context of understanding transporter and enzyme regulatory networks [63].

In this discussion, we have described the use of relatively small sets of toxicity-associated genes that have first been selected by clustering algorithms to distinguish exposed from unexposed conditions. However, the real power of the network approach is utilized if the complete dataset is used prior to clustering or classification. This preferred global dataset approach is limited by the difficulty in obtaining complete expression profiling datasets for toxicant exposure that are both statistically significant and phenotypically anchored. We hope that the developing initiatives to establish public databases of chemical and drug exposure studies in model mammalian and human systems [8,9] will alleviate this need, as we have discussed previously [13,55]. Our understanding of the mechanisms of toxicity will be significantly enhanced by the combined use of top-down and bottom-up approaches to network modeling. The establishment of a reference human interactome database will greatly facilitate comparative network analysis with the interactomes of model biological systems [85]. Such a systems toxicology-based strategy should become possible with the integration of experimentally derived proteome datasets, human-specific databases of protein–protein interactions, and improved network building algorithms [62]. As progress toward this goal, this report demonstrates that mammalian proteome networks can be generated to complement those of bacteria [86], yeast [37], and other organisms [87].

ACKNOWLEDGMENT

GeneGo is acknowledged for providing use of MetaCore™ to CNG. Ingenuity is gratefully acknowledged for providing access to Ingenuity Pathways Analysis™ for SE. CNG acknowledges research support from EHS Center Grant P30 ES06639 from the National Institute of Environmental Health Sciences.

REFERENCES

1. Barabasi A-L, Oltvai ZN. Network biology: understanding the cell's functional organization. Nat Rev Genet 2004;5:101–113.

2. Fariss MW, Chan CB, Patel M, Van Houten B, Orrenius S. Role of mitochondria in toxic oxidative stress. Mol Interv 2005;5:94–111.
3. Gaucher SP, Taylor SW, Fahy E, Zhang B, Warnock DE, Ghosh SS, Gibson BW. Expanded coverage of the human heart mitochondrial proteome using multidimensional liquid chromatography coupled with tandem mass spectrometry. J Proteome Res 2004;3:495–505.
4. Taylor SW, Fahy E, Zhang B, Glenn GM, Warnock DE, Wiley S, Murphy AN, Gaucher SP, Capaldi RA, Gibson BW, Ghosh SS. Characterization of the human heart mitochondrial proteome. Nat Biotechnol 2003;21:281–286.
5. Vo TD, Greenberg HJ, Palsson BO. Reconstruction and functional characterization of the human mitochondrial metabolic network based on proteomic and biochemical data. J Biol Chem 2004;279:39532–39540.
6. Da Cruz S, Xenarios I, Langridge J, Vilbois F, Parone PA, Martinou JC. Proteomic analysis of the mouse liver mitochondrial inner membrane. J Biol Chem 2003;278:41566–41571.
7. Jiang XS, Zhou H, Zhang L, Sheng QH, Li SJ, Li L, Hao P, Li YX, Xia QC, Wu JR, Zeng R. A high-throughput approach for subcellular proteome: identification of rat liver proteins using subcellular fractionation coupled with two-dimensional liquid chromatography tandem mass spectrometry and bioinformatic analysis. Mol Cell Proteomics 2004;3:441–455.
8. Waters M, Boorman G, Bushel P, Cunningham M, Irwin R, Merrick A, Olden K, Paules R, Selkirk J, Stasiewicz S, Weis B, Van Houten B, Walker N, Tennant R. Systems toxicology and the Chemical Effects in Biological Systems (CEBS) knowledge base. EHP Toxicogenomics 2003;111:15–28.
9. Mattes WB, Pettit SD, Sansone SA, Bushel PR, Waters MD. Database development in toxicogenomics: issues and efforts. Environ Health Perspect 2004;112:495–505.
10. Eisen MB, Spellman PT, Brown PO, Botstein D. Cluster analysis and display of genome-wide expression patterns. Proc Natl Acad Sci U S A 1998;95:14863–14868.
11. Nikolsky Y, Ekins S, Nikolskaya T, Bugrim A. A novel method for generation of signature networks as biomarkers from complex high throughput data. Toxicol Lett 2005;158:20–29.
12. Ekins S, Kirillov E, Rakhmatulin E, Nikolskaya T. A novel method for visualizing nuclear hormone receptor networks relevant to drug metabolism. Drug Metab Dispos 2005;33:474–481.
13. Ekins S, Giroux C. Computers and systems biology for pharmaceutical research and development. In: *Computer Applications in Pharmaceutical Research and Development*, edited by Ekins S, pp. 139–65. Hoboken: John Wiley and Sons, 2006.
14. Toyoshiba H, Yamanaka T, Sone H, Parham FM, Walker NJ, Martinez J, Portier CJ. Gene interaction network suggests dioxin induces a significant linkage between aryl hydrocarbon receptor and retinoic acid receptor beta. Environ Health Perspect 2004;112:1217–1224.
15. Shannon P, Markiel A, Ozier O, Baliga NS, Wang JT, Ramage D, Amin N, Schwikowski B, Ideker T. Cytoscape: a software environment for integrated models of biomolecular interaction networks. Genome Res 2003;13:2498–2504.

16. Cerami EG, Bader GD, Gross BE, Sander C. cPath: open source software for collecting, storing, and querying biological pathways. BMC Bioinformatics 2006;7:497.
17. Herrgard MJ, Covert MW, Palsson BO. Reconciling gene expression data with known genome-scale regulatory network structures. Genome Res 2003;13: 2423–2434.
18. Sharan R, Ideker T. Modeling cellular machinery through biological network comparison. Nat Biotechnol 2006;24:427–433.
19. Stuart JM, Segal E, Koller D, Kim SK. A gene-coexpression network for global discovery of conserved genetic modules. Science 2003;302:249–255.
20. Grigoryev DN, Ma S-F, Irizarry RA, Ye SQ, Quakenbush J, Garcia JGN. Orthologous gene-expression profiling in multi-species models: search for candidate genes. Genome Biol 2004;5:R34.
21. Scandalios JG. Oxidative stress: molecular perception and transduction of signals triggering antioxidant gene defenses. Braz J Med Biol Res 2005;38:995–1014.
22. Hartwell LH, Hopfield JJ, Leibler S, Murray AW. From molecular to modular cell biology. Nature 1999;402:C47–C52.
23. Lee TI, Rinaldi NJ, Robert F, Odom DT, Bar-Joseph Z, Gerber GK, Hannett NM, Harbison CT, Thompson CM, Simon I, Zeitlinger J, Jennings EG, Murray HL, Gordon DB, Ren B, Wyrick JJ, Tagne JB, Volkert TL, Fraenkel E, Gifford DK, Young RA. Transcriptional regulatory networks in *Saccharomyces cerevisiae*. Science 2002;298:799–804.
24. Wu LF, Hughes TR, Davierwala AP, Robinson MD, Stoughton R, Altschuler SJ. Large-scale prediction of Saccharomyces cerevisiae gene function using overlapping transcriptional clusters. Nat Genet 2002;31:255–265.
25. Qi Y, Rolfe A, MacIsaac KD, Gerber GK, Pokholok D, Zeitlinger J, Danford T, Dowell RD, Fraenkel E, Jaakkola TS, Young RA, Gifford DK. High-resolution computational models of genome binding events. Nat Biotechnol 2006;24: 963–970.
26. Qi Y, Rolfe A, Macisaac KD, Gerber GK, Pokholok D, Zeitlinger J, Danford T, Dowell RD, Fraenkel E, Jaakkola TS, Young RA, Gifford DK. Erratum: high-resolution computational models of genome binding events. Nat Biotechnol 2006;24:1293.
27. Segal E, Shapira M, Regev A, Pe'er D, Botstein D, Koller D, Freidman N. Module networks: identifying regulatory modules and their condition-specific regulators from gene expression data. Nat Genet 2003;34:166–176.
28. Spirin V, Mirny LA. Protein complexes and functional modules in molecular networks. Proc Natl Acad Sci U S A 2003;100:12123–12128.
29. Rives AW, Galitski T. Modular organization of cellular networks. Proc Natl Acad Sci U S A 2003;100:1128–1133.
30. Pereira-Leal JB, Enright AJ, Ouzounis CA. Detection of functional modules from protein interaction networks. Proteins 2004;54:49–57.
31. Milo R, Shen-Orr S, Itzkovitz S, Kashtan N, Chklovskii D, Alon U. Network motifs: simple building blocks of complex networks. Science 2002;298:824–827.
32. Yeger-Lotem E, Sattath S, Kashtan N, Itzkovitz S, Milo R, Pinter RY, Alon U, Margalit H. Network motifs in integrated cellular networks of transcription-

regulation and protein–protein interaction. Proc Natl Acad Sci U S A 2004;101: 5934–5939.

33. Shen-Orr SS, Milo R, Mangan S, Alon U. Network motifs in the transcriptional regulation network of *Escherichia coli*. Nat Genet 2002;31:64–68.
34. Dobrin R, Beg QK, Barabasi AL, Oltvai ZN. Aggregation of topological motifs in the *Escherichia coli* transcriptional regulatory network. BMC Bioinformatics 2004;5:10.
35. Wuchty S, Oltvai ZN, Barabasi AL. Evolutionary conservation of motif constituents in the yeast protein interaction network. Nat Genet 2003;35:176–179.
36. Albert R, Jeong H, Barabasi AL. Error and attack tolerance of complex networks. Nature 2000;406:378–382.
37. Han JD, Bertin N, Hao T, Goldberg DS, Berriz GF, Zhang LV, Dupuy D, Walhout AJ, Cusick ME, Roth FP, Vidal M. Evidence for dynamically organized modularity in the yeast protein–protein interaction network. Nature 2004;430:88–93.
38. Begley TJ, Rosenbach AS, Ideker T, Samson LD. Hot spots for modulating toxicity identified by genomic phenotyping and localization mapping. Mol Cell 2004; 16:117–125.
39. Weiss A, Delproposto J, Giroux CN. High-throughput phenotypic profiling of gene-environment interactions by quantitative growth curve analysis in *Saccharomyces cerevisiae*. Anal Biochem 2004;327:23–34.
40. Murray JI, Whitfield ML, Trinklein ND, Myers RM, Brown PO, Botstein D. Diverse and specific gene expression responses to stresses in cultured human cells. Mol Biol Cell 2004;15:2361–2374.
41. West JD, Marnett LJ. Alterations in gene expression induced by the lipid peroxidation product, 4-hydroxy-2-nonenal. Chem Res Toxicol 2005;18:1642–1653.
42. Goswami S, Sheets NL, Zavadil J, Chauhan BK, Bottinger EP, Reddy VN, Kantorow M, Cvekl A. Spectrum and range of oxidative stress responses of human lens epithelial cells to H2O2 insult. Invest Ophthalmol Vis Sci 2003;44:2084–2093.
43. Spector A, Li D, Ma W, Sun F, Pavlidis P. Differential amplification of gene expression in lens cell lines conditioned to survive peroxide stress. Invest Ophthalmol Vis Sci 2002;43:3251–3264.
44. Morgan KT, Ni H, Brown HR, Yoon L, Qualls CW, Jr., Crosby LM, Reynolds R, Gaskill B, Anderson SP, Kepler TB, Brainard T, Liv N, Easton M, Merrill C, Creech D, Sprenger D, Conner G, Johnson PR, Fox T, Sartor M, Richard E, Kuruvilla S, Casey W, Benavides G. Application of cDNA microarray technology to in vitro toxicology and the selection of genes for a real-time RT-PCR-based screen for oxidative stress in Hep-G2 cells. Toxicol Pathol 2002;30:435–451.
45. Bedard K, MacDonald N, Collins J, Cribb A. Cytoprotection following endoplasmic reticulum stress protein induction in continuous cell lines. Basic Clin Pharmacol Toxicol 2004;94:124–131.
46. Bae I, Fan S, Meng Q, Rih JK, Kim HJ, Kang HJ, Xu J, Goldberg ID, Jaiswal AK, Rosen EM. BRCA1 induces antioxidant gene expression and resistance to oxidative stress. Cancer Res 2004;64:7893–7909.
47. Efferth T, Oesch F. Oxidative stress response of tumor cells: microarray-based comparison between artemisinins and anthracyclines. Biochem Pharmacol 2004; 68:3–10.

48. Heijne WH, Slitt AL, van Bladeren PJ, Groten JP, Klaassen CD, Stierum RH, van Ommen B. Bromobenzene-induced hepatotoxicity at the transcriptome level. Toxicol Sci 2004;79:411–422.

49. Heijne WH, Stierum RH, Slijper M, van Bladeren PJ, van Ommen B. Toxicogenomics of bromobenzene hepatotoxicity: a combined transcriptomics and proteomics approach. Biochem Pharmacol 2003;65:857–875.

50. Waring JF, Jolly RA, Ciurlionis R, Lum PY, Praestgaard JT, Morfitt DC, Buratto B, Roberts C, Schadt E, Ulrich RG. Clustering of hepatotoxins based on mechanism of toxicity using gene expression profiles. Toxicol Appl Pharmacol 2001; 175:28–42.

51. Gerhold D, Lu M, Xu J, Austin C, Caskey CT, Rushmore T. Monitoring expression of genes involved in drug metabolism and toxicology using DNA microarrays. Physiol Genomics 2001;5:161–170.

52. Fountoulakis M, de Vera MC, Crameri F, Boess F, Gasser R, Albertini S, Suter L. Modulation of gene and protein expression by carbon tetrachloride in the rat liver. Toxicol Appl Pharmacol 2002;183:71–80.

53. Huang Q, Jin X, Gaillard ET, Knight BL, Pack FD, Stoltz JH, Jayadev S, Blanchard KT. Gene expression profiling reveals multiple toxicity endpoints induced by hepatotoxicants. Mutat Res 2004;549:147–167.

54. McMillian M, Nie A, Parker JB, Leone A, Kemmerer M, Bryant S, Herlich J, Yieh L, Bittner A, Liu X, Wan J, Johnson MD, Lord P. Drug-induced oxidative stress in rat liver from a toxicogenomics perspective. Toxicol Appl Pharmacol 2005; 207:171–178.

55. Ekins S. Systems-ADME/Tox: resources and network approaches. J Pharmacol Toxicol Methods 2006;53:38–66.

56. Toyoshiba H, Sone H, Yamanaka T, Parham FM, Irwin RD, Boorman GA, Portier CJ. Gene interaction network analysis suggests differences between high and low doses of acetaminophen. Toxicol Appl Pharmacol 2006;215:306–316.

57. Giroux C, Ekins S, Abdullah I, Nikolsky Y, Bugrim A, Nikolskaya T. A genetic network approach to comparative toxicogenomics and risk assessment: the oxidative stress response. Toxicol Sci 2005;84.

58. Ekins S, Andreyev S, Rybadov A, Kirillov E, Rakhmatulin EA, Sorokina S, Bugrim A, Nikolskaya T. A combined approach to drug metabolism and toxicity assessment. Drug Metab Dispos 2006;34:495–503.

59. Ekins S, Bugrim A, Brovold L, Kirillov E, Nikolsky Y, Rakhmatulin EA, Sorokina S, Ryabov A, Serebryiskaya T, Melnikov A, Metz J, Nikolskaya T. Algorithms for network analysis in systems-ADME/Tox using the MetaCore and MetaDrug platforms. Xenobiotica 2006;36:877–901.

60. Ekins S, Giroux CN, Nikolsky Y, Bugrim A, Nikolskaya T. A signature gene network approach to toxicity. The Toxicologist 2005;84.

61. Ekins S, Nikolsky Y, Bugrim A, Kirillov E, Nikolskaya T. Pathway mapping tools for analysis of high content data. In: *High Content Screening: A Powerful Approach to Systems Cell Biology and Drug Discovery*, edited by Taylor DL, Haskins JA, Giuliano KA, pp. 319–350. Totowa, NJ: Humana Press, 2006.

62. Ekins S, Nikolsky Y, Nikolskaya T. Techniques: application of systems biology to absorption, distribution, metabolism, excretion, and toxicity. Trends Pharmacol Sci 2005;26:202–209.

63. Ekins S, Shimada J, Chang C. Application of data mining approaches to drug delivery. Adv Drug Del Rev 2006;58:1409–1430.
64. Dennis G, Jr., Sherman BT, Hosack DA, Yang J, Gao W, Lane HC, Lempicki RA. DAVID: Database for Annotation, Visualization, and Integrated Discovery. Genome Biol 2003;4:P3.
65. Orrenius S, Gogvadze V, Zhivotovsky B. Mitochondrial oxidative stress: implications for cell death. Annu Rev Pharmacol Toxicol 2007;47:143–183.
66. Gibson BW. The human mitochondrial proteome: oxidative stress, protein modifications and oxidative phosphorylation. Int J Biochem Cell Biol 2005;37:927–934.
67. Liguori MJ, Anderson LM, Bukofzer S, McKim J, Pregenzer JF, Retief J, Spear BB, Waring JF. Microarray analysis in human hepatocytes suggests a mechanism for hepatotoxicity induced by trovafloxacin. Hepatology 2005;41:177–186.
68. Mackay IR, Whittingham S, Fida S, Myers M, Ikuno N, Gershwin ME, Rowley MJ. The peculiar autoimmunity of primary biliary cirrhosis. Immunol Rev 2000; 174:226–237.
69. Stockwell BR. The biological magic behind the bullets. Nat Biotechnol 2004;22:37–38.
70. Parsons AB, Brost RI, Ding H, Li Z, Zhang C, Sheikh B, Brown GW, Kane PM, Hughes TR, Boone C. Integration of chemical-genetic and genetic interaction data links bioactive compounds to cellular target pathways. Nat Biotechnol 2004; 22:62–69.
71. Schwimmer C, Rak M, Lefebvre-Legendre L, Duvezin-Caubet S, Plane G, di Rago JP. Yeast models of human mitochondrial diseases: from molecular mechanisms to drug screening. Biotechnol J 2006;1:270–281.
72. Giroux C. Signatures of exposure and pathways of susceptibility: a yeast toolkit for biomarker discovery and prediction of mechanisms of cellular toxicity. Toxicol Sci 2006;90:228.
73. Heinloth AN, Shackelford RE, Innes CL, Bennett L, Li L, Amin RP, Sieber SO, Flores KG, Bushel PR, Paules RS. Identification of distinct and common gene expression changes after oxidative stress and gamma and ultraviolet radiation. Mol Carcinog 2003;37:65–82.
74. Heinloth AN, Irwin RD, Boorman GA, Nettesheim P, Fannin RD, Sieber SO, Snell ML, Tucker CJ, Li L, Travlos GS, Vansant G, Blackshear PE, Tennant RW, Cunningham ML, Paules RS. Gene expression profiling of rat livers reveals indicators of potential adverse effects. Toxicol Sci 2004;80:193–202.
75. Reid AB, Kurten RC, McCullough SS, Brock RW, Hinson JA. Mechanisms of acetaminophen-induced hepatotoxicity: role of oxidative stress and mitochondrial permeability transition in freshly isolated mouse hepatocytes. J Pharmacol Exp Ther 2005;312:509–516.
76. Yoon B, Li G-X, Kitada K, Kawasaki Y, Katsuhide I, Kodama Y, Inoue T, Kobayashi K, Kanno J, Kim D-Y, Inoue T, Hirabayashi Y. Mechanisms of benzene-induced hematotoxicity and leukemogenicity: cDNA microarray analyses using mouse bone marrow tissue. Env Health Perspect 2003;111:1411–1420.
77. Faiola B, Fuller ES, Wong VA, Recio L. Gene expression profile in bone marrow and hematopoietic stem cells in mice exposed to inhaled benzene. Mutat Res 2004;549:195–212.

78. Young MB, DiSilvestro MR, Sendera TJ, Freund J, Kriete A, Magnuson SR. Analysis of gene expression in carbon tetrachloride-treated rat livers using a novel bioarray technology. Pharmacogenomics J 2003;3:41–52.
79. Huang Q, Dunn RT, II, Jayadev S, DiSorbo O, Pack FD, Farr SB, Stoll RE, Blanchard KT. Assessment of cisplatin-induced nephrotoxicity by microarray technology. Toxicol Sci 2001;63:196–207.
80. Davies KJ. The broad spectrum of responses to oxidants in proliferating cells: a new paradigm for oxidative stress. IUBMB Life 1999;48:41–47.
81. Scandalios JG. Oxidative stress responses—what have genome-scale studies taught us? Genome Biol 2002;3:REVIEWS1019.
82. Ogata H, Fujibuchi W, Goto S, Kanehisa MA. Heuristic graph comparison algorithm and its application to detect functionally related enzyme clusters. Nucleic Acids Res 2000;28:4021–4028.
83. Nakaya A, Goto S, Kanehisa MA. Extraction of correlated gene clusters by multiple graph comparison. Genome Inform Ser Workshop Genome Inform 2001;12:44–53.
84. Kelley BP, Sharan R, Karp RM, Sittler T, Root DE, Stockwell BR, Ideker T. Conserved pathways within bacteria and yeast as revealed by global protein network alignment. Proc Natl Acad Sci U S A 2003;100:11394–11399.
85. Gandhi TK, Zhong J, Mathivanan S, Karthick L, Chandrika KN, Mohan SS, Sharma S, Pinkert S, Nagaraju S, Periaswamy B, Mishra G, Nandakumar K, Shen B, Deshpande N, Nayak R, Sarker M, Boeke JD, Parmigiani G, Schultz J, Bader JS, Pandey A. Analysis of the human protein interactome and comparison with yeast, worm and fly interaction datasets. Nat Genet 2006;38:285–293.
86. Covert MW, Knight EM, Reed JL, Herrgard MJ, Palsson BO. Integrating high-throughput and computational data elucidates bacterial networks. Nature 2004;429:92–96.
87. Li S, Armstrong CM, Bertin N, Ge H, Milstein S, Boxem M, Vidalain PO, Han JD, Chesneau A, Hao T, Goldberg DS, Li N, Martinez M, Rual JF, Lamesch P, Xu L, Tewari M, Wong SL, Zhang LV, Berriz GF, Jacotot L, Vaglio P, Reboul J, Hirozane-Kishikawa T, Li Q, Gabel HW, Elewa A, Baumgartner B, Rose DJ, Yu H, Bosak S, Sequerra R, Fraser A, Mango SE, Saxton WM, Strome S, Van Den Heuvel S, Piano F, Vandenhaute J, Sardet C, Gerstein M, Doucette-Stamm L, Gunsalus KC, Harper JW, Cusick ME, Roth FP, Hill DE, Vidal M. A map of the interactome network of the metazoan *C. elegans*. Science 2004;303:540–543.

9

UNRAVELING MECHANISMS OF TOXICITY WITH THE POWER OF PATHWAYS: TOXWIZ TOOL AS AN ILLUSTRATIVE EXAMPLE

MARK P. KÜHNEL, BOJANA COSOVIC, GORAN MEDIC, ROBERT B. RUSSELL, AND GORDANA APIC

Table of Contents

Pathway Analysis for Drug Discovery, Edited by Anton Yuryev

9.1 INTRODUCTION

Increasing concerns about general chemical safety for humans and the environment has recently created a need for better understanding of mechanisms underlying toxic effects and safety issues. Mechanisms can often identify toxicity biomarkers, making prediction of toxic effects easier. Regulatory requirements for drugs, agrochemicals, cosmetics, and a wide range of other chemicals are changing and becoming more concerned about the mechanisms related to the potential safety issues. In order to create safe chemicals and to minimize surprises with safety issues later in the development or in marketed products, understanding the mechanisms of toxicities, as well as those of desired action, can be crucial. The knowledge of molecular mechanisms underlying biological phenomena in general, such as disease mechanisms, has accumulated as biological research has become more molecular over the last few decades. Toxicology was not traditionally required to explain the molecular basis of observed toxicity, meaning that currently comparably little is known about mechanisms of toxicity. New molecular biology data make it now possible to try to improve our understanding of mechanisms of toxicity. As the molecular biology developed in the course of 20th century, the mechanisms of cellular processes have been depicted as biological pathways such as metabolic and signaling pathways. Knowledge bases such as ToxWiz capture a wide spectrum of mechanistic hypothesis and pathways for toxic effects derived from meticulous expert analysis of millions of articles in the scientific literature, drawing parallels between mechanism of disease, pathology and toxic endpoints, and depicting them in a form of biological pathways for underlying toxic endpoints. ToxWiz also contains over 1,000 organ, tissue and pathology-specific toxic endpoints (e.g. liver hyperplasia, kidney hypertrophy, etc.) and over 3 million expert-curated literature data points to support the collected hypothesis. Additionally, software tools based on systems biology principles allow one to analyze novel compounds for toxicity. Software tools, such as ToxWiz provide an unprecedented insight for assessing safety and toxicology of novel compounds and understanding the molecular mechanism of toxicity (Figure 9.1). By describing biological pathways underlying the toxic effects, biological pathway tools promise to help design safer chemicals. This chapter illustrates how

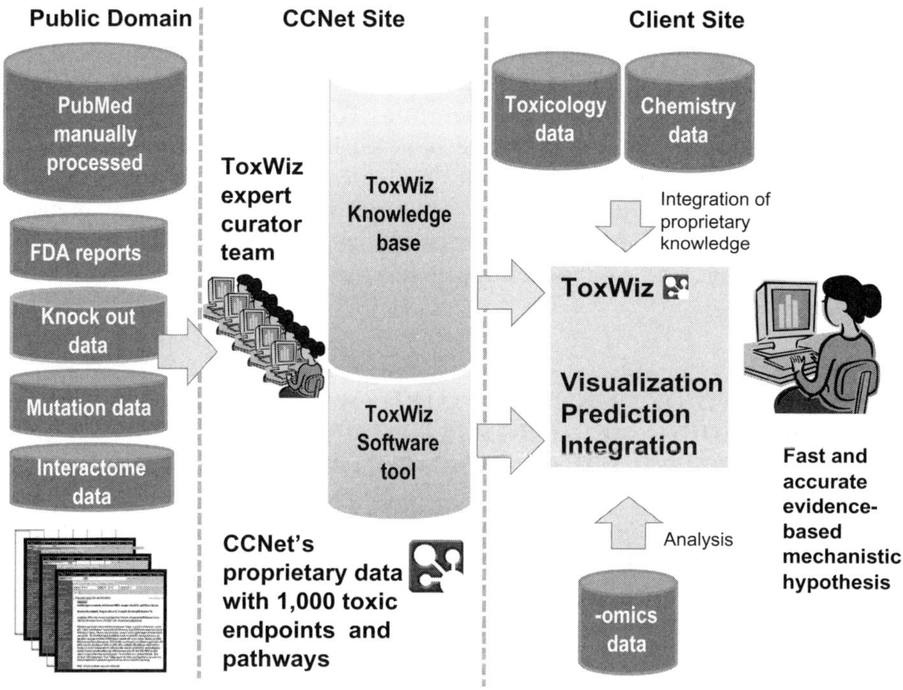

Figure 9.1 The process of elucidating mechanisms of toxicity from left to right. ToxWiz Knowledge base comprises over 100 years of manual expert curation of over 17 million scientific articles from PubMed and processing of all relevant data including FDA post-marketing reports, the entire human interactome (protein–protein interaction), etc. The curators use a text search software with over 300,000 ontologies and other systems biology software tools such as chemical structure searches in order to help them identify the relevant information faster. These data are then manually processed, interpreted, and collated via a ToxWiz software tool in a unique ToxWiz Knowledge base describing over 1,000 toxic endpoints with their mechanistic hypothesis and with direct links to evidence with over 3 million expert-curated articles supporting the hypothesis (toxicology pathways). On the client side, research can easily assess the knowledge from a ToxWiz Knowledge base in order to explain the mechanisms of observed toxicity. Additionally, the ToxWiz software tool allows creating mechanistic hypothesis based on the chemical structure features of the compound, as well as complementary to the analysis of time- and dose-dependent -omics experiments and to see them in the light of mechanistic hypothesis.

The systems biology principles make it possible to use ToxWiz software tool to integrate client's proprietary data on chemical compounds, toxicity reports or clinical data, and thus to bring the proprietary knowledge accumulated in the client's site into predictive context.

expert-curated literature data captured in a ToxWiz database, a catalogue of biological pathways underlying almost 1,000 toxic endpoints associated with over 30,000 chemical structures, can help understand the mechanisms of observed toxicity. It also describes how ToxWiz enables prediction of more than one toxicity based on mechanistic knowledge. We will provide insights of how knowledge about biological pathways and pathway analysis tools can be used in both retrospective (explaining mechanisms of observed toxicity) and prospective ways (predicting toxicity) for better safety assessment and understanding of the molecular mechanisms of toxic effects.

9.2 THE WIND OF CHANGE IN TOXICOLOGY: TOWARD UNDERSTANDING MECHANISMS OF TOXICITY AND PREDICTING TOXICITIES FOR NOVEL COMPOUNDS

Toxicology is the single most expensive aspect of preclinical drug discovery, costing roughly as much as all other preclinical operations put together. This is due to the need for rigorous testing in animals on any drug candidate prior to any contact with human subjects. It is also a major expense in chemical and agrochemical industries, since strict testing is also required on any ingredient in regular contact with humans. Animal tests are unfortunately neither scalable nor amenable to miniaturization as are other aspects of drug discovery (e.g. high-throughput screening). It has thus become a bigger and bigger bottleneck, having to cope with an increase in drug candidates with a throughput that has changed little in the last decades (e.g. Koppal [1]). As a result, there is a growing pressure to increase throughput and cut costs in this still expensive discipline.

To make matters worse, new pressure on the industry has come from initiatives such as the European Union's REACH directive (e.g. Sauer [2]), which requires retesting on thousands of chemicals already on the market, and many moves to decrease the number of animals tested in the development of new compounds. There is thus a burning need in the pharmaceutical and chemical industries to find suitable replacements or technologies to explain molecular mechanism of toxicity, predict toxic outcomes, and avoid or prioritize experiments.

It has long been known that similar chemical structures tend to induce, bind, and be metabolized by similar macromolecules in the cell. This has spurred efforts to identify *toxicity biomarkers*, which have been a major focus in predictive toxicology. Biomarkers can take the form of changes in the expression of genes as measured by toxicogenomics (e.g. Fielden et al. [3]), metabolites produced possibly owing to effects on normal cellular metabolism (e.g. Heinje et al. [4]), or in some cases proteins indicative of particular toxic responses, such as cytokines in the case of inflammation (e.g. Tsutsui et al. [5]). Biomarkers come into being as a result of the overall biological response to a xenobiotic, which is nearly always the result of the chemical, or one of its metabolites binding to a number of proteins in the cell. Binding of compounds to several proteins has long been recognized as indicators of toxic events. For instance, binding to the androgen receptor is often an indication of problems in sex-organ development relating to interference with testosterone function (e.g. Clegg et al. [6]) related to, or binding to the ether-a-go-go or ERG channel is very often associated with QT prolongation associated with cardiac problems (e.g. Sanguinetti et al. [7]).

Although it is possible to assess protein-chemical binding using a variety of biophysical techniques (e.g. Surface Plasmon Resonance), the experimental requirements in terms of the amounts of purified protein or chemical needed make it prohibitively expensive to perform on a battery of proteins. Nevertheless, there is a certain amount of binding information buried in the scientific literature, and focused efforts to collect them, like the one in creating the ToxWiz database, promise to bridge this gap of information. Software tools such as ToxWiz are thus a useful alternative to prohibitively expensive binding studies, as they allow one to see the binding partners of all similar chemical groups from the processed scientific literature and can infer other potential binding to the proteins, by combining network context algorithms with chemoinformatics. The advantage of such *in silico* predictions of toxicity over traditional QSAR methods, is that they immediately provide a mechanistic hypothesis for the predicted toxicity, as well as a set of candidate receptors or other binding events possibly related to the predicted or observed toxicity, which are easy to test in standardized receptor profiling or gene expression experiments.

During the last 50–60 years of molecular biology research, many chemicals have been used as tool compounds and many have been reported in the literature to bind to certain receptor or to cause changes in the gene expression. Some of these molecules have been already identified as toxicity biomarkers. Molecular biology also produced many ways to perturb a molecular target and to observe a produced phenotype (often pathology) in ways similar to a toxic compound perturbing a target or other off-target proteins. However, little is known and reported about mechanisms of toxicity. Mechanistic knowledge buried in the scientific literature requires expert analysis for it to be related to molecular biology and to catalogue it systematically in a database. In the case of the ToxWiz Knowledge base, a multidisciplinary team of molecular toxicologist, biochemists, chemists, pharmacologists, and medical doctors/veterinarians accumulated over 60 man-years of expert literature curation associated with toxicity biomarkers, biomarker candidates, and molecules associated with toxic endpoints (Figure 9.1). A systems biology approach made it possible for the first time to collate information on the basic set of biological pathways and to prepare literature supported mechanistic hypothesis for over 1,000 toxic endpoints. All the manually expert-curated information was put in a special systems biology framework within the underlying ToxWiz tool (Figure 9.2). This system now enables systematic analysis of toxic effects, -omics experiments or predictions of toxic endpoints.

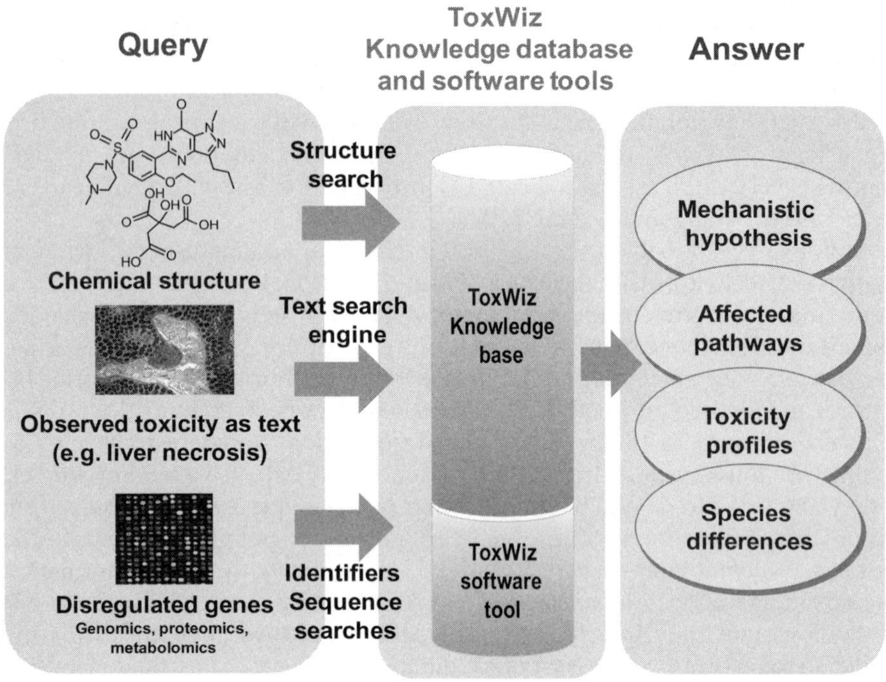

Chemicals: 40,000 biologically active chemicals (structures, other data)

- Drugs (1,200)
- Tool compounds (5,000)
- Metabolites (600)
- Industrial chemicals (400)

Proteins and Genes

- Metabolizing enzymes
- (54 CYPs, 22 UGTs, 13 GSTs, etc.)
- Nuclear receptors (51)

800 Toxic Endpoints with elaborate mechanistic hypotheses

Species differences (17 species)

Complete Human Interactome

For each chemical:

Information about reported toxicities
All metabolizing enzymes it interact with
All nuclear hormone receptors it has effect on
All diseases

Synonyms/Ontologies 2 000 000

4 Million manually expert-curated articles (PubMed, FDA, clinical, patents)

Over 10 million relationships

Figure 9.2 Predictions of toxic endpoints and mechanistic hypothesis can be generated by querying with chemical structures only, or combined with text searches for toxicities, pathologies, or with -omics data. The different ways of searching the ToxWiz Knowledge base (structure search, text search with synonyms, identifier, or sequence search) are integrated into the ToxWiz software tools.

For example, starting with a novel chemical structure causing a liver necrosis, one can easily access all the toxic effects for similar structures and the general mechanisms for liver necrosis, creating effectively a first hypothesis for how the novel chemical might be causing liver necrosis, and allowing comparison between species, as well as identifying potential toxicity biomarkers. Below the diagram are listed some of the contents of the ToxWiz Knowledge base that make associations with mechanistic knowledge possible.

9.3 SYSTEMS BIOLOGY APPROACH ENABLES UNPRECEDENTED INSIGHTS INTO MOLECULAR MECHANISMS OF TOXICOLOGY

Mechanistic knowledge accumulated in the literature as a result of molecular biology research, genome sequencing projects, and other high-throughput experimental technologies, has made biology a data-rich discipline. One of the biggest challenges scientists face today is how to interpret these large datasets. Systems biology approaches generate high-throughput data but more importantly, they allow interpretations of these data in a meaningful manner. Instead of an in-depth focus on a limited number of molecular components, these approaches adopt a comprehensive look at hundreds or thousands of interconnected components and how these combine to reveal definable phenotypes, such as disease or toxic responses.

ToxWiz contains a rich database of over 3 million of expert-curated data points that form a holistic "biological context network," where pathways are series of interactions within a cell displayed as a network of nodes (i.e. genes, proteins, chemicals, or effects) connected by edges (i.e. the lines joining molecules or effects together) and clusters, which are groups of protein/genes and chemicals associated to toxic endpoints, disease, or compound type. ToxWiz integrates the information about chemicals, genes, drug targets, metabolites, and pathologies describing their relationships or interactions. All these data, coupled with specific algorithms that allow prediction of possible toxic endpoint effects from the network context, ultimately suggest molecular mechanisms of toxicity. A collection of completely sequenced genomes and genomics information enables direct comparison of isoforms of metabolizing enzymes involved in processing of the chemical (e.g. CYP3A4) and fast cross-species validation of data across 17 species, including all toxicologically relevant species, e.g. dog, mouse, rat, etc.

This approach offers an unprecedented insight into understanding the connections between chemical structures, gene expression, protein networks, and toxic or pharmacological endpoints. Better understanding of mechanisms allows us to make a more informed assessment of risk or benefit of a drug/chemical. Combining chemical structures, together with genomics, proteomics, and a vast collection of annotated data from the scientific literature in the system are able to make accurate predictions of toxic endpoints or other aspects of pharmacology, and possibly associated mechanistic explanations, which can be easily visualized and explored further (Figure 9.1).

The captured data represent a high-level integrated holistic summary of interactions inferred from many biological contexts. However, representing the interactome as a static biological network is akin to a long-exposure photograph that can mask context-specific patterns of activation across multiple processes, cellular locations, and time. Conclusions drawn from the full network's topology may be compromised by these inherent limitations.

A central goal of systems biology research is to elucidate the underlying patterns of interaction in an effort to obtain more realistic and predictive models of the cell and eventually the organism. This has prompted the development of a broad range of graphical representations coupled with mathematical equations intended to model cellular dynamics.

The overall goal is to make toxicology more a predictive science than a diagnostic as it used to be. Toxicology is an ideal place to apply this kind of technology, for the simple reason that there is a pressing need to integrate a lot of information in order to improve efficiency.

Knowledge bases and tools such as ToxWiz allow one to rationalize toxicity findings by understanding mechanisms and their significance, and predict toxicity by linking molecular structures and effects through knowledge of pathways. It also helps in making decisions about the likely properties of chemicals in biological systems and the potential for adverse health effects.

ToxWiz contains a large, mainly manually curated database of protein–protein, protein–chemical interactions, pathways, and information about biological function, toxicology and other data from the literature. The software allows one to query this database with one or more protein or gene sequences, one or more chemicals, or text queries that are processed with the latest text-mining technologies and synthesize complex information fast. As a result, one can get specific answers such as where a xenobiotic perturbs a particular system, what other pathways might be affected, and how one can link chemical structures to a complex network of interactions (Figure 9.2).

It is often important to predict and understand more than one toxicity in more than one organ and rank the significance of the predictions and mechanistic hypothesis. The ToxWiz *mapping* system allows one to rank-order possible effected pathways or toxic-endpoint clusters. This system derives a score for each pathway/cluster inside the system considering genes/proteins that overlap within a set of query genes (e.g. from a microarray dataset), in addition to those that interact with members of a pathway/cluster (i.e. indirect associations). Several case studies performed with ToxWiz with well-established chemical toxicities are given below, and show how this technique is able to tease true signals out of very noisy data, predicting nonobvious associations that might be missed by more conventional means.

9.4 WHY NOW AND NOT 30 YEARS AGO?

An understanding of molecular mechanisms is not only important when developing new drugs but also for safety of chemicals in a workplace, industrial pollutants, and naturally occurring hazardous compounds found in food and drinking water. A new legislative directive from the European Union, REACH, was approved in summer 2007. REACH requires many commonly used chemicals to be re-registered and often retested for safety. Testing tens of thousands of chemicals in animals is extremely expensive and alternatives to animal testing are thus needed more than ever before. At the moment, there are insufficient *in vitro* alternatives available and there is a pressing need to make the best out of *in vitro* and *in silico* systems to gain a better mechanistic understanding of chemical actions and to define mechanism-based biomarkers of toxicity.

Significant progress in genomics has provided tools for the investigation and interpretation of important biochemical events. By combining *in vitro* and *in silico* systems, we should be better able to define biomarkers of toxicity, which will in turn allow better translation of effects across species and add specificity to predictions of toxic response.

9.5 PREDICTING TOXICITY AND MAKING A MECHANISTIC HYPOTHESIS BASED ONLY ON CHEMICAL STRUCTURE: A CASE STUDY WITH A PHTHALATES

Phthalates are common ingredients in plastics and are thus omnipresent in our environment. The toxic effects they can cause largely became known only once they were in frequent use. Today, certain phthalates are known to cause a range of reproductive toxicities in certain mammals (see Third National Report on Human Exposure to Environmental Chemicals, U.S. CDC, July 2005), though comparatively little is known about the mechanism. Here we will describe how a chemical structure of the phthalate is investigated inside ToxWiz, in order to make a mechanistic hypothesis about the toxic effects of diethylhexyl phthalate and about phthalates in general.

9.6 CHEMICAL STRUCTURE SEARCH: SEARCHING FOR PHTHALATE COMPOUNDS

A chemical substructure search for a phthalic acid fragment (Figure 9.3) inside a larger molecule identifies 29 closely related chemical structures, out of the roughly 40,000 biologically active chemicals inside the system (i.e. those

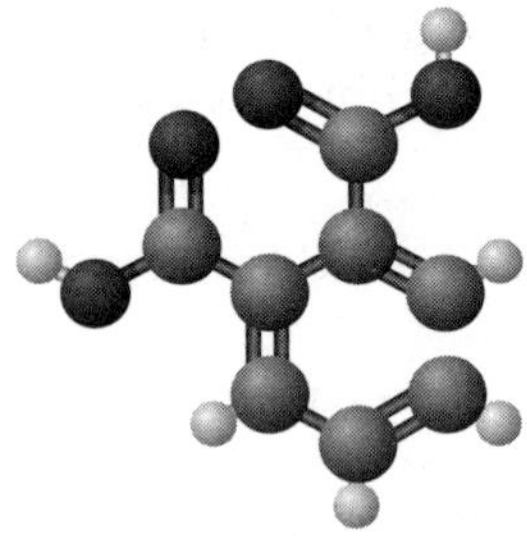

Figure 9.3 Phthalic acid.

chemicals with some molecular information about their biological action). These 29 similar chemicals are linked to over 300 proteins, genes, or biological effects in the system in the ToxWiz Version 2.1. These data allow us now to map out a mechanistic hypothesis of biological pathways activated by a very similar group of chemicals. The recent advances in systems biology and the prior knowledge have made it possible to make use of structural information describing a compound in order to predict pathways and exact sets of genes that will be affected with the chemical compound. The chemical structure can thus offer a first mechanistic guess. It can be done before any—omics experiment has been done and can therefore be used to guide, for instance, microarray or proteomics experiment design.

9.7 FINDING TOXICITY IN TESTIS BY ANALYZING EXPRESSION STUDY OF LIVER TISSUE FROM MICE TREATED WITH PHTHALIC ACID

Toxicogenomics studies are commonly used to identify a set of disregulated genes when an animal has been treated with a chemical of interest. Microarray studies allow not only chart disregulated genes, but also to observe changes over time and/or dosage. The sets of disregulated genes can help to obtain a more in-depth understanding about mechanisms of toxicity and eventually to identify toxicity biomarkers. Here, we show an analysis of whole-genome microarray experiment of a diethylhexyl phthalate-treated mice. A liver microarray dataset of mice treated with diethylhexyl phthalate over the four different doses was analyzed (the data are deposited in the Environment, Drugs and Gene Expression [EDGE] database). ToxWiz allows us to simultaneously load multiple experiments.

Though nearly all the mouse genes will be recognized by the system, it makes sense to further analyze for mechanism only those genes that have something known about them, either in terms of their function or their interactions with proteins and chemicals.

The ToxWiz database is human centric but it also contains data for 17 other species, including higher mammals such as chimp, dog, and mouse, as well as simpler organisms like zebra fish, nematode, and yeast. When microarray data of an animal data are imported into ToxWiz, the tool determines how many of the genes in the animal are equivalent in humans.

Typically, between 50% and 70% of all the genes in the dataset will be considered in the mechanistic analysis with a tool such as ToxWiz. It is possible to identify distinctively disregulated genes where nothing is known about the function of the gene, and that these can become biomarkers. Though useful as biomarkers, these genes cannot tell us much about mechanism. When importing a set of disregulated genes, we can also specify thresholds for the genes to include in the analysis. Here only genes with a fold change of 1.5 or higher are taken into account.

9.8 MAPPING THE MICROARRAY DATA ONTO TOXIC ENDPOINTS

A quick and easy way to see what a set of disregulated genes might do in a biological system is to map them onto biological pathways or toxicity clusters in the system. A cluster is a set of genes or chemicals associated with a biological phenomenon such as a toxic endpoint or a disease. Genes, proteins, and chemicals within a toxicity cluster can be viewed as a mechanistic hypothesis for a toxic endpoint. In this case, we will consider only toxic endpoints (Figure 9.4). The toxicity pathways or clusters (on the right) are ranked according to our statistical mapping system, which considers how common genes are in the system, in addition to connections between genes. The third ranked toxic endpoint in this case is testis atrophy. This is a well-known problem with phthalate molecules, which cause genital and germ-line development problems.

9.9 EXPLORING TOXIC ENDPOINTS

A toxic endpoint cluster is a set of genes, proteins, and chemicals associated somehow with a particular endpoint. Figure 9.5 shows part of the testis atrophy cluster. There are many molecules linked to the testis atrophy toxic endpoint, suggesting that this pathology has been extensively studied. The genes from our microarray dataset that are inside this cluster appear in blue, with bars indicating the degree of up- or down-regulation. Hovering over these bars displays the numbers from the original data file.

It is also possible to see how other genes in our dataset are indirectly connected to this toxic endpoint. If we consider all types of connections, then several other genes from our dataset are added to the cluster (Figure 9.6). These connections contributed to the score that we got in our previous mapping (Figure 9.4). These powerful and unique features of ToxWiz allow us to relate genes to toxic endpoints that have not yet been reported in the literature to

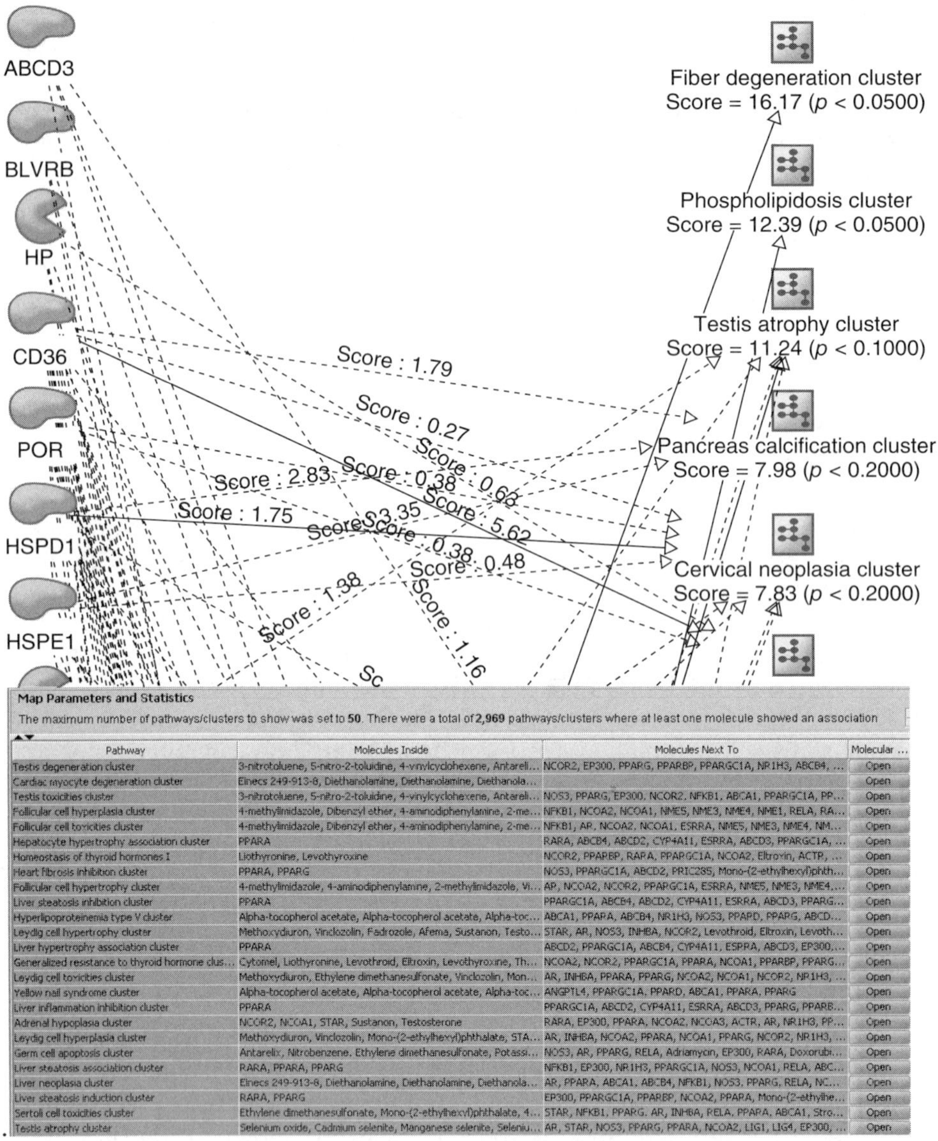

Map Parameters and Statistics

The maximum number of pathways/clusters to show was set to **50**. There were a total of **2,969** pathways/clusters where at least one molecule showed an association

Pathway	Molecules Inside	Molecules Next To	Molecular ...
Testis degeneration cluster	3-nitrotoluene, 5-nitro-2-toluidine, 4-vinylcyclohexene, Antareli...	NCOR2, EP300, PPARG, PPARBP, PPARGC1A, NR1H3, ABCB4, ...	Open
Cardiac myocyte degeneration cluster	Einecs 249-913-8, Diethanolamine, Diethanolamine, Diethanola...		Open
Testis toxicities cluster	3-nitrotoluene, 5-nitro-2-toluidine, 4-vinylcyclohexene, Antareli...	NOS3, PPARG, EP300, NCOR2, NFKB1, ABCA1, PPARGC1A, PP...	Open
Follicular cell hyperplasia cluster	4-methylimidazole, Dibenzyl ether, 4-aminodiphenylamine, 2-me...	NFKB1, NCOA2, NCOA1, NME5, NME3, NME4, NME1, RELA, RA...	Open
Follicular cell toxicities cluster	4-methylimidazole, Dibenzyl ether, 4-aminodiphenylamine, 2-me...	NFKB1, AR, NCOA2, NCOA1, ESRRA, NME5, NME3, NME4, NM...	Open
Hepatocyte hypertrophy association cluster	PPARA	RARA, ABCB4, ABCD2, CYP4A11, ESRRA, ABCD3, PPARGC1A, ...	Open
Homeostasis of thyroid hormones I	Liothyronine, Levothyroxine	NCOR2, PPARBP, RARA, PPARGC1A, NCOA2, Eltroxin, ACTP, ...	Open
Heart fibrosis inhibition cluster	PPARA, PPARG	NOS3, PPARGC1A, ABCD2, PRIC285, Mono-(2-ethylhexyl)phth...	Open
Follicular cell hypertrophy cluster	4-methylimidazole, 4-aminodiphenylamine, 2-methylimidazole, Vi...	AP, NCOA2, NCOR2, PPARGC1A, ESRRA, NME5, NME3, NME4,...	Open
Liver steatosis inhibition cluster	PPARA	PPARGC1A, ABCB4, ABCD2, CYP4A11, ESRRA, ABCD3, PPARG...	Open
Hyperlipoproteinemia type V cluster	Alpha-tocopherol acetate, Alpha-tocopherol acetate, Alpha-toc...	ABCA1, PPARA, ABCB4, NR1H3, NOS3, PPAPD, PPARG, ABCD...	Open
Leydig cell hypertrophy cluster	Methoxydiuron, Vinclozolin, Fadrozole, Afema, Sustanon, Testo...	STAR, AR, NOS3, INHBA, NCOR2, Levothroid, Eltroxin, Levoth...	Open
Liver hypertrophy association cluster	PPARA	ABCD2, PPARGC1A, ABCB4, CYP4A11, ESPRA, ABCD3, EP300,...	Open
Generalized resistance to thyroid hormone clus...	Cytomel, Liothyronine, Levothroid, Eltroxin, Levothyroxine, Th...	NCOA2, NCOR2, PPARGC1A, PPARA, NCOA1, PPARBP, PPARG...	Open
Leydig cell toxicities cluster	Methoxydiuron, Ethylene dimethanesulfonate, Vinclozolin, Mon...	AR, INHBA, PPARA, PPARG, NCOA2, NCOA1, NCOR2, NR1H3, ...	Open
Yellow nail syndrome cluster	Alpha-tocopherol acetate, Alpha-tocopherol acetate, Alpha-toc...	ANGPTL4, PPARGC1A, PPARD, ABCA1, PPARA, PPARG	Open
Liver inflammation inhibition cluster	PPARA	PPARGC1A, ABCD2, CYP4A11, ESRRA, ABCD3, PPARG, PPARB...	Open
Adrenal hypoplasia cluster	NCOR2, NCOA1, STAR, Sustanon, Testosterone	RARA, EP300, PPARA, NCOA2, NCOA3, ACTR, AR, NR1H3, PP...	Open
Leydig cell hyperplasia cluster	Methoxydiuron, Vinclozolin, Mono-(2-ethylhexyl)phthalate, STA...	AR, INHBA, NCOA2, PPARA, NCOA1, PPARG, NCOR2, NR1H3, ...	Open
Germ cell apoptosis cluster	Antarelix, Nitrobenzene, Ethylene dimethanesulfonate, Potassi...	NOS3, AR, PPARG, RELA, Adriamycin, EP300, RARA, Doxorubi...	Open
Liver steatosis association cluster	RARA, PPARA, PPARG	NFKB1, EP300, NR1H3, PPARGC1A, NOS3, NCOA1, RELA, ABC...	Open
Liver neoplasia cluster	Einecs 249-913-8, Diethanolamine, Diethanolamine, Diethanola...	AR, PPARA, ABCA1, ABCB4, NFKB1, NOS3, PPARG, RELA, NC...	Open
Liver steatosis induction cluster	RARA, PPARG	EP300, PPARGC1A, PPARBP, NCOA2, PPARA, Mono-(2-ethylhe...	Open
Sertoli cell toxicities cluster	Ethylene dimethanesulfonate, Mono-(2-ethylhexyl)phthalate, 4...	STAR, NFKB1, PPARG, AR, INHBA, RELA, PPARA, ABCA1, Stro...	Open
Testis atrophy cluster	Selenium oxide, Cadmium selenite, Manganese selenite, Seleniu...	AR, STAR, NOS3, PPARG, PPARA, NCOA2, LIG1, LIG4, EP300, ...	Open

Figure 9.4 The upper part shows a map of genes down the left side, connected to toxic endpoints on the right. Solid lines connecting genes to toxic endpoints indicate that the gene is reported in the literature to be associated with a toxic endpoint. Dashed lines indicate that the gene is interacting or influencing something that is reported to be related with the toxicity and thus highly probably plays a role in the mechanism of toxicity. Below is the alternative table version with names of endpoints in the left column, molecules associated with a toxic endpoint in the middle column, and molecules interacting or influencing a given toxicity endpoint in the right column.

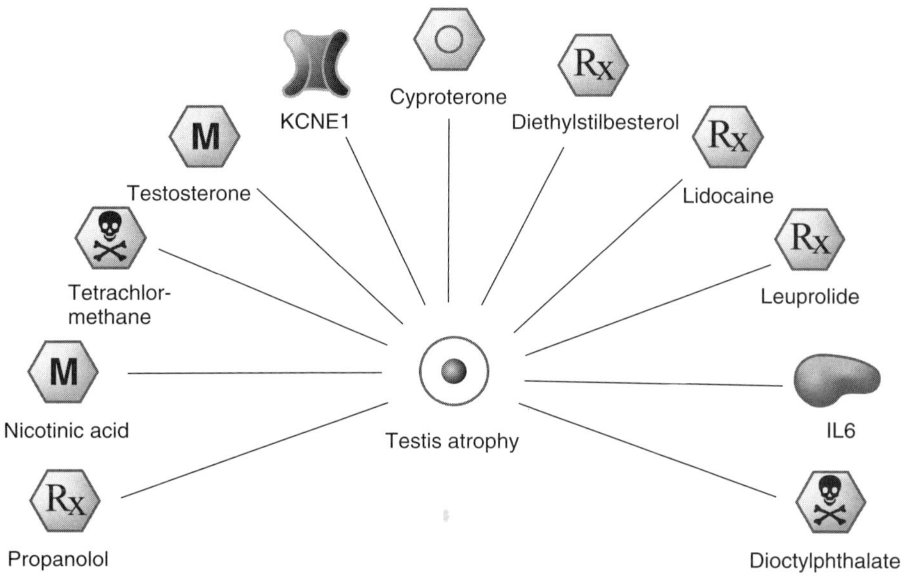

Figure 9.5 Some of genes, proteins, and chemicals (with chemical structure information) expert curated from the literature to be associated with the testis atrophy toxic endpoint.

be associated with the toxic endpoints, opening new avenues for discovering biomarkers and creating hypotheses about mechanism.

The mechanistic view of testis atrophy reveals that phthalates have previously been seen to up-regulate PEX11A2. This gene is also up-regulated in our dataset and while there is no evidence of a direct link between PEX11A [8] and testis problems in the current literature, ToxWiz has inferred such a relationship, based on the vast network of genes, proteins, and chemicals stored in the database. Other genes that are known to have an effect on testis development are also present in the cluster, such as cytochrome c (T-Cc), which causes problems when it is deleted in the mouse [8].

9.10 COMBINING THE RESULTS OF THE SUBSTRUCTURE SEARCH WITH THE MICROARRAY DATA

Combining the prediction of toxic endpoints based on the chemical structure with the microarray dataset provides an unprecedented way of creating a mechanistic hypothesis. Figure 9.7 shows part of the highest scoring cluster from the previous map, testis degeneration, with both the connected chemicals from the phthalate substructure search and the connected genes from the microarray dataset added.

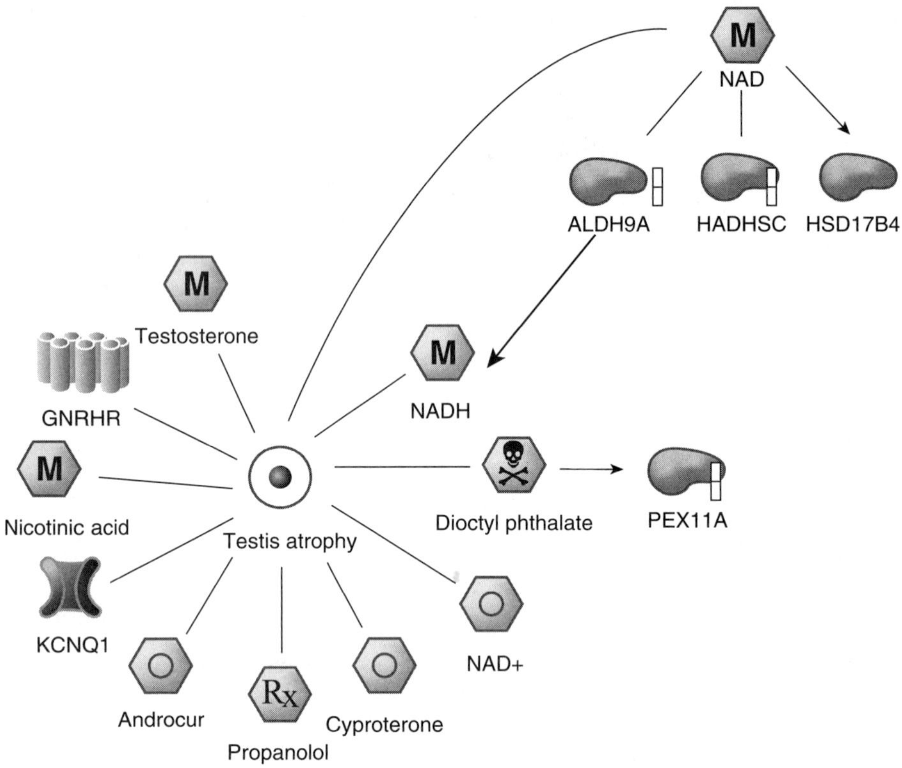

Figure 9.6 Mechanistic view of testis atrophy toxic endpoint cluster shows connected genes from the mouse liver microarray dataset. The connecting lines have expert-curated literature references linked to them, thus allowing fast access to literature supporting the mechanistic hypothesis.

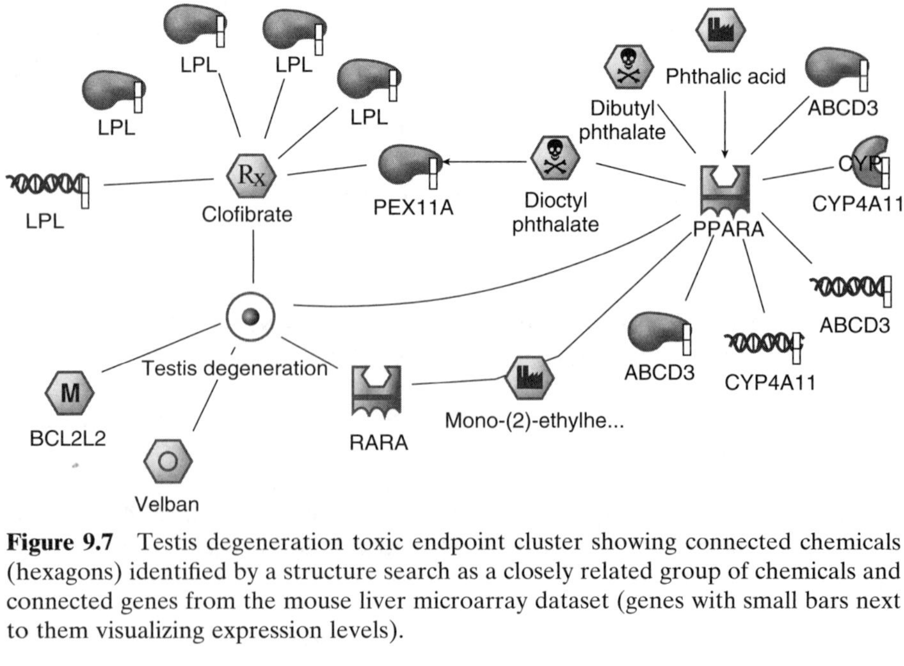

Figure 9.7 Testis degeneration toxic endpoint cluster showing connected chemicals (hexagons) identified by a structure search as a closely related group of chemicals and connected genes from the mouse liver microarray dataset (genes with small bars next to them visualizing expression levels).

This provides a better picture of what might be happening when mice are given these compounds. The figure shows that phthalates are known to activate or induce various nuclear receptors such as PPAR alpha (PPARA) and RAR alpha 2 (RARA) [8]. Moreover, several genes from our microarray dataset appear to be activated by one or more of these transcription factors, such as the transporter molecule, ABCD3.

This analysis demonstrates one of the many ways in which ToxWiz can be used to investigate toxicogenomics data. It shows how by combining microarray data with the chemical information in the system, it is possible to gain insights into possible mechanisms of toxicity.

9.11 SOLUTION TO COMMON PROBLEMS TOXICOLOGISTS FACE: FINDING A MOLECULAR RATIONALE FOR UNEXPECTED TOXIC ENDPOINTS

AstraZeneca toxicologists described an unexpected observation of testis degeneration in rats, when studying a series of compounds consisting of a "lipophilic core connected to a nicotinic acid" [9]. Subsequent work suggested that undesirable interactions with retinoic acid receptors (RARs) might be the molecular basis of this effect.

Although full details of the compounds were not disclosed because they were proprietary, it was possible to suggest a molecular rationale for the findings by using the ToxWiz Knowledge base combined with the chemical structure analysis from the ToxWiz tool. It was possible to integrate data in such a way that diverse sources of information such as gene knockouts in model organisms, chemistry, and pathology data can be combined to reveal the likely mechanism of these (and other) unexpected toxic endpoints.

The ToxWiz system works by exploiting a network of thousands of interacting genes, proteins, chemicals, pathways, and biological effects, including toxic endpoints and diseases. The neighborhood of a molecule in the system captures everything that has been observed about it in the past, either in the public domain or in company proprietary datasets, and therefore provides an excellent basis to help make decisions in toxicology and many other areas of drug discovery.

9.12 CHEMICAL SEARCH FOR NICOTINATE

The first step to perform is a substructure search of the ToxWiz database for compounds linked to biology and containing an appropriately substituted nicotinic acid group, and then find toxicology-related biological pathways in which these compounds are active. This search yielded 19 compounds containing the nicotinic group (Figure 9.8), of which 10–12 contained lipophilic

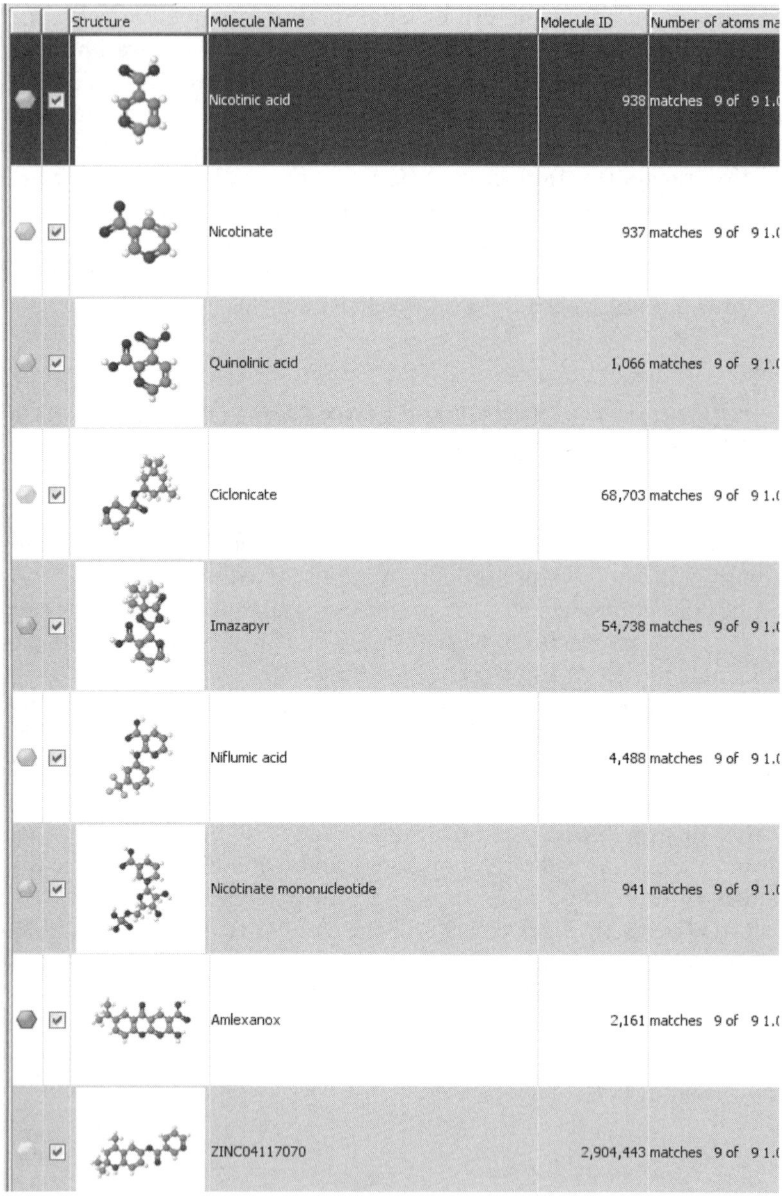

		Structure	Molecule Name	Molecule ID	Number of atoms ma
			Nicotinic acid	938	matches 9 of 9 1.(
			Nicotinate	937	matches 9 of 9 1.(
			Quinolinic acid	1,066	matches 9 of 9 1.(
			Ciclonicate	68,703	matches 9 of 9 1.(
			Imazapyr	54,738	matches 9 of 9 1.(
			Niflumic acid	4,488	matches 9 of 9 1.(
			Nicotinate mononucleotide	941	matches 9 of 9 1.(
			Amlexanox	2,161	matches 9 of 9 1.(
			ZINC04117070	2,904,443	matches 9 of 9 1.(

Figure 9.8 Some of the 19 compounds returned using a chemical substructure search with nicotinic acid.

substituents, and were thus possibly similar to the proprietary novel compound in question. Note that this filtering process also involved removing compounds that did not contain true nicotinate groups (for example, the drug Amlexenox, in which the nicotinate group is part of a larger ring system).

9.13 MAPPING THE COMPOUND SET ONTO TOXIC ENDPOINTS

This set of molecules was then mapped onto over 1,000 toxic endpoint clusters in the ToxWiz database (Figure 9.9). Each cluster contains chemicals, genes, and proteins that have previously been associated with each endpoint. The mapping process considers whether each molecule is reported in the literature to be associated with the toxic endpoints (full lines), or if it is not (yet) reported in the literature to be associated with the toxic endpoints. We found that it acts on genes or proteins that themselves have reported to be associated with the toxic endpoint and thus these chemicals are likely to be related somehow to the toxicity. The toxic endpoint pathways (clusters) are then ranked according to a statistical significance, which accounts for the abundance of proteins in biological processes and the size of pathways. For interactions it also considers the strength, or confidence, of the interaction, for example, giving a greater weight to manually curated interactions than those determined from text mining. The clusters are then listed in order of significance on the right of the map.

Testis atrophy and testis degeneration are among the highest ranked clusters (first and third, respectively). One can also see that only three of the compounds map with any significance to endpoints. Inspection shows that

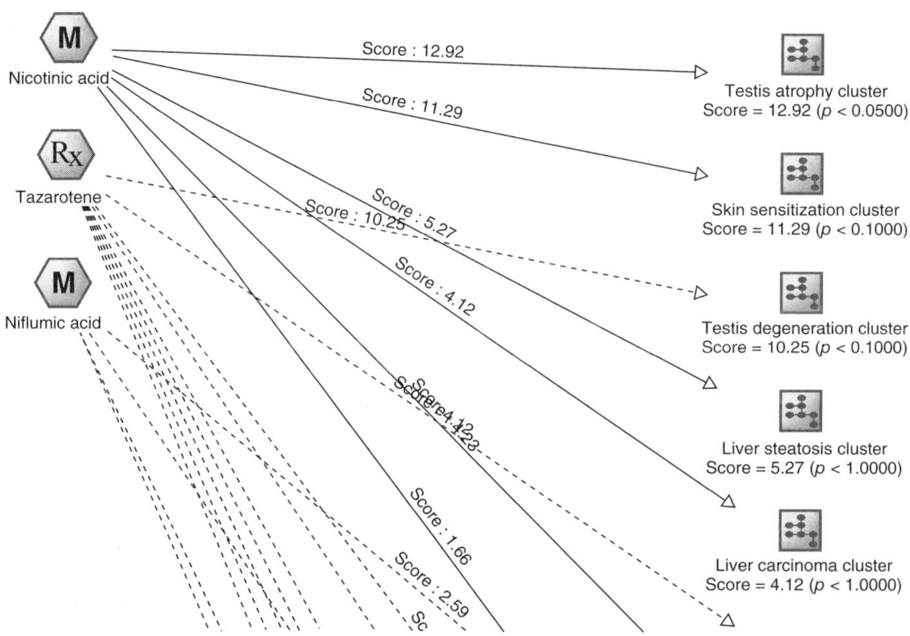

Figure 9.9 Mapping compounds containing nicotinic acid without significant polar group substituents.

several of the compounds are from high-throughput assays in PubChem and lack any literature evidence about their biological function.

The power of the network is illustrated by the map in Figure 9.7. Solid lines indicate that there is at least one previous observation of a relationship between the molecule and the endpoint. By contrast, dashed lines indicate an indirect relationship, in which the molecule interacts with something (usually a protein) inside the endpoint cluster. Such compounds might have a secondary effect. Such network-based systems and analysis tool help highlight and discover novel mechanistic associations.

9.14 PREVIOUSLY OBSERVED XENOBIOTICS THAT CAUSE TESTIS PROBLEMS

The association of nicotinic acid to problems in the testis is most likely due to processes downstream from, or related to, toxicity induced by phthalates and other similar molecules [10]. Phthlates are known to influence RAR in the course of causing testis developmental problems [11,12]. Searches within the system also identify key problems.

9.15 NICOTINATES WITH LIPOPHILIC SUBSTITUENTS ACTING ON RARα

One of the compounds containing lipophilic attachments is the drug tazarotene (Figure 9.10), an optical retinoid used for the treatment of psoriasis. This compound is more active against RAR beta or gamma than alpha, but it does show activity against RAR alpha [13]. Tazarotene acts by affecting activities of RARs. The compound itself is not associated with testis problems, only with the receptor (hence the dashed lines in Figure 9.9). Indeed, no evidence of testis problems has been found for this compound. Although, as it acts by

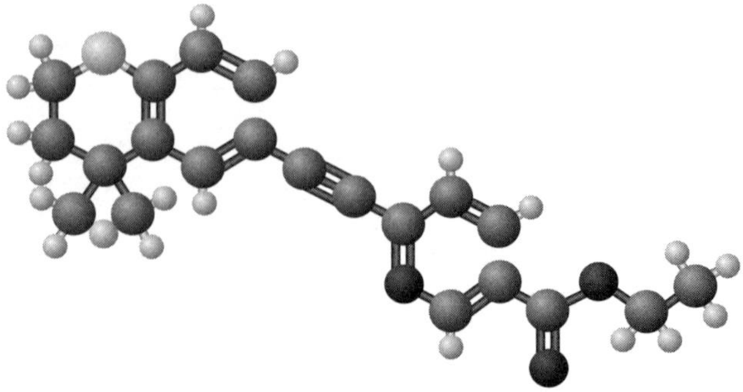

Figure 9.10 Tazarotene.

targeting this receptor, that is, perhaps, unsurprising. But because it conforms to the pattern mentioned by Scott Boyer in his presentation at ISSX, the proprietary compounds could be acting in a similar way to exert deleterious (instead of desired) effects.

9.16 THE MOLECULAR DETAILS OF THE PREDICTION

Displaying the mechanistic hypothesis view for testis degeneration and then adding to it the compounds from the search, creates a pathway-like diagram that might explain how the compounds are eliciting their effect (Figure 9.11). In particular, relationships with both retinoic acid (e.g. RAR alpha) and peroxisome proliferative activated receptors (e.g. PPAR alpha) are apparent among the compounds known to cause developmental problems in the testis.

This retrospective analysis demonstrates the great predictive potential of ToxWiz. Moreover, its ability to deal with the concept of a group of compounds rather than just one compound can be a great advantage when studying a series of molecules, or when considering the best substituents to try during lead optimization studies. Knowledge of previous studies is always a great benefit in designing new ones, and ToxWiz can be an effective way of finding this information quickly.

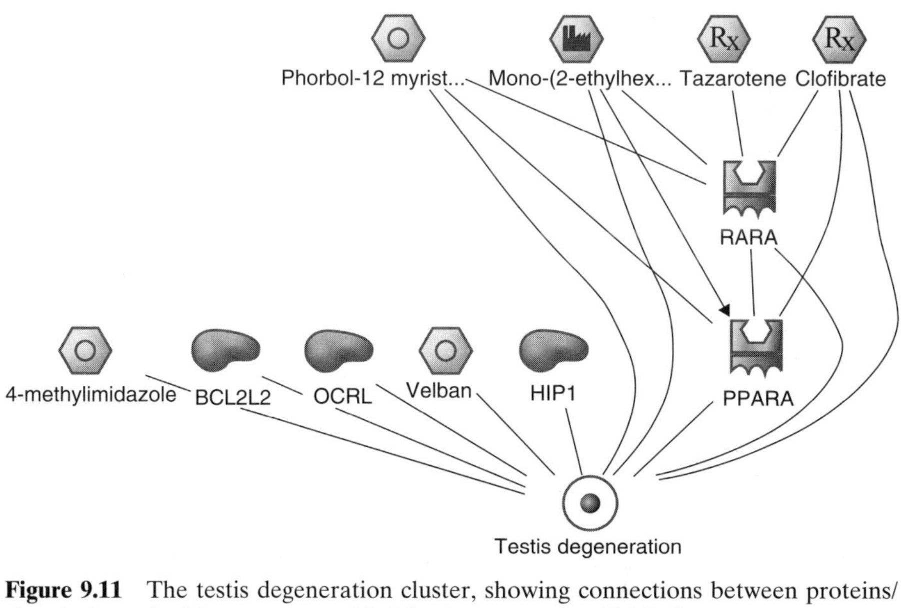

Figure 9.11 The testis degeneration cluster, showing connections between proteins/chemicals, and with tazarotene added (and connected to RARα).

9.17 EXPLORING THE MOLECULAR BASIS OF COMBINATION DRUGS USING TOXWIZ

Drug combinations represent intriguing possibilities for new therapies. The basic principle is that two active compounds can lead to effects that are more than the sum of their parts, possibly by simultaneously blocking two parts of the same pathway, or by one compound augmenting the activity of another. Borisy et al. [14], from CombinatoRx systematically screened about 120,000 combinations of reference listed drugs to find combinations that had new activities. Using ToxWiz, all combinations that came out of this study were investigated to see if the ToxWiz database and algorithms could suggest how one drug could complement another.

One of the interesting combination effects that were reported was that the antiplatelet drug dipyridamole, when combined with the glucocorticoid dexamethasone, prevented TNF-α production in response to stimulation by phorbol 12-myristate 13-acetate (PMA) or ionomycin. The best possible paths between these two drugs were sought using ToxWiz and also, additional connections between them were explored by investigating the interaction network around these two compounds (see Figure 9.12).

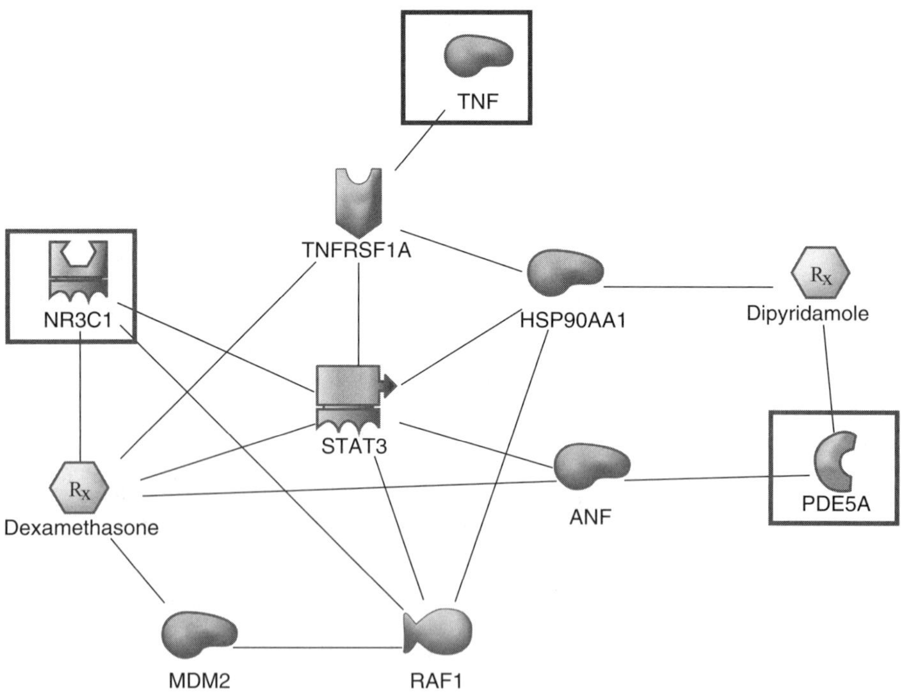

Figure 9.12 Some of the possible biological links between dexamethasone and dipyridamole suggested by ToxWiz. The targets for the two drugs are boxed left and right. TNF-alpha is boxed on the top of the picture.

There are several possibilities for how these two drugs can affect each other inside the system. As pointed out by Borisy et al. [14], the effects of steroids, like dexamethasone, on TNF are well documented [15], and anti-TNF effects of dipyridamole are thought to be due to its blockage of adenosine uptake. However, their combined mode of action could not be explained by these prior observations [14].

Interrogation of ToxWiz's vast database of protein–protein and protein–chemical associations, however, suggested other possibilities (Figure 9.10). One particularly interesting possibility arises owing to a screening result, where dipyridamole was one of 220 compounds found to inhibit HSP90 in tumor cell lysate [16]. The relationship between HSP90-like molecules, such as TRAP-1, and TNF is well understood—complexes involving HSP90 are often critical for TNF-mediated signaling (e.g. Chen et al. [17]). It is compelling to suggest that affects of TNF signaling could be due to the inhibition of HSP90, thus operating in a manner similar to the well-known HSP90 inhibitor geldenamycin, which itself can decrease TNF secretion in response to inflammatory stimulants [18].

There are several other possibilities for interactions between these two drugs suggested by the system. These possible relationships became apparent after just a few minutes of study, thus demonstrating the power of the very fast synthesis of data from diverse sources to arrive at hypotheses very quickly.

9.18 CONCLUSIONS

The topics discussed in this chapter are illustrating the importance of collating and analyzing information on biological pathways in order to gain a better insight into the mechanisms underlying safety issues, especially mechanisms of toxicity. Specialist databases of expert-curated knowledge, such as the ToxWiz Knowledge base with mechanistic information related to over 1,000 toxic endpoints, can offer a first insight into the possible molecular mechanisms of observed toxicities. The systems biology approach underlying the databases and specially designed software allows one to analyze experiments, capturing various aspects of the exposure to a toxic compound over time or dosages (e.g. such as genomics, proteomics, and metabolomics experiments) and potentially to identify toxicity biomarkers. Collated mechanistic information and network-based systems are amenable to be used not only to retrospectively (to explain observed toxicity) but also to predict pathways that are likely to be affected with a certain chemical structure class. Integrated pathway systems such as ToxWiz are able to point to more than one toxicity in more than one tissue, offering unprecedented insights into the molecular mechanisms of toxicity. With several million expert-curated articles on clinical manifestations, such expert systems hold great promise to improve other safety-related issues important in planning of clinical trials. The biological pathway tools are thus becoming an integral part not only of a drug discovery process but also in the assurance of safety of a wide variety of chemicals such as pesticides, herbicides, plastics, and detergents, which we are surrounded with in everyday life.

9.19 ABOUT CAMBRIDGE CELL NETWORKS

Cambridge Cell Networks (CCNet) based in Cambridge, UK, supplies a range of industry-leading content on biological pathways, chemistry, and toxicology, combined with an integrated pathway visualization and exploration tools to the pharmaceutical and biotechnology industries. Using cutting-edge biological and computational methods combined with knowledge management techniques, CCNet offers a novel approach to pathway analysis, providing effective target validation and predictive toxicology data, which will ensure the production of safer drugs. CCNet has facilities in three countries and is staffed by a team of expert biochemists, pharmacologists, bioinformaticians, chemists, and industrial toxicologists.

REFERENCES

1. Koppal T. Microarrays: migrating from discovery to diagnostics. Drug Discov Dev 2004;7:30–34.
2. Sauer UG. The new EU Chemicals Policy—comments of Eurogroup for Animal Welfare and the Deutscher Tierschutzbund on the EU Commission's REACH system Consultation Documents. ALTEX. 2003;20(3):225–227.
3. Fielden MR, Kolaja KL. The state-of-the-art in predictive toxicogenomics. Curr Opin Drug Discov Devel 2006;9(1):84–91.
4. Heijne WH, Kienhuis AS, van Ommen B, Stierum RH, Groten JP. Systems toxicology: applications of toxicogenomics, transcriptomics, proteomics and metabolomics in toxicology. Expert Rev Proteomics 2005;2(5):767–780.
5. Tsutsui H, Matsui, K, Okamura H, Nakanishi K. Pathophysiological roles of interleukin-18 in inflammatory liver diseases. Immunol Rev 2000;174:192–209.
6. Clegg ED, Cook JC, Chapin RE, Foster PM, Daston GP. Leydig cell hyperplasia and adenoma formation: mechanisms and relevance to humans. Reprod Toxicol 1997;11:107.
7. Sanguinetti MC, Chen J, Fernandez D, Kamiya K, Mitcheson J, Sanchez-Chapula JA. Physicochemical basis for binding and voltage-dependent block of hERG channels by structurally diverse drugs. Novartis Found Symp 2005;266:159–166.
8. Narisawa S, Hecht NB, Goldberg E, Boatright KM, Reed JC, Millán JL. Testis-specific cytochrome *c*-null mice produce functional sperm but undergo early testicular atrophy. Mol Cell Biol 2002;22(15):5554–5562.
9. Boyer S. In silico toxicity predictions—present and future tools in systems biology. International Society for the Study of Xenobiotics (ISSX), Manchester, UK, June 2006.

10. Fukuwatari T, Ohsaki S, Suzuki Y, Fukuoka S, Sasaki R, Shibata K. The effects of phthalate esters on the tryptophan-niacin metabolism. Adv Exp Med Biol 2003;527:659–664.
11. Dufour JM, Vo MN, Bhattacharya N, Okita J, Okita R, Kim KH. Peroxisome proliferators disrupt retinoic acid receptor alpha signaling in the testis. Biol Reprod 2003;68:1215–1224.
12. Lufkin T, Lohnes D, Mark M, Dierich A, Gorry P, Gaub MP, LeMeur M, Chambon P. High postnatal lethality and testis degeneration in retinoic acid receptor alpha mutant mice. Proc Natl Acad Sci U S A 1993;90:7225–7229.
13. Bianchi L, Orlandi A, Campione E, Angeloni C, Costanzo A, Spagnoli LG, Chimenti S. Topical treatment of basal cell carcinoma with tazarotene: a clinicopathological study on a large series of cases. Br J Dermatol 2004;151:148–156.
14. Borisy AA, Elliott PJ, Hurst NW, Lee MS, Lehar J, Price ER, Serbedzija G, Zimmermann GR, Foley MA, Stockwell BR, Keith CT. Systematic discovery of multicomponent therapeutics. Proc Natl Acad Sci U S A 2003;100(13): 7977–7982.
15. Joyce DA, Steer JH, Abraham LJ. Glucocorticoid modulation of human monocyte/macrophage function: control of TNF-alpha secretion. Inflamm Res 1997;46(11):447–451.
16. Rodina A, Vilenchik M, Moulick K, Aguirre J, Kim J, Chiang A, Litz J, Clement CC, Kang Y, She Y, Wu N, Felts S, Wipf P, Massague J, Jiang X, Brodsky JL, Krystal GW, Chiosis G. Selective compounds define Hsp90 as a major inhibitor of apoptosis in small-cell lung cancer. Nat Chem Biol 2007;3(8):498–507.
17. Chen G, Cao P, Goeddel DV. TNF-induced recruitment and activation of the IKK complex require Cdc37 and Hsp90. Mol Cell 2002;9(2):401–410.
18. Vega VL, De Maio A. Geldanamycin treatment ameliorates the response to LPS in murine macrophages by decreasing CD14 surface expression. Mol Biol Cell 2003;14(2):764–773.

10

IMPACT OF CHEMISTRY INFORMATION ON PATHWAY ANALYSIS

SREENIVAS DEVIDAS

Table of Contents

10.1 INTRODUCTION

The past decade has witnessed the emergence of Pathway Analysis as an essential component of Drug Discovery. In the recent past, it has moved across all aspects of drug development from discovery through to the Clinical Phases. The recent past has also witnessed a surge in the number of publications from

Pathway Analysis for Drug Discovery, Edited by Anton Yuryev

Figure 10.1 Conventional approach to Pathway Analysis.

both industry and academia. Over 300 publications have been deposited in PubMed and continue to grow at a rapid pace.

During the old days, the approach to Pathway Analysis took a "cut and paste" approach. In essence, researchers would identify and read papers of interest to them in the areas relevant to their research. They would then attempt to reconstruct a pathway on the blackboard, with a hope of trying to identify novel drug targets for the drugs under consideration.

The process though reasonably successful in the long term suffered from the following:

- The process was extremely slow, expensive, and inefficient. It took an immense amount of time to identify, summarize, and infer the data from the relevant references.
- Electronic reference repositories were few if any and slowed down the process tremendously.
- Nonstandard approaches to manuscript preparations across journals and reviews, as well as the lack of automated approaches such as Natural Language Processing, made the process even slower (Figure 10.2).

10.2 EMERGENCE OF PATHWAY TOOLS

The problems highlighted in the previous section were significantly overcome with the emergence of Pathway Databases and Pathway Analysis Tools. The last decade has witnessed the emergence of over a dozen commercial and public sector efforts. The fundamental rationale behind all these efforts was that the number of protein–protein interactions in the published and patented literature had grown to a number high enough that automated pathway reconstruction from the underlying data made sense.

The rationale was that the database would collate the known protein–protein interactions from the published and patented literature. This would

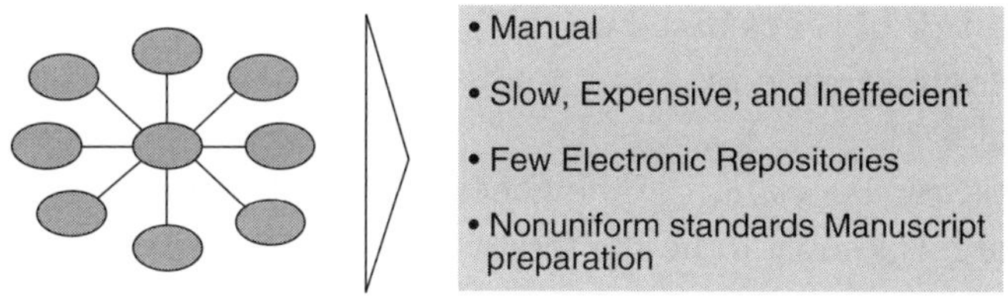

Figure 10.2 Issues with traditional pathway analysis.

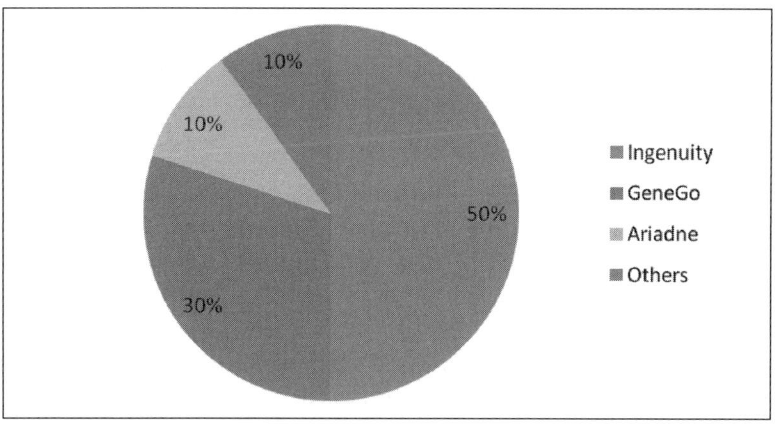

Others include: Jubilant Biosys, Keymolnet, Genomatix, Molecular Connections, and others.

Figure 10.3 Market Share of Major Pathway Database Vendors—independent survey based on number of users and dollars spent.

be accompanied by an intuitive user interface that would allow users to input gene and protein lists of interest to them. The application layer would infer all the relevant interactions associated with the input dataset and map a protein–protein interaction network based on the input data. The last couple of years has witnessed a couple of the providers taking a lion's share of the market (>90%) in both the academic and industry segments (Figure 10.3).

10.3 LIMITATIONS OF CURRENT OFFERINGS OF PATHWAY ANALYSIS TOOLS

The workflow that is supported by most of the Pathway Analysis Tools is illustrated in Figure 10.4. The end point is limited to either of the following:

a. A network of genes based on the input dataset OR
b. The resultant network filtered down by a few criteria such as (but not limited to)
   i. Disease association of the genes in the network
   ii. Targets known to be the target of one or more launched drugs
   iii. Targets known to be the target of one or more compounds in Clinical Trials
   iv. Targets reasonably known to be potential candidate or clinical biomarkers

Almost all the tools in use today are biased toward the biological side and contain little if any information on the chemical inhibitors associated with the components of a given biological network. The limitation resides in the following:

- The fact that the number of inhibitors that have been published or patented is probably several fold (probably greater than 10-fold) more than the number of published protein–protein interactions. Curation of such data is both expensive and time-consuming and could impact ROI significantly.
- The nonstandardization associated with protein and gene names makes mapping reasonably non-straightforward.

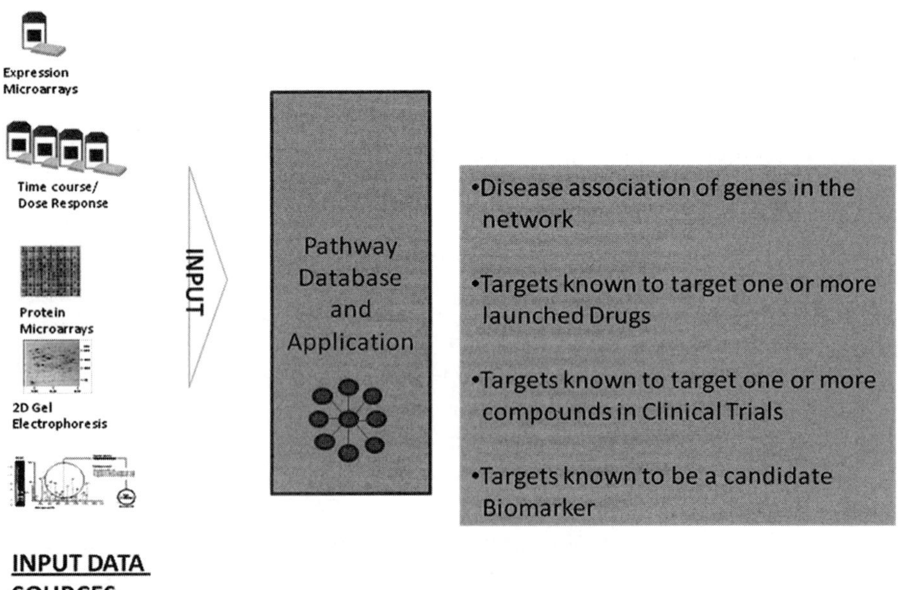

Figure 10.4 Limitations of current Pathway Tools.

- The number of patented and published inhibitors across target classes increases by over a million per year. Keeping the databases current becomes a monumental problem.
- Modifications required in the product architectures to accommodate chemistry-related content.

10.4 IMPACT OF CHEMICAL INFORMATION ON PATHWAY ANALYSIS

The remainder of this review focuses on the impact of chemical information on Pathway Analysis. The rationale is split into the following sections:

a. The content
b. The integration
c. The utility and workflow
d. Return on Investment

10.4.1 The Content

The present pathway databases (and the databases that underlie the applications) contain data on protein–protein interactions. As illustrated above, the end point of the analysis of an input dataset is a network diagram of such interactions that have been filtered by some simple means. However, the starting point of the experiment that fed the data into the application was in most cases a compound that was being tested by the investigator against an indication. Hence, once a network of interactions has been identified, one needs to identify the following:

a. Did the compound under investigation target a known druggable target?
b. Were any of the targets contained in the final network novel to the investigational compound?
c. Has the investigational compound been used to test against other targets in the published or patented literature?
d. If the compound has been tested in the literature, then what are the known SAR (structure–activity relationships), assay types, and toxicity information?
e. Side Effects-Off or Multiple Target effects—If the investigational compound has been studied and shown to demonstrate activity against multiple targets, then could it be indicative of favorable or unfavorable cross-target effects?

The key to generating answers to the aforementioned questions lies in the access to a well-collated and structured database of the relevant information. The degree of difficulty lies in the fact that such information has been:

a. Published in a vast number of published journals (possibly exceeding a thousand distinct journals)
b. Published in several languages
c. Patented in different countries

It is estimated that over half a million new inhibitors are published against the well-studied biological targets (Kinase, GPCR, Ion Channels, Transporters, and Proteases) alone every year.

There are three sources of obtaining such information (Figure 10.5):

a. Do it yourself—This is essentially a no-go right from the beginning. The magnitude of the effort requires screening and curation of data from thousands of journals and millions of patents. The likelihood of a non-comprehensive analysis is extremely high. Several pharmaceutical companies have had their library services try to attempt to collate such information and abandoned their efforts.

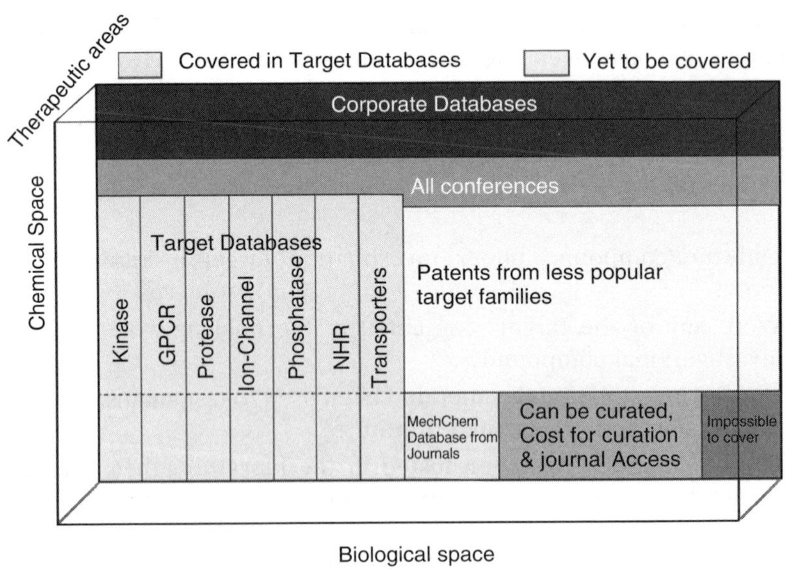

Figure 10.5 Data Coverage in Cheminformatics Databases with content relevant for Pathway Analysis.

b. Use Commercial online database sources—While the three major online data providers do have a good coverage of the desired content, the price can be fairly prohibitive. However, they seem to be the vendor of choice for most of the relevant user base.
c. Commercial Vendors with the relevant information—At least three commercial database providers (GVK BioSciences [www.gvkbio.com], Aureus Pharma [www.aureus-pharma.com], Jubilant Biosys [www.jubilantbiosys.com]) have attempted to develop target-specific databases covering the content of what is desired for pathway analysis. In fact, three of the databases of GVK BioSciences have been integrated into two popular Pathway Tools (Ingenuity and GeneGo). These vendors are capturing the following information on each compound that has been published or patented and claimed to work on a specific target:
   i. Structure: which is queryable
   ii. Activity Information: IC 50, Ki, Km, etc., which illustrates the activity of the compound against the claimed target
   iii. Assays: which illustrate the assay(s) that were used to test for the activity
   iv. Assay Type: illustrates the type of assay to include Binding, Functional, Toxicity, ADME, etc.
d. Public Sources such as PubChem: While these efforts are extensive, they do not contain current or the most relevant data. It will be a few years before they can be used for credible inferencing.

The data in the currently available databases cover mainly the inhibitors against the well-studied biological targets. These include Kinases, G Protein Coupled Receptors, Proteases, Nuclear Receptors, Transporters, and Ion Channels. The entire biological space, however, comprises several thousand targets. Most of these targets, however, contain very few patented or published inhibitor data. Hence, even the availability of data for the popular classes should add sufficient value to merit their inclusion in the Pathway Tools.

10.4.2 The Integration

Perhaps the most important issue next to the curation of the data is the integration of the data to make it accessible and relevant in the context of the pathway application (Figure 10.6). This is an issue independent of the type of application that one uses. The primary reasons for issues in integration and possible solutions are as follows:

A

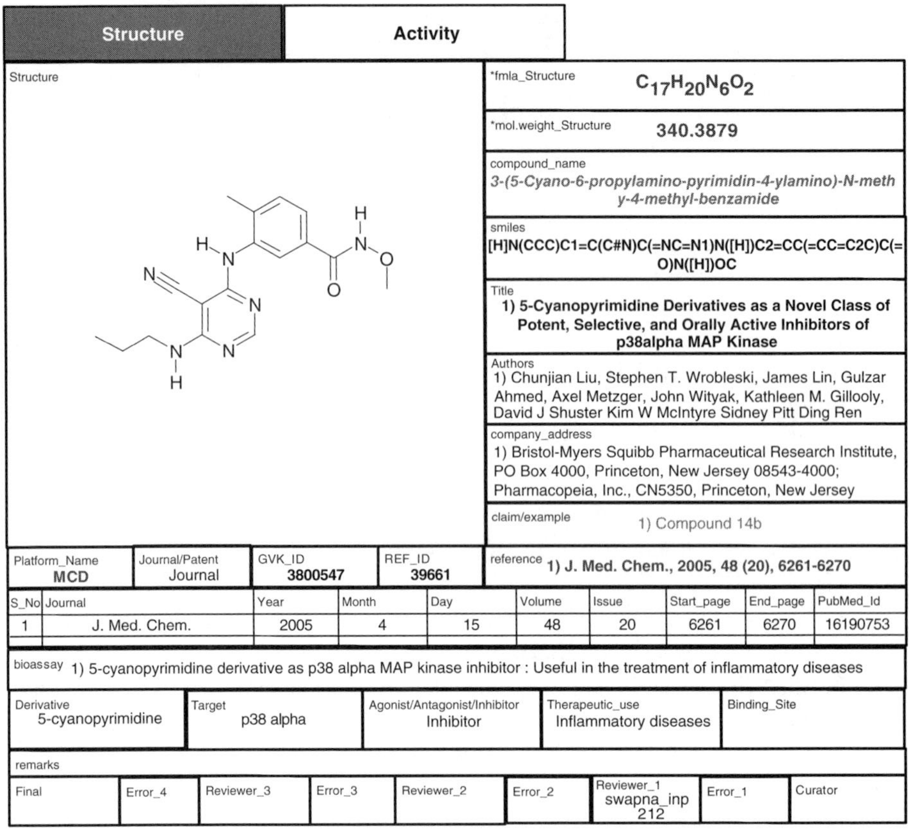

Figure 10.6 (A) Schematic of the data that are represented in Cheminformatics databases. (Source: GVK BioSciences). (B) Schematic of the data that are represented in Cheminformatics databases. (Source: GVK BioSciences).

a. Gene-Target names: The nonstandardization of names (though this is getting a lot better) makes mapping of every known target to its inhibitor a difficult task. However, some of the database vendors (GVK BioSciences for example) have overcome some aspects of this problem by providing an exhaustive Synonym Mapping.

b. Structure Integration: Most of the Pathway Tools do not integrate structure information. There is a degree of difficulty involved in embedding structure information into these tools. Both Database Size and Performance can be affected disproportionately.

c. Pricing: The need for structure cartridges (such as those from Accelrys, ChemAxxon, MDL, etc.) requires changes in the architecture of the

B

Structure	Activity

GVK_ID	reference	claim/example
3800547	1) J. Med. Chem., 2005, 48 (20), 6261-6270	1) Compound 14b

protein\cell\anima	Source_name	Source_code	official_name	Locus_ID	MultipleLoci	Locus_Ref	assay_type	Assay_no	REFER
p38 alpha	Escherichia coli	Human	MAPK14	1432			B	1	1
	PBM cell line	Human					FI	2	1

protein	Activity Type	Activity UOM	Activity Pre	Activity Value	Molarvalue	SD	enzyme/cell_assay	REFER
p38 alpha	Ki	nM	=	0.97000000 00000000	0.000000 00097000 00		Binding affinity of the compound towards human p38 alpha kinase expressed in E. coli upon incubation for 2 I hrs at RT in pH 7.4 using [gamma-33P]ATP as radioligand with compound dissolved in DMSO; n=4	1
	IC50	uM	=	158.000000 0000000000	0.000158 00000000 00		Inhibitory concentration of the compound against LPS-induced TNFalpha production in human peripheral blood mononuclear cells; n=3	1

Target_class	Family	Subfamily	Sub_subfamily	PDB_ID	Standard_name	Alias	Other_names	P/S	REFERENCE
Kinase	Ser/Thr protein kinase family	MAP kinase subfamily		1A9U, 1BL6, 1BL7, 1BMK, 1DI9, 1IAN, 1KV1, 1KV2, 1M7Q, 1OUK	Mitogen-activated protein kinase 14	CSBP1, CSBP2, CSPB1, EXIP, Mxi2, PRKM14, PRKM15, RK, SAPK2A	p38alpha	P	1

remarks

Figure 10.6 (*Continued*)

products, as well as pricing models (due to the additional cost associated with licensing these cartridges).

d. Performance: In addition, memory requirements become intensive. ASP-based applications could also slow down the overall application and affect performance.
e. Updates: Most Pathway Tools do not update in real time. The rate of growth of inhibitor data (Figure 10.7) is so fast (over half a million new inhibitors published or patented every year) that real-time updates are a must to yield the most reliable conclusions.
f. License Costs: The cost of curation of data from a vast number of additional journals (many of whom have exorbitant subscription costs), as well as the manpower required to curate the data, could result in significant additional license fees for the Pathway Tools. A better option may be to license the data, with some customization, from the vendors who have access to such structured information.

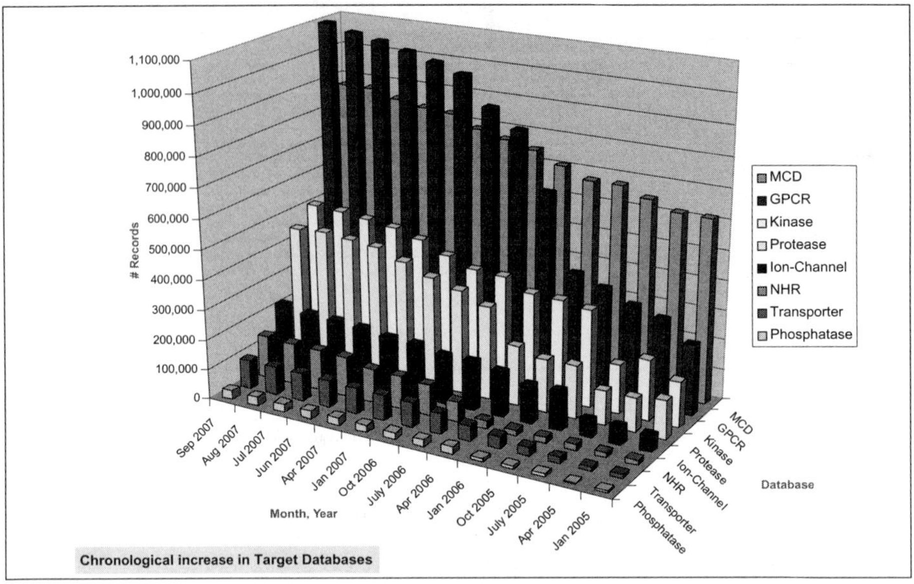

Figure 10.7 Rate of growth of content in Cheminformatics databases (Source: GVK BioSciences). See color insert.

10.4.3 Utility and Workflow

The content described above has clear utilities in the following areas:

a. To accelerate candidates through the discovery process
b. Eliminate candidates that demonstrate inactivity, cross-target effects, or toxicity effects earlier than normal resulting in significant cost savings
c. Add more significance than can be attributed at present to the results generated by Pathway Analysis Tools
d. Accelerate the acceptance of data generated from Pathway Tools by the FDA reviewers—Several Pathway Tool vendors have collaborations with the US FDA to obtain review input on incorporating workflows in their products with a view to increase their utility in the drug review process
e. Competitive Intelligence—An exhaustive set of inhibitor data can be used to analyze competitor intelligence. Since the investigator information is curated for each inhibitor, interesting matrices can be built to study the therapeutic areas, scaffold types, and targets being worked upon by the competition.

Having provided an overview of the content and utility, it is important to illustrate the utility of these products in the context of a Pathway Analysis

Tool. The next section attempts to illustrate each of the workflows that were described in brief in an earlier section on the content. Five such possible workflows are described below:

10.4.3.1 Did the Compound under Investigation Target a Known Druggable Target? Several investigators that have either launched drugs (or were marketed drugs) or compounds that either are or were in a Clinical Trial have published the target affected by the drug or candidate. Unfortunately, there do not exist too many sources of reliable data that can provide this information. Some of the Pathway Vendors have attempted to incorporate this information. However, a current limitation resides in the fact that the information provided is limited to the name of the possible "druggable target" in the context of a network diagram. To maximize utility, a lot more information, including but not limited to the following, would add significant utility to Pathway Analysis (Figure 10.8):

- Pharmacokinetic Data
- Pharmacodynamic Data
- Toxicity Data
- Clinical Parameters

10.4.3.2 Were Any of the Targets Contained in the Final Network Novel to the Investigational Compound? Given the cost of drug development, investigators wish that their compound were novel. It is a near impossible task to determine this without a database that contains Structure-Target information. Novelty cannot only add value but also allow for early termination of redundant experiments (Figure 10.9).

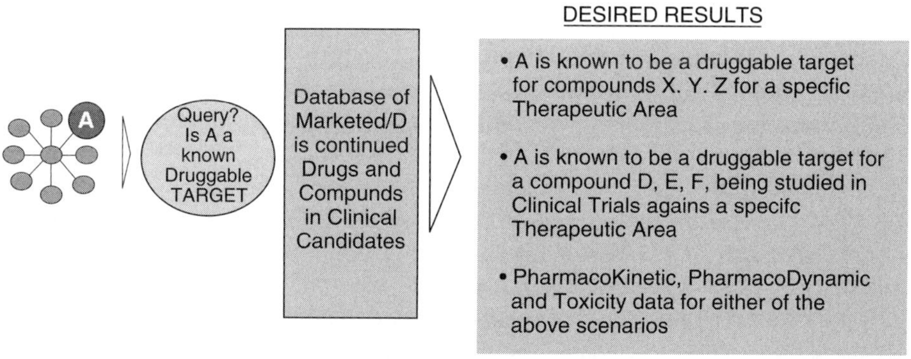

Figure 10.8 Workflow demonstrating known information about an investigational compound with respect to its "Druggability."

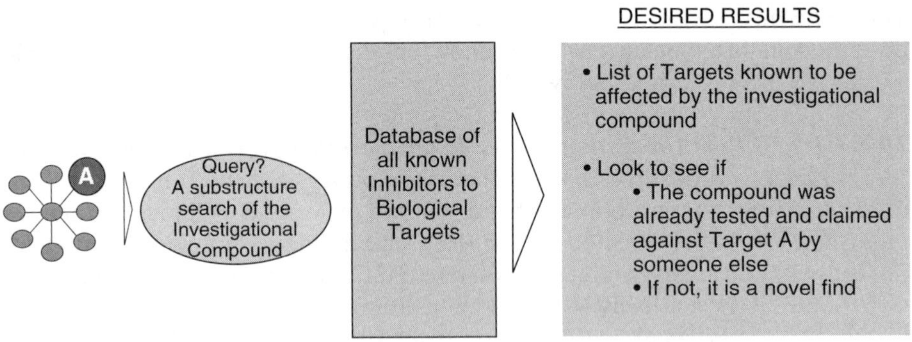

Figure 10.9 Demonstrating "Novelty" in the context of Pathway Analysis.

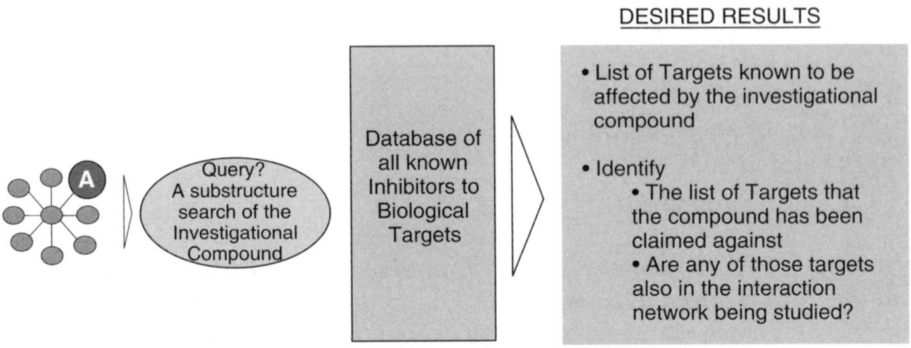

Figure 10.10 Demonstrating "Target-Novelty" in the context of Pathway Analysis.

10.4.3.3 Has the Investigational Compound Been Used to Test Against Other Targets in The Published or Patented Literature? Investigators always wish that their compound under investigation were targeting a specific target. However, the issue of cross-target effects and side effects is often due to action by the compound on multiple targets—often undesired. The ability to determine this in the context of a Pathway Tool adds tremendous value to the analysis. In the event that multiple targets are targeted by the investigational compound, it can raise an early alert to look for possible side effects (Figure 10.10).

10.4.3.4 Inhibitor Experimental Data—If the Compound Has Been Tested in the Literature, Then What Are the Known SAR (Structure–Activity Relationships), Assay Types, and Toxicity Information? It is quite likely that an investigational compound has been claimed against one or more targets. However, in many cases the "Activity" could be reported as "Inactive" or "extremely small or irrelevant activity." This can be extremely misleading to the investigator. Hence, it is important that the Chemistry content not only covers the inhibitors against the targets, but also the activities or Structure–Activity Relationships (often referred to as SAR). In addition, activity is also very dependent on the Assay that was used to measure the activity point. Hence, the content needs to cover the Assay type. Some of the popular assays classifications include Binding, Functional, Toxicity, and ADME (Figure 10.11).

10.4.3.5 Side Effects-Off or Multiple Target Effects—Is the Candidate Active against Multiple Targets? If the investigational compound has been studied and shown to demonstrate activity against multiple targets, then could it be indicative of favorable or unfavorable cross-target effects (Figure 10.12)?

10.4.3.6 Competitor Intelligence—How Extensively Was the Target, Candidate, or Scaffold of Interest Being Worked upon by Others? An exhaustive

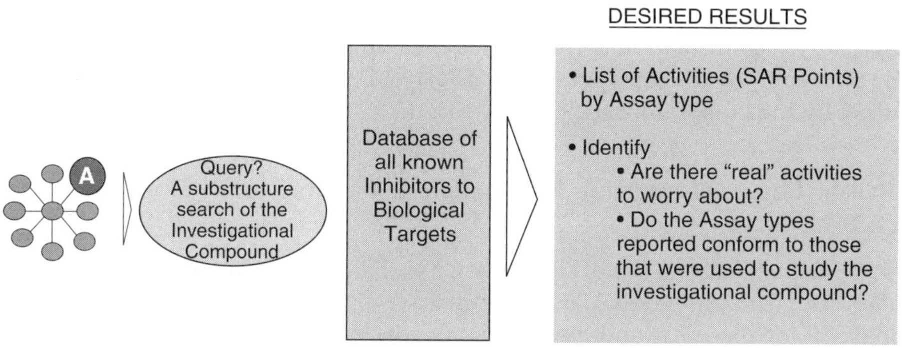

Figure 10.11 Demonstrating "SAR-Novelty" in the context of Pathway Analysis.

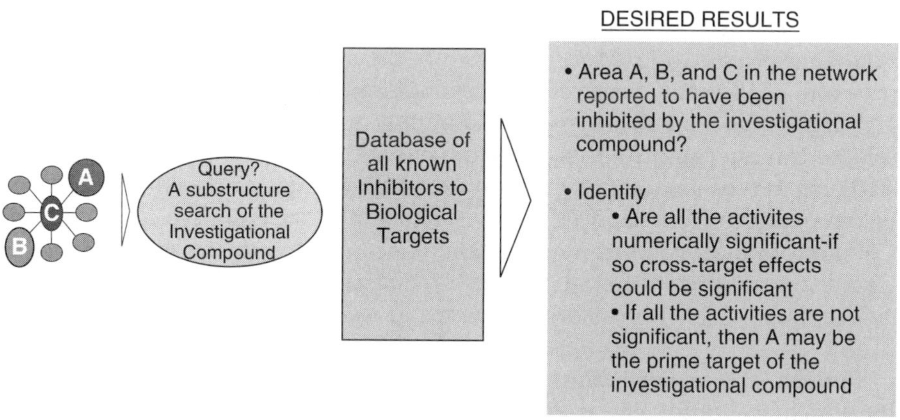

Figure 10.12 Demonstrating "Cross-Target effects" in the context of Pathway Analysis.

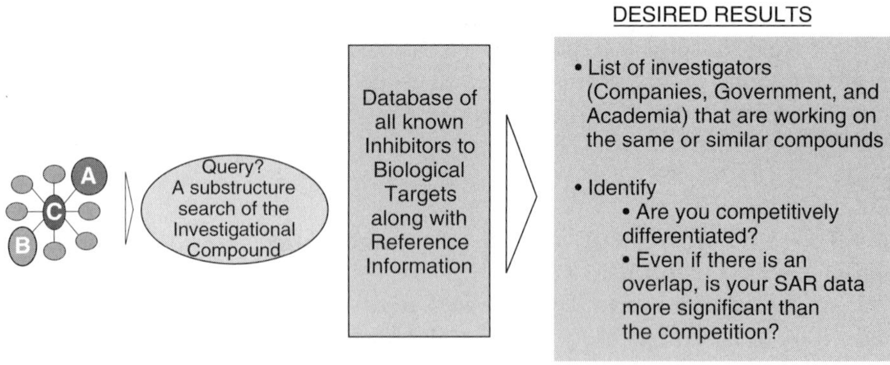

Figure 10.13 Demonstrating "Competitive Differentiation" in the context of Pathway Analysis.

set of inhibitor data can be used to analyze competitor intelligence. Since the investigator information is curated for each inhibitor, interesting matrices can be built to study the therapeutic areas, scaffold types, and targets being worked upon by the competition (Figure 10.13).

10.4.4 Impact of Chemical Information on Pathway Analysis—Return on Investment

Several publications have highlighted the Return on Investment (ROI) of Pathway Analysis Tools. The findings have been reasonably validated by a majority of the users and also by the exponential increase in the number of publications featuring results generated using Pathway Analysis Tools.

The section below attempts to highlight the ROI of adding Chemistry content to the Pathway Analysis Tools. The data have been generated after discussions with several key decision makers in the industry, government, and academia (Table 10.1). More importantly the questions were asked to scientists who are actually involved in relevant data analysis on a daily basis.

Early estimates indicate that the addition of Chemistry-Cheminformatics-related content can further increase the impact of standalone Pathway Analysis Tools. It is estimated that an additional 10% savings across different phases of discovery can be achieved by the inclusion of such content. In addition, Cheminformatics content have a standalone utility with the Computational Chemistry groups within the industry and academia. This could further "spread" the cost within an organization and result in an additional 5% savings (Table 10.2).

Further, the use of Pathway Analysis Tools has just begun to increase in the Preclinical and Clinical Phases of drug development. It is estimated that the cost of development from the PreClinical through Phase 3 is close to 500 million dollars. Pathway Analysis Tools are estimated to reduce the cost of the relevant phases by about 8%. It is estimated that the addition of Chemistry

TABLE 10.1 Independent Survey on Utility of Chemistry-Related Information on Pathway Analysis

Question?	Response
Do you currently use Pathway Analysis Tools and Databases?	Over 90% said YES
Do you currently use any "nonbiological" content within these tools?	Over 95% said NO
Do the commercially available tools presently contain any useful "nonbiological" information?	Over 80% said NO
Would adding Chemistry-related data (of the type discussed in this review) add value to your analysis workflows?	Over 90% said YES
Would addition of such data:	
• Speed up Analysis	• Overwhelming Majority said YES
• Speed up Decision Making	• Overwhelming Majority said YES
• Increase Utility	• Overwhelming Majority said YES
• Help Reduce Pipeline Attrition	• Overwhelming Majority said YES
• Increase Productivity	• Overwhelming Majority said YES

TABLE 10.2 Independent Survey (Numerical Data from Several Published Market Reports) on Potential Savings from the Inclusion of Chemistry-Related Information in Pathway Analysis Tools

Phase	Average Cost of Phase $M	Percentage of Savings	Cost Benefit using only Pathway Analysis Tools $M	Additional Cost Benefit using Chemistry Content into Pathway Tools
Target Identification, Quantification, and Prioritization	185	10	18.5	18.5
Target Validation	225	10	22.5	22.5
Compound Screening	60	10	6	6
Lead Optimization	145	10	14.5	14.5
Total	**615**		**61.5**	**61.5**

content, especially Pharmacodynamic, Pharmacokinetic, and Toxicity content, will further reduce the cost by an additional 5% to 8%.

10.5 AN ALTERNATE WORKFLOW—BRIDGING CHEMINFORMATICS TO PATHWAYS

Currently, the usage of Pathway Tools is restricted primarily to the Discovery Informatics groups. The input data tend to be (see Figure 10.2) gene-protein datasets and associated numerical data (for example, Microarray Data).

An alternate workflow is to map the Chemical Structure to identify the Pathway or Interaction Map affected:

The interaction map so generated can highlight the following:

a. Which targets in the patented and published literature have been known to have been affected by the Query compound?
b. What types of Activities have been observed? Is it a structure-specific effect or cross-target effect?
c. Novelty: The query compound may
   i. Not find any hits against any of the targets in the database being queried. This could suggest a novel find or an "unknown target."
   ii. May hit against a target that has not been claimed before and thus demonstrate novelty.

10.6 CONCLUSION

The review suggests that the extension of Pathway Analysis to include Chemistry- and Cheminformatics-related content could increase ROI significantly (Figure 10.14). More importantly, it can increase the collaboration within different groups in an organization and could result in reduced attrition and improved productivity. Several challenges exist with respect to the proposed integration (Table 10.3). These include, but are not limited to:

- The availability of the relevant content in a structured format
- Keeping the content current
- Issues of Integration, Structure Visualization, and Searching
- Impact on licensing costs
- Product Architecture modifications that may be necessitated by inclusion of such content

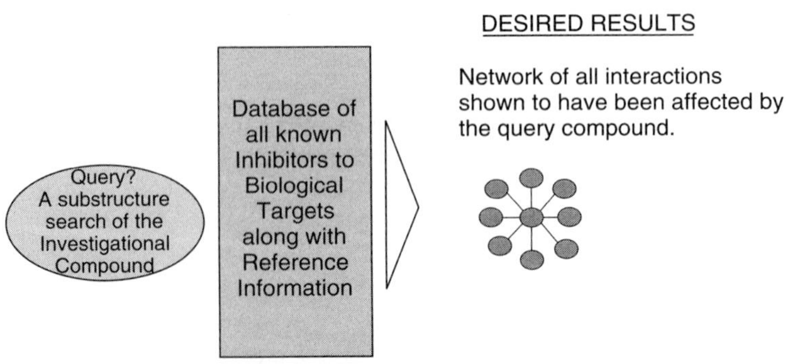

Figure 10.14 A cheminformatics to pathway workflow.

TABLE 10.3 Challenges in Developing a Pathway Database and Application

	Challenge
DATA MODEL	• Nonstandardization of Ontology • A problem of multiple dimensionality
CONTENT	• Enormous volume of knowledge • Thousands of Disparate Journals • Contradictory results • Lack of common naming standard
APPLICATION LAYER	• Large networks generated for small input datasets • Filtering, visualization, and interpretation
INTEGRATION	• Integration to multiple data sources • Integration with the "data generators"—the software packages that do the initial data analysis
PRODUCT LICENSING MODELS	• Local with access to the underlying database • Remote with no or limited access to the underlying database
UPDATES	• Large amount of data being published on a daily basis • Even a single miss could result in misinterpretation

Given the weakness of the pipeline of most pharmaceutical and biotechnology companies and the increased attrition rate, this new paradigm might result in an increased productivity.

ACKNOWLEDGMENTS

The author wishes to acknowledge the help of Nikhil Tamhankar, Dr. Sarma Jagarlapudi, and Manni Kantipudi at GVK BioSciences. The author also wishes the inputs received from several companies working in the pathway analysis space and scientists and personnel in business development at both pharmaceutical and biotechnology companies.

REFERENCES

1. Southan C, Várkonyi P, Muresan S. Complementarity between public and commercial databases: new opportunities in medicinal chemistry informatics. Curr Top Med Chem 2007;7(15):1502–1508.
2. Hopkins AL, Polinsky A. Knowledge and intelligence in drug design. Annu Rep Med Chem 2006;41:425–437.
3. American Chemical Society. Chemical Abstract Database (CAS): Scifinder. CAS web site Available at http://www.cas.org/SCIFINDER/
4. Leeson PD, Springthorpe B. The influence of drug-like concepts on decision-making in medicinal chemistry. Nat Rev Drug Discov 2007;6(11):881–890.

5. Ingenuity Systems. Available at www.ingenuity.com
6. Ariadne Genomics. Available at www.ariadnegenomics.com
7. Genomatix. Available at www.genomatix.de
8. BioBase. Available at www.biobase.de

11

PROPAGATION OF CONCENTRATION PERTURBATIONS IN EQUILIBRIUM PROTEIN BINDING NETWORKS

SERGEI MASLOV AND IAROSLAV ISPOLATOV

Table of Contents

Pathway Analysis for Drug Discovery, Edited by Anton Yuryev

11.1 INTRODUCTION

Recent high-throughput experiments performed in a wide variety of organisms revealed networks of protein–protein physical interactions (PPI) that are interconnected on a genome-wide scale. In such "small-world" PPI networks, most pairs of nodes can be linked to each other by relatively short chains of interactions, involving just a few intermediate proteins [1]. While globally connected architecture facilitates biological signaling and possibly ensures a robust functioning of the cell following a random failure of its components [2], it also presents a potential problem by providing a conduit for propagation of undesirable cross talk between individual functional modules and pathways. Indeed, large (several-fold) changes in proteins' levels in the course of activation or repression of a certain functional module affect bound concentrations of their immediate interaction partners. These changes have a potential to cascade down a small-world PPI network, affecting the equilibrium between bound and unbound concentrations of progressively more distant neighbors including those in other functional modules. Most often such indiscriminate propagation would represent an undesirable effect, which has to be either tolerated or corrected by the cell. On the other hand, a controlled transduction of reversible concentration changes along specific conduits may be used for biologically meaningful signaling and regulation. A routine and well-known example of such regulation is inactivation of a protein by sequestration with its strong binding partner.

In this study, we quantitatively investigate how large concentration changes propagate in the PPI network of yeast *Saccharomyces cerevisiae*. We focus on the non-catalytic or reversible binding interactions whose equilibrium is governed by the Law of Mass Action (LMA) and do not consider irreversible, catalytic processes such as protein phosphorylation and dephosphorylation, proteolytic cleavage, etc. While such catalytic interactions constitute the most common and best-studied mechanism of intracellular signaling, they represent only a rather small minority of all protein–protein physical interactions (for example, only ~5% links in the yeast network used in our study involve a kinase).

Furthermore, the balance between free and bound concentrations of proteins matters even for irreversible (catalytic) interactions. For example, the rate of a phosphorylation reaction depends on the availability of free kinases and substrate proteins, which are both controlled by the LMA equilibrium calculated here. Thus, perturbations of equilibrium concentrations considered in this study could be spread even further by other mechanisms such as transcriptional and translational regulation, and irreversible post-translational protein modifications.

11.2 RESULTS

11.2.1 The PPI Network and Protein Abundance Data

To illustrate general principles on a concrete example, in this study we used a highly curated genome-wide network of protein–protein physical interactions in yeast (*S. cerevisiae*), which, according to the BIOGRID database [3], were independently confirmed in at least two publications. We combined this network with a genome-wide dataset of protein abundances in the log-phase growth in rich medium, measured by the TAP-tagged Western blot technique [4]. Average protein concentrations in this dataset range between 50 and 1,000,000 molecules/cell with the median value around 3,000 molecules/cell. After keeping only the interactions between proteins with known concentrations, we were left with 4,185 binding interactions among 1,740 proteins. The BIOGRID database [3] lists all interactions as pairwise and thus lacks information about multi-protein complexes larger than dimers. Thus in the main part of this study, we consider only homo- and heterodimers and ignore the formation of higher-order complexes. However, we show that the reliable data on multi-protein complexes can be easily incorporated into our analysis. Furthermore, we demonstrate that taking into account such complexes leaves our results virtually unchanged (see Figure 11.1).

11.2.2 The Assignment of Dissociation Constants K_{ij}

The state-of-the-art genome-wide PPI datasets lack information on dissociation constants K_{ij} of individual interactions. The only implicit assumption is that the binding is sufficiently strong to be detectable by a particular experimental technique (some tentative bounds on dissociation constants detectable by different techniques were reported recently [5]). A rough estimate of the average binding strength in functional protein–protein interactions could be obtained from the PINT database [6]. This database contains about 400 experimentally measured dissociation constants between wild-type proteins from a variety of organisms. In agreement with predictions of Lancet et al. [7] and Deeds et al. [8], the histogram of these dissociation constants has an approximately log-normal shape. The average relevant for our calculations is that of

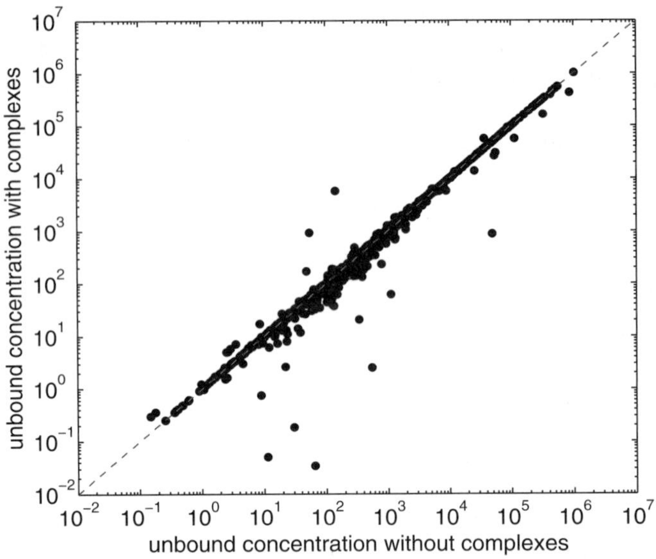

Figure 11.1 The scatter plot of calculated free (unbound) concentrations of individual proteins with (y-axis) and without (x-axis) multi-protein complexes listed in the MIPS CYGD database [20,21]. The dissociation constant of all interactions in 4,185 dimers and 81 multi-protein complexes in this plot is K_{ij} = const = 10 nM.

the *association* constant $\langle 1/K_{ij} \rangle = 1/(5\,\text{nM})$. Common sense dictates that the dissociation constant of a functional binding between a pair of proteins should increase with their abundances. The majority of specific physical interactions between proteins are neither too weak (to ensure a considerable number of bound complexes) nor unnecessarily strong. Indeed, there is little evolutionary sense in increasing the binding strength between a pair of proteins beyond the point when both proteins (or at least the rate limiting one) spend most of their time in the bound state. The balance between these two opposing requirements is achieved by the value of dissociation constant K_{ij} equal to a fixed fraction of the largest of the two abundances C_i and C_j of interacting proteins. In our simulations, we used $K_{ij} = \max(C_i, C_j)/20$ in which case the average association constant nicely agrees with its empirical value (1/[5 nM]) observed in the PINT database [6]. In addition to this, perhaps, more realistic assignment of dissociation constants we also simulated binding networks in which dissociation constants of all 4,185 edges in our network are *equal to each other* and given by 1 nM, 10 nM, 100 nM, and 1 μM.

11.2.3 Numerical Calculation of Bound and Free (Unbound) Equilibrium Concentrations

The Law of Mass Action (LMA) relates the free (unbound) concentration F_i of a protein to its total (bound and unbound) concentration C_i as

$$F_i = \frac{C_i}{1+\sum_j F_j / K_{ij}}. \tag{11.1}$$

Here, the sum is over all specific binding partners of the protein i with free concentrations F_j and dissociation constants K_{ij}. While in the general case these nonlinear equations do not allow for an analytical solution for F_i, they are readily solved numerically, e.g. by successive iterations.

11.2.4 Concentration-Coupled Proteins

To investigate how large changes in abundances of individual protein affect the equilibrium throughout the PPI network, we performed a systematic numerical study in which we recalculated the equilibrium free concentrations of all protein nodes following a twofold increase in the total concentration of just one of them: $C_i \to 2C_i$. This was repeated for the source of twofold perturbation spanning the set of all 1,740 of proteins in our network [9]. The magnitude of the initial perturbation was selected to be representative of a typical shift in gene expression levels or protein abundances, following a change in external or internal conditions. Thus here we simulate the propagation of functionally relevant changes in protein concentrations and not that of background stochastic fluctuations. A change in the free concentration F_j of another protein was deemed to be significant if it exceeded the 20% level, which according to Newman et al. [10] is the average magnitude of cell-to-cell variability of protein abundances in yeast. We refer to such protein pairs $i \to j$ as *concentration-coupled*. The detection threshold could be raised simultaneously with the magnitude of the initial perturbation. For example, we found that the list of concentration-coupled pairs changes very little if instead of twofold (+100%) perturbation and the 20% detection threshold, one applies a sixfold (+500%) initial perturbation and twofold (100%) detection threshold.

In general, we found that lists of concentration-coupled proteins calculated for different assignments of dissociation constants strongly overlap with each other. For example, more than 80% of concentration-coupled pairs observed for the variable $K_{ij} = \max(C_i, C_j)/20$ assignment described above were also detected for the uniform assignment $K_{ij} = \text{const} = 10\,\text{nM}$ (for more details see Table 11.1). This relative robustness of our results allowed us to use the latter conceptually simplest case to illustrate our findings in the rest of the chapter.

11.3 CENTRAL OBSERVATIONS

We found that:

- *On average*, the magnitude of cascading changes in equilibrium free concentrations exponentially decays with the distance from the source of

TABLE 11.1 Numbers of Concentration-Coupled Proteins as a Function of Network Distance L Calculated Using Linear Response Formalism [12]

L	Constant $K_{ij} = 5\,\text{nM}$	Variable $K_{ij} = \max(C_i, C_j)/20$ (Inverse Mean 5 nM)	Variable K_{ij} Reshuffled	Overlap Between Columns 2 and 1	Overlap Between Columns 2 and 3	Overlap Between Columns 1 and 3
1	2,155	2,099	2,192	84	73	77
2	842	456	749	78	64	63
3	120	42	140	59	59	65
4	8	0	19	0	0	25
5	0	0	0	0	0	0

Data in columns 2–4 were obtained using identical network topology and protein abundances but different assignments of dissociation constants (with mean of 5 nM). In column 2, all edges were assigned the same dissociation constant $K_{ij} = 5\,\text{nM}$, in column 3 we used $K_{ij} = \max(C_i, C_j)/20$, and in column 4 (a null-model control) the K_{ij} values from column 3 were randomly reshuffled among all the edges. Columns 5–7 show the percentage of overlap between lists of concentration-coupled

a perturbation. This explains why, despite a globally connected topology, individual modules in such networks are able to function fairly independently.

- Nevertheless, specific favorable conditions identified in our study cause perturbations to selectively affect proteins at considerable network distances (sometimes as far as four steps away from the source). This indicates that in general, such cascading changes *could not be neglected* when evaluating the consequences of systematic changes in protein levels, e.g. in response to environmental factors, or in gene knockout experiments. Conditions favorable for propagation of perturbations combine high yet monotonically decreasing concentrations of all heterodimers along the path with low free (Unbound) concentrations of intermediate proteins. While reversible protein binding links are symmetric, the propagation of concentration changes is usually asymmetric with the preferential direction pointing down the gradient in the total concentrations of proteins.

11.3.1 Examples of Multistep Cascading Changes

In Figure 11.2A,B, we illustrate these observations using two examples. In each of these cases, the twofold increase in the abundance of just one protein (marked with the yellow circle in the center of each panel) has significantly (>20%) affected equilibrium free concentrations of a whole cluster of proteins some as far as four steps away from the source of the perturbation. However, the propagation beyond immediate neighbors is rather specific. For example, in the case of SUP35 (Figure 11.2A) only 1 out of 169 of its third nearest neighbors was affected above the 20% level. Note that changes in free

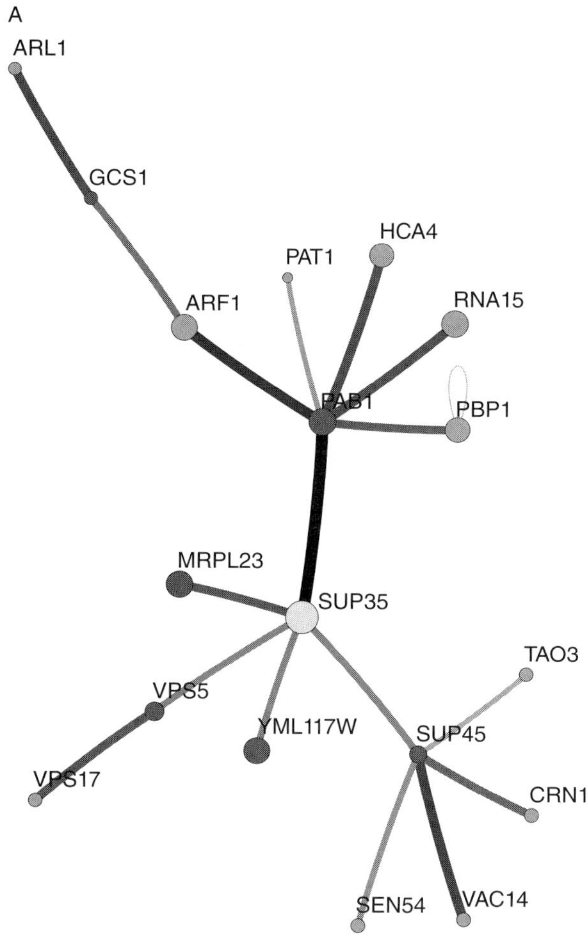

Figure 11.2 Two cases of propagation of large concentration changes in the yeast protein binding network. The total (bound + unbound) concentration of the protein marked with the yellow circle (the SUP35 protein (A), the SEC27 protein (B)) was increased twofold from its wild-type value in the rich growth medium [4]. Red and green circles mark all other proteins whose equilibrium free (unbound) concentrations have increased (green) or decreased (red) by more than 20%. The area of each circle is proportional to the logarithm of the change in free concentration. Edges show all physical interactions among this group of proteins with the shade of gray proportional to the logarithm of the equilibrium concentration of the corresponding dimer calculated for K_{ij} = const = 10 nM.

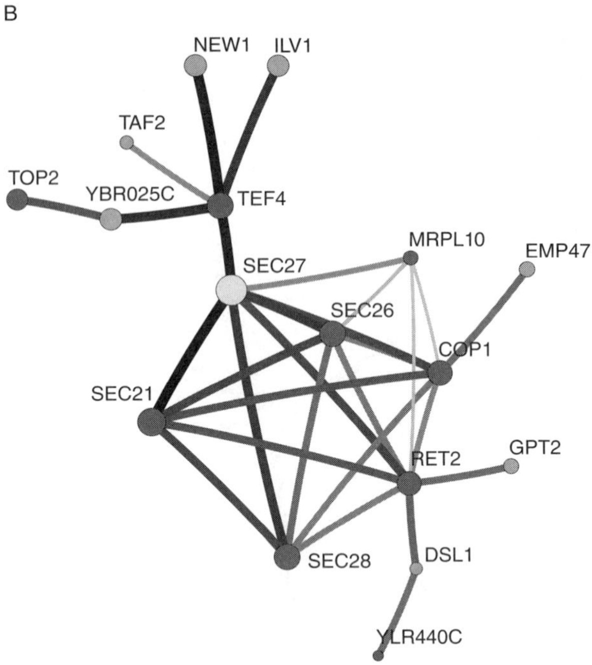

Figure 11.2 (*Continued*)

concentrations generally sign-alternate with the network distance from the source. Indeed, free concentrations of immediate binding partners of the perturbed protein usually drop as more of them become bound in heterodimers with it. This, in turn, lowers concentrations of the next-nearest heterodimers and thus *increases* free concentrations of proteins at distance 2 from the source of perturbation, and so on.

11.3.2 Exponential Decay with the Network Distance

The results of our quantitative network-wide analysis of these effects are summarized in Figure 11.3 and Table 11.2. From Figure 11.3, one concludes that the fraction of proteins with significantly affected free concentrations rapidly (exponentially) decays with the length L of the shortest path (network distance) from the perturbed protein. The same statement holds true for bound concentrations if the distance is measured as the shortest path from the perturbed protein to any of the two proteins forming a heterodimer. Thus, on average, the propagation of concentration changes along the PPI network is indeed considerably dampened. On the other hand, from Table 11.2, one concludes that the total number of multistep chains along which concentration changes propagate with little attenuation remains significant for all but the largest values of the dissociation constant. These two observations do not

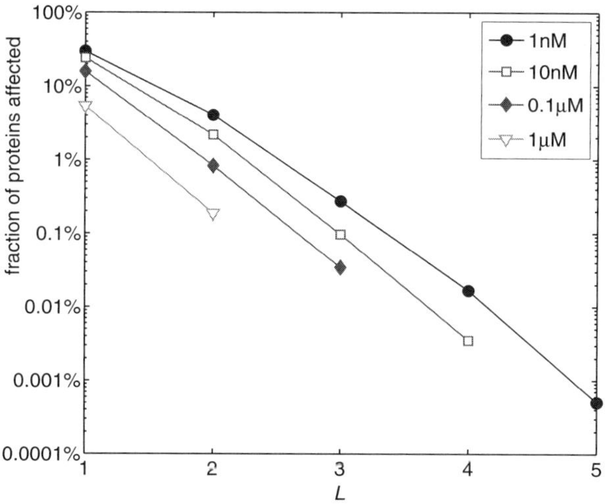

Figure 11.3 Indiscriminate propagation of concentration perturbations is exponentially suppressed. The fraction of proteins with free concentrations affected by more than 20% among all proteins at network distance L from the perturbed protein. Different curves correspond to simulations with K_{ij} = const = 1 nM (solid circles), 10 nM (empty squares), 0.1 μM (solid diamonds), and 1 μM (empty triangles).

TABLE 11.2 The Number of Concentration-Coupled Pairs of Yeast Proteins Separated by Network Distance L

L	var. 5 nM	1 nM	10 nM	0.1 μM	1 μM	All
1	2,003	2,469	1,915	1,184	387	8,168
2	415	1,195	653	206	71	29,880
3	15	159	49	8	0	87,772
4	2	60	19	0	0	228,026
5	0	3	0	0	0	396,608

Numerical simulations (twofold initial perturbation, 20% detection threshold) were performed for different assignment of dissociation constants: $K_{ij} = \max(C_i, C_j)/20$ (column 2), K_{ij} = const = 1 nM, 10 nM, 0.1 μM, 1 μM (columns 3–6). Column 7 lists the total number of protein pairs at distance L.

contradict each other since the number of proteins separated by distance L (the last column in Table 11.2) rapidly grows with L.

11.3.3 Conditions Favoring the Multistep Propagation of Perturbations

What conditions favor the multistep propagation of perturbations along particular channels? In Figure 11.4A, we show a group of highly abundant proteins along with all binding interactions between them. Then on panel B of the same figure, we show only those interactions that according to our LMA

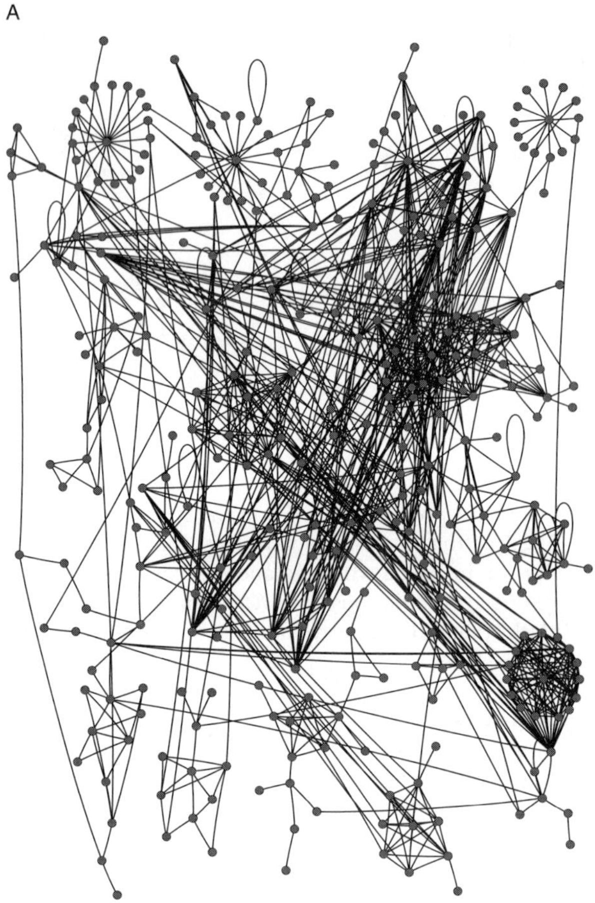

Figure 11.4 (A) All binding links between a subset of 312 highly abundant proteins. (B) Binding links characterized by high concentration of heterodimers (>1,000 molecules/cell). The level of gray of binding links scales with the logarithm of concentration of the corresponding heterodimer. (C) Concentration-coupled proteins A $\rightarrow$ B with the property that a twofold increase in the abundance A reduces free concentration of its immediate binding partner B by 20% or more. Note that links roughly coincide with highly abundant dimers shown in panel B. Arrows reveal the preferential direction of propagation of perturbations.

calculation give rise to highly abundant heterodimers (equilibrium concentration >1,000 per cell). This breaks the densely interconnected subnetwork drawn in panel A into 10 mutually isolated clusters. Some of these clusters contain pronounced linear chains, which serve as conduits for propagation of concentration perturbations. The fact that perturbations indeed tend to propagate via highly abundant heterodimers is illustrated in the next panel (Figure 11.4C) where red arrows correspond to concentration-coupled nearest neigh-

B

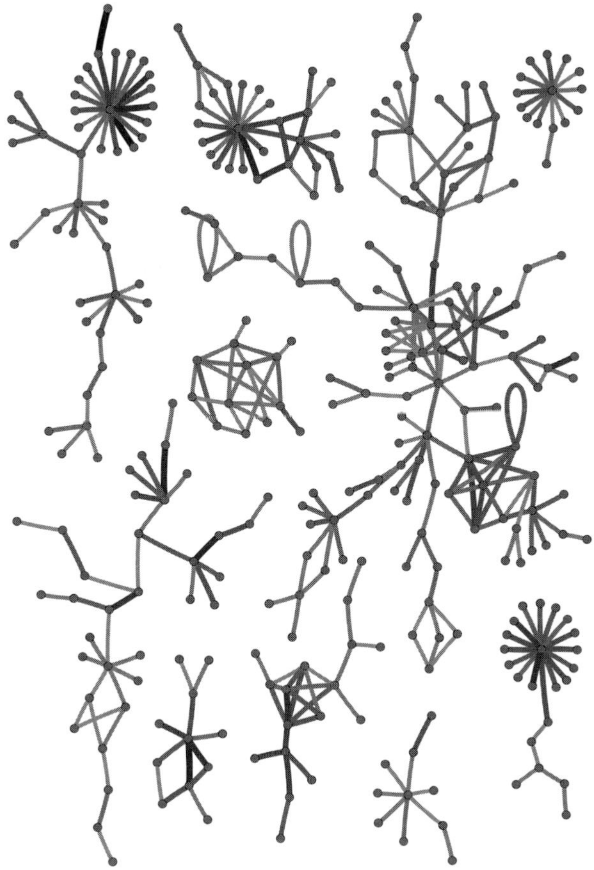

Figure 11.4 (*Continued*)

bors A → B. Evidently, the edges in panels B and C largely (but not completely) coincide. Additionally, panel C defines the preferred direction of propagation of perturbations from a more abundant protein to its less abundant binding partners.

To further investigate what causes concentration changes to propagate along particular channels, we took a closer look at eight three-step chains $A \rightarrow A_1 \rightarrow A_2 \rightarrow B$ With the largest magnitude of perturbation of the last protein B (twofold detection threshold following a twofold initial perturbation). The identification of intermediate proteins A_1 and A_2 was made by a simple optimization algorithm, searching for the largest overall magnitude of intermediate perturbations along all possible paths connecting A and B.

Inspection of the parameters of these chains shown in Figure 11.5 allows one to conjecture that for a successful transduction of concentration changes, the following conditions should be satisfied:

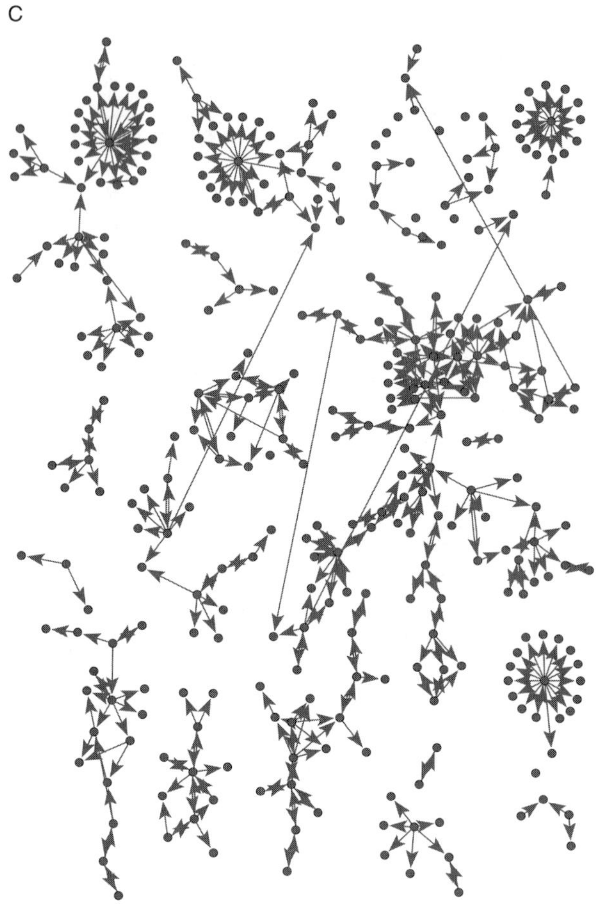

Figure 11.4 (*Continued*)

- Heterodimers along the whole path have to be of sufficiently high concentration D_{ij}.
- Intermediate proteins have to be highly sequestered. That is to say, in order to reduce buffering effects, free-to-total concentration ratios F_i/C_i should be sufficiently low for all but the last protein in the chain.
- Total concentrations C_i should gradually decrease in the direction of propagation. Thus, propagation of perturbations along virtually all of these long conduits is unidirectional and follows the gradient of concentration changes (a related concept of a "gradient network" was proposed for technological networks by Toroczkai and Bassler [11]).
- Free concentrations F_i should alternate between relatively high and relatively low values in such a way that free concentrations of proteins at steps 2 and 4 have enough "room" to go down. The two apparent excep-

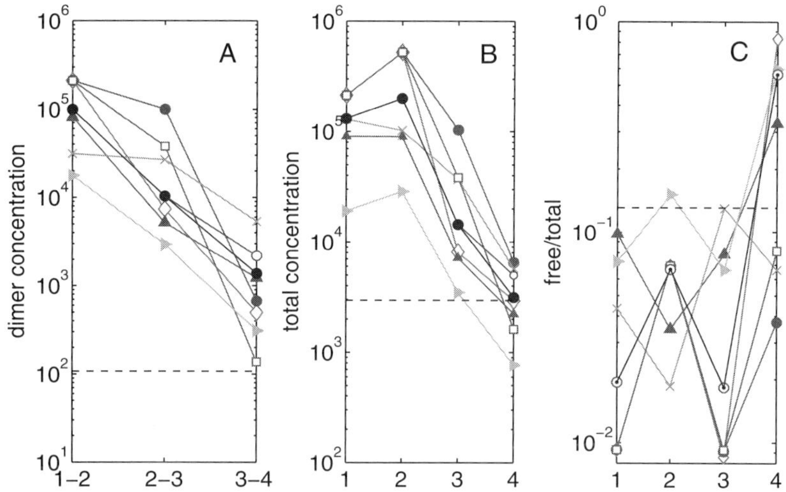

Figure 11.5 Parameters of the eight three-step chains that exhibit the best transduction of concentration changes: Heterodimer concentrations D_{ij} (A) for three binding links along the chain. Total concentrations C_i (B) and free-to-total concentration ratios F_i/C_i (C) of the four proteins involved in these chains. Dashed lines correspond to network-wide geometric averages of the corresponding quantities: $\langle D_{ij} \rangle \sim 100$ copies/cell, $\langle C_i \rangle \sim 3{,}000$ copies/cell, and $\langle F_i/C_i \rangle = 13\%$.

tions to this rule visible in Figure 11.5 may be optimized to respond to a drop (instead of increase) in the level of the first protein.

These findings are in agreement with our more detailed numerical and analytical analysis of propagation of fluctuations presented in an earlier study [12]. Previously, we demonstrated that the linear response of the LMA equilibrium to *small* changes in protein abundances could be approximately mapped to a current flow in the resistor network in which heterodimer concentrations play the role of conductivities (which need to be large for a good transmission) while high F_i/C_i ratios result in the net loss of the perturbation "current" on such nodes and thus need to be minimized [12].

11.4 DISCUSSION

11.4.1 Robustness with Respect to Assignment of Dissociation Constants

It has been often conjectured that the qualitative dynamical properties of biological networks are to a large extent determined by their topology rather than by quantitative parameters of individual interactions such as their kinetic or equilibrium constants (for a classic success story, see for example vonDassow et al. [13]). Our results generally support this conjecture, yet go one step further: we observe that the response of reversible protein–protein binding networks to large changes in concentrations strongly depends not only on topology but also on abundances of participating proteins. Indeed, perturbations tend to preferentially propagate via paths in the network in which abundances of intermediate proteins monotonically decrease along the path (see Figure 11.4). Thus by varying protein abundances while strictly preserving the topology of the underlying network, one can select different conduits for propagation of perturbations.

On the other hand, our results indicate that these conduits are to a certain degree insensitive to the choice of dissociation constants. In particular, we found (see Figure 11.6) that equilibrium concentrations of dimers and the remaining free (unbound) concentrations of individual proteins calculated for two different K_{ij} assignments (K_{ij} = const = 5 nM and $K_{ij} = \max(C_i, C_j)/20$ with the inverse mean of 5 nM) had a high Spearman rank correlation coefficient of 0.89 and even higher linear Pearson correlation coefficient of 0.98. The agreement was especially impressive in the upper part of the range of dimer concentrations (see Figure 11.6). For example, the typical difference between dimer concentrations above 1,000 molecules/cell was measured to be as low as 40%. As we demonstrated above, it is exactly these highly abundant heterodimers that form the backbone for propagation of concentration perturbations. Thus, it should come as no surprise that sets of concentration-coupled protein pairs observed for different K_{ij} assignments also have a large (~70–80%) overlap with each other. Such degree of robustness with respect to quantitative parameters of interactions can be partially explained by the following observation: proteins whose abundance is higher than the sum of abundances of all of their binding partners cannot be fully sequestered into heterodimers for any assignment of dissociation constants. As we argued above, such proteins with substantial unbound concentrations considerably dampen the propagation of perturbations, and thus cannot participate in highly conductive chains. Another argument in favor of this apparent robustness is based on extreme heterogeneity of wild-type protein abundances (in the dataset of Ghaemmaghami et al. [4], they span 5 orders of magnitude). In this case, concentrations of heterodimers depend more on relative abundances of two constituent proteins than on the corresponding dissociation constant (within a certain range).

In a separate numerical control experiment, we verified that the main results of this study are not particularly sensitive to false positives and false negatives in the network topology inevitably present even in the best-curated large-scale data. The percentage of concentration-coupled pairs surviving a random removal or addition of 20% of links in the network generally ranges between 60% and 80% (see Table 11.3).

11.4.2 Genetic Interactions

The effects of concentration perturbations discussed above could explain some of the genetic interactions between proteins. Consider for example a

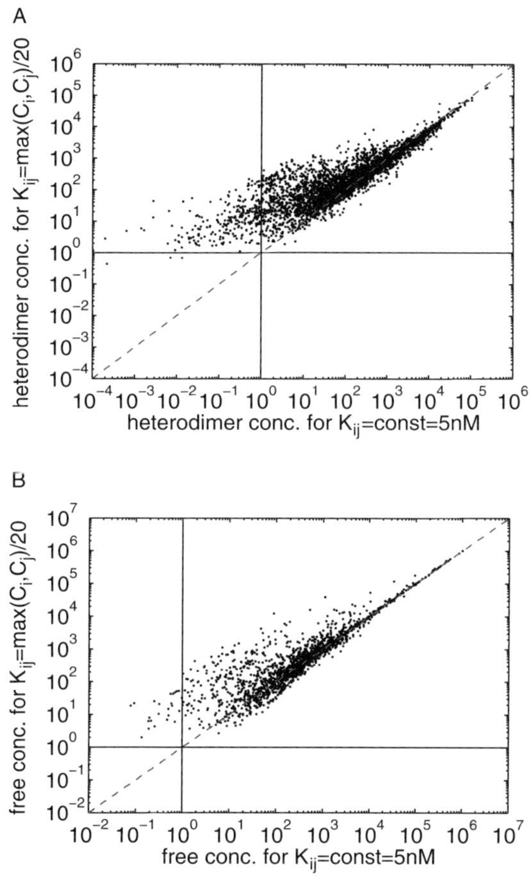

Figure 11.6 The scatter plot of 4,185 bound concentrations D_{ij} (panel A) and 1,740 free concentrations F_i (panel B) calculated for two different assignments of dissociation constants to links in the PPI network. The x-axis was computed for the homogeneous assignment K_{ij} = const = 5 nM, while the y-axis was computed for the heterogeneous assignment $K_{ij} = \max(C_i, C_j)/20$ with the same average strength. The dashed lines along the diagonals are drawn at $x = y$, while the horizontal and vertical solid lines denote the concentration of 1 molecule/cell. Note that equilibrium concentrations in the upper part of their range (e.g. above 1,000 molecules/cell) are nearly independent of the choice of K_{ij}. Also, our choice of heterogeneous assignment nearly eliminates free or bound concentrations in a biologically unreasonable range <1 molecule/cell.

"dosage rescue" of a protein A by a protein B, or the correction of an abnormal phenotype caused by deletion or other type of inactivation of A by overexpression of B. One possible mechanism behind this effect is that the knockout of A and overexpression of B affect the LMA equilibrium in opposite directions and to some extent cancel one another. In order for this mechanism to be applicable (albeit tentatively), concentrations of both A and B must be simultaneously coupled (in the sense used throughout this work) to at least

TABLE 11.3 Fraction of Concentration-Coupled Protein Pairs that Survive a Random Addition or Removal

L	Addition of 20% of Links (%)	Removal of 20% of Links (%)
1	79.2	79.3
2	69.5	62.5
3	60.6	55.9
4	43.8	31.3
New	39.5	30.7

Note: The fraction of concentration-coupled protein pairs in the original network (K_d = 10 nM, 20% cutoff) that survive a random addition (column 2) or removal (column 3) of 20% of links as a function of network distance. The last row is the percentage of new pairs in modified networks.

one crucial protein *C* whose free or bound concentration has to be maintained at or close to wild-type levels. To assess this hypothesis, we analyzed the set of 772 dosage rescue pairs involving proteins from the PPI network used in this study of 2,531 dosage rescue pairs listed in the BIOGRID database [3]. For 136 pairs (or 18% of all dosage rescue pairs), we were able to identify one or more putative "rescued" protein whose free concentration was considerably (by >20%) affected by changes in abundances of both *A* and *B*. This overlap is highly statistically significant, having the Fisher's exact test *p*-value around 10^{-216}. Even more convincing evidence that perturbations to the LMA equilibrium state cause some of genetic interactions is presented in Figure 11.7. It plots the fraction of protein pairs at distance *L* from each other in the PPI network that are known to dosage rescue each other. From this figure, one concludes that proteins separated by distances 1, 2, and 3 are significantly more likely to genetically interact with each other than one expects by pure chance alone (the expected background level is marked with a dashed line or better yet visible as a plateau for large values of *L*). Furthermore, the slope of the exponential decay in the fraction of dosage rescue pairs as a function of *L* is roughly consistent with that shown in Figure 11.3 for the fraction of concentration-coupled pairs.

11.4.3 Possibility of Functional Signaling and Regulation Mediated by Multistep Reversible Protein Interactions

Another intriguing possibility raised by our findings is that multistep chains of reversible protein–protein bindings might in principle be involved in meaningful intracellular signaling and regulation. There are many well-documented cases in which one-step "chains" are used to reversibly deactivate individual proteins by the virtue of sequestration with their binding partner(s). An

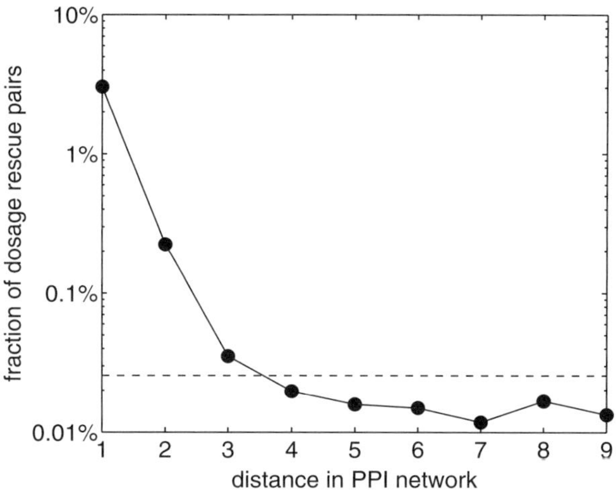

Figure 11.7 The fraction of dosage rescue protein pairs separated by distance L in the PPI network. Note that pairs at distances 1, 2, and 3 are significantly over-represented over the background level marked with dashed line ($772/1{,}740^2$) or visible as a plateau at large distances L. The exponential decay constant at low values of L is consistent with that in Figure 11.3.

example involving a longer regulatory chain of this type is the control of activity of condition-specific sigma factors in bacteria. In its biologically active state, a given sigma victor is bound to the RNA polymerase complex. Under normal conditions, it is commonly kept in an inactive form by the virtue of a strong binding with its specific anti-sigma factor (anti-sigma factors are reviewed by Hughes and Mathee [14]). In several known cases, the concentration of the anti-sigma factor in turn is controlled by its binding with the specific anti-anti sigma factor [14]. The existence of such experimentally confirmed three-step regulatory chains in bacteria hints at the possibility that at least some of the longer conduits we detected in yeast could be used in a similar way.

11.4.4 Application to Microarray Data Analysis

In order to unequivocally detect cascading perturbations, in our simulations we always modified the total concentration of just one protein at a time. In more realistic situations, expression levels of a whole cluster of genes change, for example, in response to a shift in environmental conditions. Our general methods could be easily extended to incorporate this scenario. With the caveat that changes in expression levels of genes reflect changes in overall abundances of corresponding proteins, our algorithm allows one to calculate the impact of an external or internal stimulus measured in a microarray on free and bound concentrations of all proteins in the cell. Including such indirectly perturbed targets could considerably extend the list of proteins affected by a given shift in environmental conditions. Simultaneous shifts in expression levels of several genes may amplify changes of free concentrations of some proteins and/or mutually inhibit changes of others.

11.4.5 Effects of Intracellular Noise

Another implication of our findings is for intracellular noise, or small random changes in total concentrations C_i of a large number of proteins. The randomness, smaller magnitude, and the sheer number of the involved proteins characterize the differences between such noise and systematic several-fold changes in the total concentration of one or several proteins considered above. Our methods allow one to decompose the experimentally measured [10] noise in total abundances of proteins into biologically meaningful components (free concentrations and bound concentrations within individual protein complexes). Given a fairly small magnitude of fluctuations in protein abundances (on average around 20% [10]), one could safely employ a computationally efficient linear response algorithm (see Maslov et al. [12]). Several recent studies [10,15,16] distinguish between the so-called extrinsic and intrinsic noise. The extrinsic noise corresponds to synchronous or correlated shifts in abundance of multiple proteins which, among other things, could be attributed to variation in cell sizes and their overall mRNA and protein production or degradation rates. Conversely, the intrinsic noise is due to stochastic fluctuations in production and degradation, and thus lacks correlation between different proteins. We found that extrinsic and intrinsic noise affect equilibrium concentrations of proteins in profoundly different ways. In particular, while multiple sources of the extrinsic noise partially (yet not completely) cancel each other, intrinsic noise contributions from several sources can sometimes add up and cause considerable fluctuations in equilibrium free and bound concentrations of particular proteins (see Figure 11.8).

11.4.6 Limitations of the Current Approach and Directions for Further Studies

In our study, we used a number of fundamental approximations and idealizations including the assumption of spatially uniform concentrations of proteins, the neglect of temporal dynamics, or equivalently, the assumption that all concentrations have sufficient time to reach their equilibrium values, the continuum approximation neglecting the discrete nature of proteins and their bound complexes, etc. Another set of approximations was mostly due to the lack of reliable large-scale data quantifying these effects. They include not taking into account the effects of cooperative binding within multi-protein complexes, using a relatively small number (81) of well-curated multi-protein complexes used in our study (Table 11.4), neglecting systematic changes in protein abundances in the course of the cell cycle, etc. We do not expect these

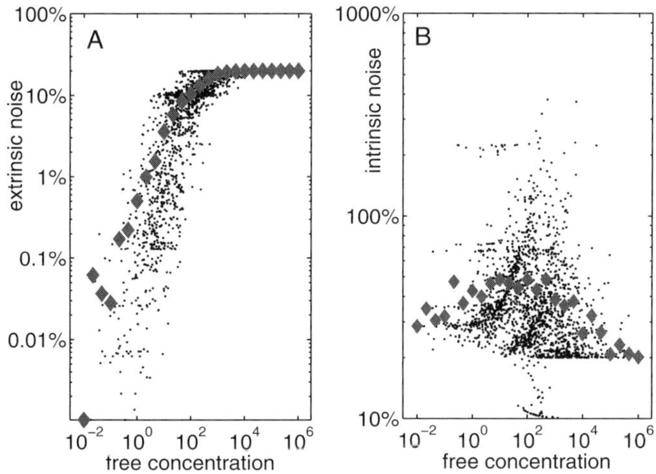

Figure 11.8 The magnitude of extrinsic (panel A) and intrinsic noise in free concentrations F_i of proteins when their total concentrations C_i fluctuate by 20%. In this plot, we used K_{ij} = const = 1 nM. One can see that while the extrinsic noise is suppressed in the low concentrations region, the intrinsic one is uniformly high and reaches as much as >300% in the mid-F_i range.

TABLE 11.4 Fraction of Concentration-Coupled Pairs that Survive Addition of 81 Multi-Protein Complexes

L	% Pairs
1	97.4
2	91.2
3	65.9
4	12.5
New	21.2

The fraction of concentration-coupled pairs in the original network (K_{ij} = const = 10 nM) that survive the addition of 81 multi-protein complexes curated from MIPS CYGD database [20,21] (column 2). The last row is the percentage of new pairs in the network after these complexes were added.

effects to significantly alter our main qualitative conclusions, namely, the exponential decay of the amplitude of changes in equilibrium concentrations, the existence of three- to four-step chains that nevertheless successfully propagate concentration changes, and the general conditions that enhance or inhibit such propagation.

In the future we plan to extend our study of fluctuations in equilibrium concentrations by incorporating the effects of protein diffusion (nonuniform spatial concentration) and kinetic effects. Another interesting avenue for further research is to apply the concept of "potential energy landscape" (for definitions see Ao [17] and references therein) to reversible processes governed by the law of mass action, e.g. the equilibrium in protein binding networks. In the past, this concept was applied to processes involving catalytic, irreversible protein–protein interactions such as phosphorylation by kinases or regulation by transcription factors. In this case, it helped to reveal the robustness of regulatory networks in the cell cycle [18] and in a simple two-protein toggle switch [19].

11.5 METHODS

11.5.1 Source of Interaction and Concentration Data

The curated PPI network data used in our study are based on the 2.020 release of the BIOGRID database [3]. We kept only pairs of physically interacting proteins that were reported in at least two publications using the following experimental techniques: Affinity Capture-MS (28,172 pairs), Affinity Capture-RNA (55 pairs), Affinity Capture-Western (5,710 pairs), Co-crystal Structure (107 pairs), FRET (43 pairs), Far Western (41 pairs), and Two-hybrid (11,935 pairs). That left us with 5,798 nonredundant interacting pairs. Further restriction for both proteins to have experimentally measured total abundance [4] narrowed it down to 4,185 distinct interactions among 1,740 yeast proteins.

The list of manually curated yeast protein complexes was obtained from the latest release (May 2006) of the MIPS CYGD database [20,21]. The database contains 1,205 putative protein complexes, 326 of which are not coming from systemic analysis studies (high-throughput MS experiments). In the spirit of using only the confirmed PPI data, we limited our study to these manually curated 326 complexes. For 99 of these complexes, the MIPS database lists three or more constituent proteins. After elimination of proteins with unknown total concentrations, we were left with 81 multi-protein complexes.

Genetic interactions of dosage rescue type were also obtained from the BIOGRID database. There are 772 pairs of dosage rescue interactions among 1,740 proteins participating in our PPI network (the full list contains 2,531 dosage rescue pairs).

11.5.2 Sensitivity to False Positives and Negatives in Network Topology

The list of concentration-coupled protein pairs that we identify in our calculations is relatively insensitive to false positives and false negatives in the network topology. Table 11.3 shows the fraction of concentration-coupled pairs (K_d = 10 nM) that survive a random addition (column 2) or removal (column 3) of 20% of all protein–protein interactions. This confirms the robustness of our observations with respect to incompleteness and errors that are inevitable even for well-curated PPI networks in the best-studied model organisms.

11.5.3 Numerical Algorithms

The numerical algorithm calculating all free concentrations F_i, given the set of total concentrations C_i and the matrix of dissociation constants K_{ij}, was implemented in MATLAB 7.1 and is available for downloading on http://www.cmth.bnl.gov/~maslov/programs.htm. It consists of iterating equation 11.1 starting with $F_i = C_i$. Iterations stop once relative change of free concentration on every node in the course of one iteration step becomes smaller than 10^{-8}, which for networks used in our study takes less than a minute on a desktop computer. When necessary, multi-protein complexes are incorporated into this algorithm as described below.

The effects of large concentration perturbations were calculated by recalculating free concentrations, following a twofold increase in abundance of a given perturbed protein. The effects of small perturbations such as those of concentration fluctuations were calculated using the faster linear response matrix formalism described elsewhere [12].

11.5.4 Rigorous Results for Simple Lattices

To illustrate and rigorously validate our observations of the exponential decay of concentration perturbation, we analytically investigate a simple example of a network, the Bethe lattice, where each vertex has the same number of interaction partners (degree) $d_i = d$. In addition, we assume that all dissociation constants are equal, $k_{ij} = k$. When total concentrations of all proteins are also identical $C_i = C$, the equilibrium concentrations of all monomers F_i, and heterodimers D_{ij} are given by

$$F_i = F = \frac{k}{2d}\left(\sqrt{1+\frac{4dC}{k}}-1\right)$$
$$D_{ij} = D = \frac{k}{2d}\left(\frac{2C}{k}+\frac{1}{d}-\frac{1}{d}\sqrt{1+\frac{4dC}{k}}\right). \qquad (11.2)$$

For arbitrary concentrations C_i, using the mass conservation and LMA, it is also simple to derive the following recurrent in the lattice index l relation for free concentrations,

$$C_l - F_l = \frac{F_l}{k}[(d-1)F_{l+1} + F_{l-1}], \qquad (11.3)$$

Assuming that the total concentration is perturbed from its universal for all network value C at node 0, and the deviation F_l of free concentrations from their equilibrium value given by equation (11.2) are small, equation (11.3) yields

$$-\left(d+\frac{k}{F}\right)\underset{\cdot}{F}_l=(d-1)\underset{\cdot}{F}_{l+1}+\underset{\cdot}{F}_{l-1}. \tag{11.4}$$

It has an exponentially decaying solution $\underset{\cdot}{F}_l=\underset{\cdot}{F}_0 l^l$, where

$$l=-\frac{1}{d-1}\left[\frac{d+k/F}{2}-\sqrt{\left(\frac{d+k/F}{2}\right)^2-(d-1)}\right]. \tag{11.5}$$

As expected, $-1 < l < 0$ which means that perturbations sign alternate and exponentially decay as a function of l. In a strong binding limit, the combination of equations (11.5) and (11.2) yields

$$l=-\frac{1}{d-1}\left(1-\frac{1}{d-2}\sqrt{\frac{dk}{C}}\right)+\mathcal{O}\left(\frac{dk}{C}\right). \tag{11.6}$$

This confirms our qualitative prediction that in the "ideal" scenario when free concentrations vanish, the perturbation still decays exponentially due to branching of the "perturbation current" at each node. For a linear chain of proteins ($d = 2$), the complete solution in terms of C and k looks particularly elegant:

$$l_{d=2}=-\frac{(1+8C/k)^{1/4}-1}{(1+8C/k)^{1/4}+1}. \tag{11.7}$$

As one expects heuristically, in the limit of strong binding, a perturbation in a linear chain propagates indefinitely, $|l_{d=2}| \to 1$.

To explore the effect of nonideal concentration setup on propagation of perturbation, we solve for the decay exponent in the linear chain ($d = 2$) with oscillating total concentrations,

$$C_i=C\left[1+(-1)^i a\right]. \tag{11.8}$$

Response to perturbation of the even- and odd-numbered vertices has different amplitudes A_{2i} and A_{2i+1} yet decays with the same exponential coefficient $l_{1D\pm}$:

$$\begin{aligned}\underset{\cdot}{F}_{2i}&=A_{2i}l^{2i}\\ \underset{\cdot}{F}_{2i+1}&=A_{2i+1}l^{2i+1}\end{aligned} \tag{11.9}$$

Substitution of equation (11.9) into linearized around the equilibrium concentration recursion relation (11.3) yields the system of two equations for the relative amplitude A_{2i}/A_{2i+1} and l with the solutions

$$\frac{A_{2i}}{A_{2i+1}} = \sqrt{\frac{F_{2i}(k+2F_{2i})}{F_{2i+1}(k+2F_{2i+1})}} \tag{11.10}$$

$$l_{\pm} = -\frac{4\chi\sqrt{1-a^2} - \sqrt{2f(1+4\chi)-2f^2}}{1+4\chi-f}, \tag{11.11}$$

where $\chi = C/k$ and $f = \sqrt{16a^2\chi^2 + 8\chi + 1}$. Evidently, $|l_{\pm}(C/k,a)| \leq |l_{d=2}(C/k)|$ with equality being achieved only for $a = 0$. For example, for $k/C \to 0$ and a small,

$$l_{\pm} \to -\left(1-\sqrt{2a}\right)$$

Thus, as it was discussed above, any variation in C_i/d_i, which results in a larger average unbound concentrations F_i, leads to a faster decay of perturbations.

11.5.5 Complexes of Three and More Proteins

Our approach can be easily generalized to take into consideration the formation of complexes with more than two constituent proteins. Consider for example the case of a three-protein complex, consisting of proteins 1, 2, and 3 in which protein 1 interacts with proteins 2 and 3 with the same dissociation constant K_d and proteins 2 and 3 are not in direct contact with each other. In this case, concentrations of heterodimers (partially formed complexes) 1 – 2 and 1 – 3 are still given by F_1F_2/K_d and F_1F_3/K_d correspondingly, while the concentration of the fully formed complex is simply $F_1F_2F_3/K_d^2$. The conservation of mass dictates that $C_1 = F_1 + F_1F_2/K_d + F_1F_3/K_d + F_1F_2F_3/K_d^2$ or

$$F_1 = \frac{C_1}{1 + F_2/K_d + F_3/K_d + F_2F_3/K_d^2}.$$

This expression can be readily modified to describe larger complexes and more complicated subsets of intra-complex bindings.

We incorporated multi-protein complexes into a version of our numerical algorithm and calculated free and bound concentrations of all proteins, as well as the linear response of the LMA equilibrium to small perturbations. As for heterodimers before, we assume all dissociation constants attaching each of the constituent proteins to a complex to be equal to each other and thus, the equilibrium concentration of an n-protein complex to be given by $F_1F_2 \ldots F_n/K_d^{n-1}$. In the absence of more detailed knowledge of the architecture intra-complex interactions, we also chose to omit concentrations of partially formed complexes except for the ones formed by the known pairs of directly binding proteins (heterodimers) within a complex. The list of manually curated yeast protein complexes was obtained from the latest release

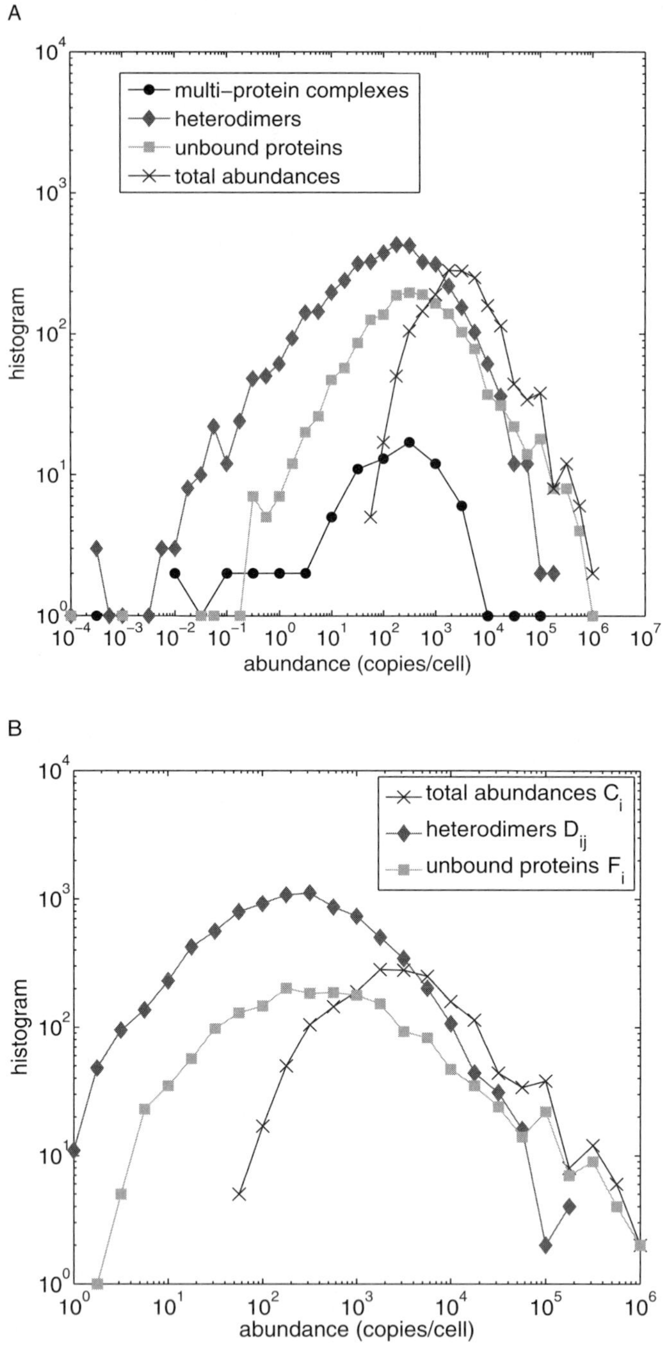

Figure 11.9 (A) The histogram of calculated equilibrium concentrations of 81 multi-protein complexes [20] (black circles), 4,185 two-protein complexes (dimers) (red diamonds), and the free (unbound) concentrations of 1,740 individual proteins F_i (green squares) given the experimentally measured total proteins concentrations C_i from Ghaemmaghami et al. [4] (blue crosses) and K_{ij} = const = 10 nM. (B) The same histograms (minus that of multi-protein complexes) recalculated for the evolutionary-motivated assignment of dissociation constants: $K_{ij} = \max(C_i, C_j)/20$. Note the lack of unreasonably low free or bound concentrations (those below 1 molecule/cell) in this case.

(May 2006) of the MIPS CYGD database [20,21]. In the spirit of using only the confirmed PPI data, we limited our study to these manually curated 326 complexes.

For 99 of these complexes, the MIPS database lists three or more constituent proteins. After elimination of proteins with unknown total concentrations, we were left with 81 multi-protein complexes. The histogram of their equilibrium concentrations (along with equilibrium concentrations of dimers and monomers), calculated using our algorithm at K_{ij} = const = 10 nM, is plotted in Figure 11.9A using black circles.

ACKNOWLEDGMENTS

We thank Kim Sneppen for valuable discussions and contributions in early phases of this project. This work was supported by a National Institute of General Medical Sciences Grant 1 R01 GM068954-01. Work at Brookhaven National Laboratory was carried out under Division of Material Science, U.S. Department of Energy Contract DE-AC02-98CH10886. S.M.A.'s visit to the Kavli Institute for Theoretical Physics, where part of this work was accomplished, was supported by a National Science Foundation Grant PHY05-51164.

REFERENCES

1. Wagner A. The yeast protein interaction network evolves rapidly and contains few redundant duplicate genes. Mol Biol Evol 2001;18:1283–1292.
2. Jeong H, Mason SP, Barabasi AL, Oltvai ZN. Lethality and centrality in protein networks. Nature 2001;411(6833):41–42.
3. Stark C, Breitkreutz B-J, Reguly T, Boucher L, Breitkreutz A, Tyers M. BioGRID: a general repository for interaction datasets. Nucleic Acids Res 2006;34(Suppl 1):D535–D539.
4. Ghaemmaghami S, Huh WK, Bower K, Howson RW, Belle A, Dephoure N, O'Shea EK, Weissman JS. Global analysis of protein expression in yeast. Nature 2003;425:737–741.
5. Piehler J. New methodologies for measuring protein interactions in vivo and in vitro. Curr Opin Struct Biol 2005;15:4–14.
6. Kumar MD, Gromiha MM. PINT: Protein-protein Interactions Thermodynamic Database. Nucleic Acids Res 2006;34:D195A–D198.
7. Lancet D, Sadovsky E, Seidemann E. Probability model for molecular recognition in biological receptor repertoires: significance to the olfactory system. Proc Natl Acad Sci U S A 1993;90(8):3715–3719.
8. Deeds EJ, Ashenberg O, Shakhnovich EI. A simple physical model for scaling in protein-protein interaction networks. Proc Natl Acad Sci U S A 2006;103(2):311–316.

9. As an alternative to this computationally expensive approach, we also tried the linear response matrix formalism [12] relating small changes in F_j to the ones in C_i. We found the linear response algorithm to be much less computationally expensive, while still providing remarkably good approximation to directly computed results even for large changes in protein levels.

10. Newman JRS, Ghaemmaghami S, Ihmels J, Breslow DK, Noble M, De Risi JL, Weissman JS. Single-cell proteomic analysis of S. cerevisiae reveals the architecture of biological noise. Nature 2006;441:840–846.
11. Toroczkai Z, Bassler KE. Jamming is limited in scale-free systems. Nature 2004;428(716):170.
12. Maslov S, Sneppen K, Ispolatov I. Spreading out of perturbations in reversible reaction networks. New J Phys 2007;9(8):273.
13. von Dassow G, Meir E, Munro EM, Odell GM. The segment polarity network is a robust development modeule. Nature 2000;406(6792):188–192.
14. Hughes KT, Mathee K. The anti-sigma factors. Annu Rev Microbiol 1998;52(1): 231–286.
15. Elowitz MB, Levine AJ, Siggia ED, Swain PS. Stochastic gene expression in a single cell. Science 2002;297(5584):1183.
16. Raser JM, O'Shea EK. Noise in gene expression: origins, consequences, and control. Science 2005;309(5743):2010–2013.
17. Ao P. Potential in stochastic differential equations: novel construction. J Phys A: Math Gen 2004;37(3):L25–L30.
18. Wang J, Huang B, Xia X, Sun Z. Funneled landscape leads to robustness of cell networks: yeast cell cycle. PLoS Comput Biol 2006;2(11):e147, 1385.
19. Kim K, Wang J. Potential energy landscape and robustness of a gene regulatory network: toggle switch. PLoS Comput Biol 2007;3(e60):0656.
20. Guldener U, Munsterkotter M, Kastenmuller G, Strack N, van Helden J, Lemer C, Richelles J, Wodak SJ, Garcia-Martinez J, Perez-Ortin JE, et al. CYGD: the Comprehensive Yeast Genome Database. Nucleic Acids Res 2005;33:364–368.
21. Mewes HW, Heumann K, Kaps A, Mayer K, Pfeiffer F, Stocker S, Frishman D. MIPS: a database for genomes and protein sequences. Nucleic Acids Res 1999;27(1):44–48.

12

AN ADAPTIVE SYSTEM MODEL OF THE YEAST GLUCOSE SENSOR SYSTEM

TODOR VUJASINOVIC AND ANDRÉ SINIŠA ŽAMPERA

Table of Contents

Pathway Analysis for Drug Discovery, Edited by Anton Yuryev

12.1 INTRODUCTION

Most morbidity and mortality in industrialized countries is associated with multifactorial diseases for which current treatment is insufficient, for example, cancers, cardiovascular diseases, neurological diseases, and autoimmune diseases. The identification and validation of novel, more effective treatment strategies are therefore required. Large-scale observations and treatment failures have clearly demonstrated the complexity of the mechanisms involved: in general, more than 100 genes or proteins, together with all the intermediate products and their modifications, are involved in a given cellular physiological function. For example, the query "induction of apoptosis" retrieves 97 human gene products in the Gene Ontology database [10]. This complexity is observed on every level of the physiological/biochemical processes, from the topology of signaling and transcriptional networks (recurrent and heterogeneous) to their highly nonlinear and complex dynamic behaviors, reflecting the complex mechanisms of organization and regulation of living organisms. Research in human physiopathology, one of the main aims of which is to identify new therapeutic targets, must take this complexity into account, at best by the use of dedicated integrative models.

12.1.1 Strategies for the Large-Scale Modeling of Human Physiopathology

Modeling, in the generally understood sense of creating a faithful representation of reality, can today only be applied to a few well-delimited and well-described physiological phenomena involving no more than a few tens of genes or proteins. Among others, such models have been developed for the cell cycle [4], the osmotic shock response [5], the extracellular-signal-regulated kinase (ERR) signaling [6], and the glucose enzymatic metabolism [7,8]. Many other annotated dynamic models can be found in the BioModels database of the European Bioinformatics Institute [9]. Essentially, most of these models have been developed to gain a better understanding of the generic properties of biological phenomena, such as homeostatic feedback [5], bi-stability [4,5], oscillations [7,8], hysteresis [4], and others. These models generally represent simplified signaling/metabolic modules but in fine detail to provide accurate global and local simulations. For the review of methods and results on the modeling and simulation of genetic regulatory networks, see for example De Jong [33] or Smolen et al. [34].

However, this detailed picture cannot be easily scaled up to a complexity realistically reflecting most eukaryotic physiological functions, because the laws regulating elementary chemical processes are generally unknown, even qualitatively (permissive, additive, cooperative effects, etc.). This makes it unrealistic to calculate the precise kinetic constants of the elementary chemical reactions. Our knowledge of the networks of interactions between molecules also remains incomplete. In this context, the need to integrate extensive networks has led us to an alternative modeling strategy: instead of describing the biological system at the elementary chemical reaction level (low level symbolism: intermediate chemical states such as phosphorylations), we describe it in terms of the relative amount or activity of the molecules (mid-level symbolism), in the logic of signal propagation. Data are effectively available at this level.

The lack of knowledge also implies that large-scale modeling strategies may be seen as credible only if they include an adaptive or data mining approach. These approaches are designed to obtain knowledge from large amounts of data, structured to various extents, using automatic methods sometimes called learning methods. Generally speaking, they involve the construction of a "flexible" model, in our case a differentiable dynamic system. The flexibility is reflected by the presence of parameters to be determined by optimization methods. Rather than aiming at the mimetic reproduction of the process modeled itself, these approaches aim at the resolution of concrete, specific problems—generating hypotheses concerning potential therapeutic targets in our case. Therefore, our modeling strategy aims more at predicting than explaining, it is qualitative rather than quantitative, and focuses on the design of a decision support tool capable of producing credible hypotheses.

The adaptive aspect of our strategy lies in "learning" the constants of the model—determination of the set of parameters, minimizing the distance between the simulated kinetics on the one hand, and temporal experimental data on the other. Learning of kinetics does not in itself impose constraints on the dynamic behavior to be learned; however, the incomplete nature of the interaction networks and their formalization, and the insufficient accuracy of expression data suggest that we should limit ourselves to the simplest types of behavior—steady states, which are the lasting regimes of functioning of living organisms, as opposed to other dynamic regimes such as periodic, quasi periodic, and chaotic behaviors. In this context, the kinetics represent transient states that the system learns to follow between stable states.

Under the hypothesis that the network represents the principal mechanisms operating in these transitions and that these stable states provide an exhaustive description of the geometry of all stable states, at least in a limited region of the state space, the model parameterized in this manner should be able to classify manipulations (typically inactivation or overexpression of genes) by simulations according to their capacity to direct the system to one of these states, and thus identify the key elements required to induce or to prevent a given type of response.

We applied this classification strategy to the signal transduction network of the glucose repression/derepression system (GRDS, including the glucose sensor system) of the model organism *Saccharomyces cerevisiae* (baker's and brewing yeast), with the aim of identifying which genes in this network are essential for good growth in the absence of glucose. We present here the corresponding medium-sized quantitative dynamic model that integrates both the topography and dynamic responses of the GRDS, and its classifying efficiency.

12.2 IMPLEMENTATION OF THE MODEL

12.2.1 Mathematical Formalism

The available data describing topology and behavior of large signaling networks include: (1) data concerning molecular interactions (annotated according to the activating or inhibitory nature of the action, its compilation can be used to reconstitute signaling and metabolic networks) and (2) high-throughput expression screening data in the form of kinetics.

We model a network of symbolic entities that represent both the genes and the corresponding proteins. The network of interactions is formalized as a signed directed graph. We describe the dynamics of the system by means of a system of differential equations inspired by standard models of signal transduction. This type of model could be described as a heteronomous dynamic system in which the kinetic constants are inferred from time-course data of gene expression using standard adaptive system techniques. We assumed that molecules are permanently synthesized and degraded, and that the rates of synthesis and degradation compensate each other in equilibrium state (e.g. here in glucose-rich medium prior to the diauxic shift). The degradation is not considered subjected to regulation and is modeled by the classic linear decay term. We also assumed that the expression level of every molecule in the network can be saturated and thus the synthesis rate is modeled by a sigmoid-shape transformation of the total regulatory input, which is in turn a linear superposition of elementary regulatory inputs. As in the case of the diauxic shift where the modeled biological response is globally synchronous, we did not introduce time delays into the model. The mathematical formalism is detailed in the Material and Methods section.

12.2.2 The Glucose Repressor/Derepressor System: Definition and Construction of the Signaling Network

In *Saccharomyces cerevisiae*, the expression of glucose metabolism enzymes is dynamically regulated by the availability of extracellular glucose, and their expression pattern follows accurately the biochemical direction of the metabolite flow that depends on the nature of the carbon source [2]. This dynamic response is driven by specific glucose-sensing, -signaling, and -metabolizing pathways of which the intracellular qualitative topography has been relatively well described [3]. These pathways interact and regulate each other, forming a complex, nonlinear, and relatively large network of about 100 genes/proteins, as is often observed for signal transduction pathways in eukaryotic cells.

Yeast cultured in the presence of high extracellular concentrations of glucose breaks down this glucose by glycolysis to generate energy (in the form of ATP) and ethanol (by fermentation). If glucose is not supplemented in the medium, this consumption of glucose decreases the extracellular glucose concentration, leading to the so-called diauxic shift, in which the directions of synthesis and metabolic degradation are modified (and partly reversed) so that the yeast can use the ethanol it has synthesized to produce energy and glucose (gluconeogenesis). This transition is required for growth in the absence of glucose and allows the yeast to use other carbon sources.

The signal propagation logic made us to represent these signaling pathways by directly connecting in the graph the signaling genes/proteins to the genes/proteins that they regulate directly (or semi-directly through second messengers), regardless of the chemical mechanisms involved. The few cases where the product of a metabolic enzyme directly regulates the activity of another metabolic enzyme were also considered to propagate a signal and treated this way. The principle of such a symbolism is quite similar to the widely used 2D graphical displays of signaling pathways.

The representation of the metabolic component of the network was more problematic for two reasons. First, it was not possible to directly connect the signaling component to some of the metabolic enzymes due to insufficient knowledge in the literature, whereas it was important to avoid gaps within the network that would have stopped the signal propagation. The straightforward solution to this, and relatively classical from a biological point of view, would have been to take advantage of the good knowledge of the cascades of the metabolic reactions, which have no solution of continuity by themselves, by explicitly representing them and connecting to them the signaling module as described above, in other words to interconnect the flows of signals with the flows of metabolites. However, the second difficulty was that the amounts of the substrates/products of the metabolic enzymatic reactions were not available, so that explicitly representing them in the model would have been incompatible with the adaptive aspect of our strategy. This imposed to alternatively represent the metabolic component of the network. Thus, we decided to take advantage of the fact that in this very case of glucose signaling and metabolism, the cells do not regulate the expression of enzymes independently from each other, due to common regulatory elements in their promoters (references cited in Uemura et al. [1]), so that their expression regulation during the diauxic shift is finely tuned [2] in a coherent way with regard to metabolic requirements of the cells and chemical reactions. Therefore, an acceptable solution was to directly connect the enzymes when they are engaged in consecutive chemical reactions. This symbolism provides the coherence of the global network in terms of signal propagation, while preserving the topology of the substrate/product biochemical network (metabolic pathways). It is also a way to symbolize the enzyme regulations that are not identified yet, and is compatible with our objective to use models as data mining/decision support tools. Though this formal representation results in the metabolic enzymes regulating each other (which is not the case in cells), the simulation results show that it does not significantly weaken the selectivity-specificity even though many genes in which invalidation impairs the fitness of yeast to ethanol medium are metabolic enzymes (see Results).

To build the network, we compiled the available information on the glucose sensor system, the glucose repression/derepression system, and the glucose metabolism from 144 scientific articles and from additional data from *Saccharomyces* Genome Database [11]. The final modeled network is shown in Figure 12.1. Its complete description and the principles driving its compilation are detailed in the Material and Methods section. It is represented in table format and certain additional assumptions are described in the additional data available on Helios Biosciences website (www.heliosbiosciences.com).

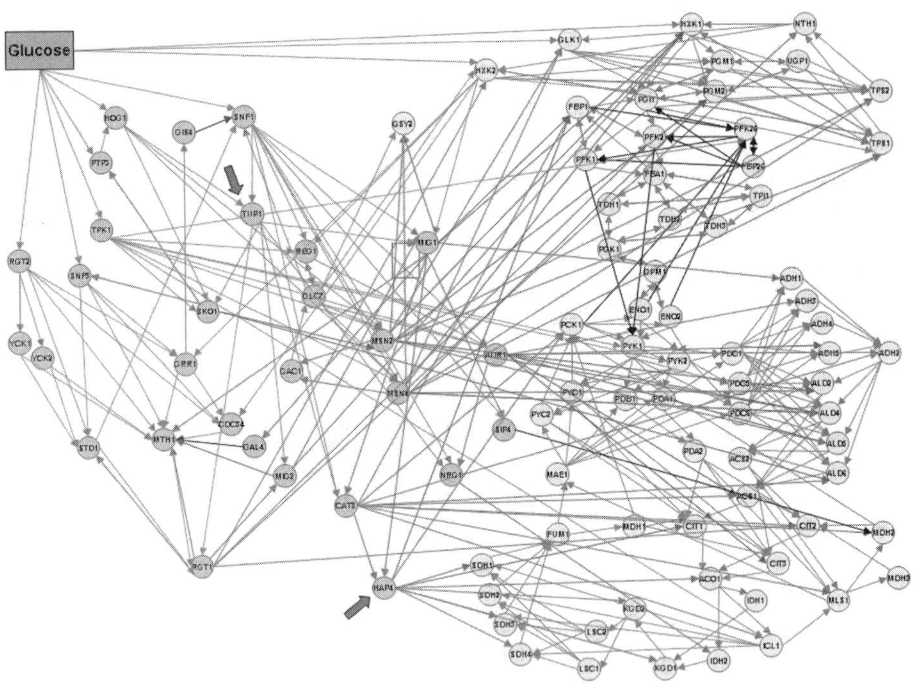

Figure 12.1 The glucose repressor/derepressor system. Green circles: genes/proteins transducing the glucose signal; yellow circles: genes/proteins metabolizing glucose; blue arrows: regulations at the protein activity level; green arrows: regulations at the gene transcription level; orange arrows: metabolic reactions; arrows are directed according either to the propagation of the signal or to the direction of the metabolic reactions; genes/proteins are identified by their standard names, the correspondences between gene standard names and systematic names according to the *Saccharomyces* Genome Database [11] and the list of interactions with bibliographic references are provided in additional material. See color insert.

12.3 RESULTS

12.3.1 Fitting the Model to Training Data

The parameters for the model were calculated to fit the gene expression time-course data of the diauxic shift (during 690 minutes) by using a trajectory learning procedure known as backpropagation through time (BPTT) (see Material and Methods for more details). The learning procedure was carried out several times with different random initial weights over the interval $\langle -1, 1 \rangle$. The initial time constants were set to 60 minutes.

The overall correlation between the weights and between the time constants for different sets of parameters was >0.95. Therefore, the results from the simulations carried out with these different sets of parameters were very similar.

The relative error with respect to the shift trajectory was ≈20, 4%. This was satisfactory enough to perform simulations, considering we are not dealing with an artificial network, but with a real network (with interaction sign constraints) and noisy data (error estimated at 15% to 20%), as well as the evolution of the network being completely derived from only one source (glucose input).

Figure 12.2 shows the clustering of the experimental time course and how they were reproduced by the model.

12.3.2 Simulations/Extrapolations

12.3.2.1 Knockout/Overexpression Experiments. We first observed the dynamics of simulations after systematically deleting every molecule in the network under initial state conditions (high constant glucose concentration) by setting the state variable of the deleted gene to 0 and replacing the glucose reduction input by a maximum (=0.75) constant input representing a stable glucose-rich medium. We obtained no complex dynamic patterns (limit cycles, strange attractors, etc.), with all state variables relaxing toward steady states after 690 minutes.

A stringent test of the predictive power of the model is its ability to simulate knockout/overexpression experiments. We would expect knockout/overexpression simulations to predict the evolution of the system with enough accuracy to discriminate the deleted/overexpressed molecule if the discrimination criterion is how close the simulated quantities are to the corresponding experimentally measured states.

Although there are several published quantitative datasets of gene expression results for knockout or overexpression of yeast genes, to be suitable as controls they need to concern genes and culture conditions that correspond to our model. We found only two datasets meeting these criteria: the knockout of the TUP1 gene [31] and the overexpression of the HAP4 gene [32]. Both datasets were produced in glucose-rich culture conditions, so the

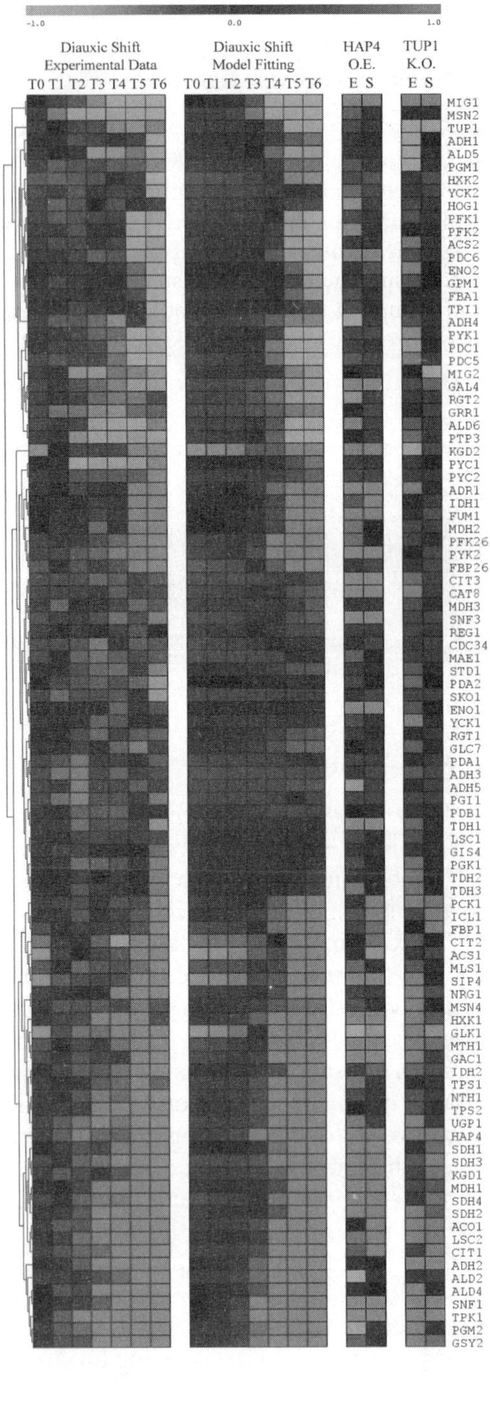
-1.0
0.0
1.0
Diauxic Shift Experimental Data
T0 T1 T2 T3 T4 T5 T6
Diauxic Shift Model Fitting
T0 T1 T2 T3 T4 T5 T6
HAP4 O.E.
E S
TUP1 K.O.
E S
MIG1
MSN2
TUP1
ADH1
ALD5
PGM1
HXK2
YCK2
HOG1
PFK1
PFK2
ACS2
PDC6
ENO2
GPM1
FBA1
TPI1
ADH4
PYK1
PDC1
PDC5
MIG2
GAL4
RGT2
GRR1
ALD6
PTP3
KGD2
PYC1
PYC2
ADR1
IDH1
FUM1
MDH2
PFK26
PYK2
FBP26
CIT3
CAT8
MDH3
SNF3
REG1
CDC34
MAE1
STD1
PDA2
SKO1
ENO1
YCK1
RGT1
GLC7
PDA1
ADH3
ADH5
PGI1
PDB1
TDH1
LSC1
GIS4
PGK1
TDH2
TDH3
PCK1
ICL1
FBP1
CIT2
ACS1
MLS1
SIP4
NRG1
MSN4
HXK1
GLK1
MTH1
GAC1
IDH2
TPS1
NTH1
TPS2
UGP1
HAP4
SDH1
SDH3
KGD1
MDH1
SDH4
SDH2
ACO1
LSC2
CIT1
ADH2
ALD2
ALD4
SNF1
TPK1
PGM2
GSY2

Figure 12.2 Model fitting and simulation results: Diauxic shift experimental data, simulated data (both at times T0 = 0, T1 = 150, T2 = 240, T3 = 360, T4 = 450, T5 = 570, T6 = 690 minutes), HAP4 overexpression data (O.E.) and TUP1 knockout data (K.O.). E = experimental, S = simulated. The gene expression ratios to a T0 experimental value corresponding to a similar independent experiment are log2-transformed. The resulting data were imported in the TIGR MeV MultiExperiment Viewer 3.1 software from the Institute for Genomic Research [37] to produce the images; hierarchical clustering of the genes/proteins according to their expression behavior (gene tree only) was performed using the Euclidean distance, with complete linkage clustering, 100 iterations, no resampling of samples, and all other parameters set to default values. Gray color = experimental data not available. See color insert.

corresponding simulations were carried out using a high constant glucose concentration input.

The TUP1 gene product (Tup1p) associates with Cyc8p to form a general corepressor that acts as a cofactor for several transcription factors [35]. It is involved in the initial repression of several genes in our network under initial high glucose culture conditions, among which are Mig1p targets [36]. Thus, it plays an important role in the biological mechanism we have modeled. The experimental data show that the TUP1 knockout partially mimics the diauxic shift by triggering the derepression of a subset of genes in the network. However, the up-regulation of genes during the shift also involves other mechanisms, such as the direct inhibition of repressor transcription factors [36]. Thus, we also expected the inhibition of other genes in the network to partially mimic the TUP1 knockout. Although Tup1p corepressor function is also involved in other physiological mechanisms/pathways [2], we assumed that this would not have a major impact on glucose repression/derepression. We tested whether we could identify TUP1 as being the gene that, when inhibited *in silico*, would best reproduce the TUP1 knockout experimental data.

For every single knockout simulation, we compared the simulated expression levels of all the molecules in the network with those from the experimental TUP1 knockout data (expression level of TUP1 = 0.1). We first normalized the experimental data by the same factor as for the training data, and as the maximum simulated rate of any product in our model is 1, any values exceeding 1 were set to 1 to avoid "overweighting" highly expressed genes in the error function. The error function was the standard Euclidean distance in the state variable space. We classified all molecules with respect to the distance between the knockout simulation result (after 690 minutes) and the experimental TUP1 knockout data. Figure 12.3 shows that the simulated deletion of TUP1 approached the system to the experimental data more than the deletion of any other gene. Specificity-sensitivity analysis of the up- or down-regulation ratios showed that the simulated deletion of TUP1 gives the highest selectivity-specificity trade-off (for the level of variation of 10%). Thus, our model efficiently discriminated TUP1 as the deleted gene.

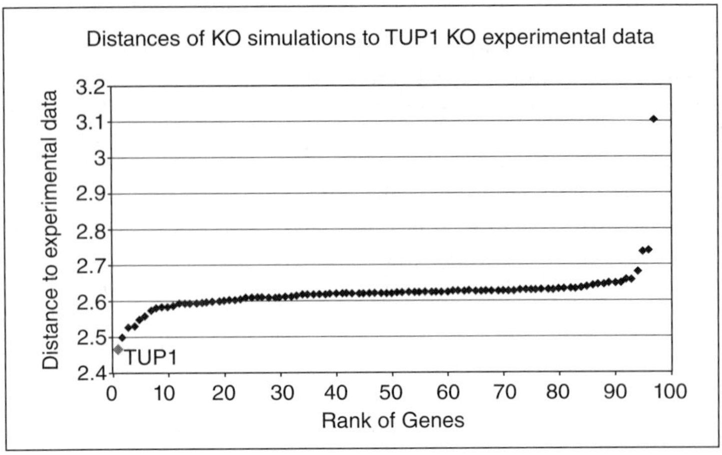

Figure 12.3 Simulation results of TUP1 deletion. Each dot represents a gene deletion simulation with its distance to the TUP1 KO experimental data on *y*-axis and its rank in increasing order with respect to this distance on *x*-axis.

The up-regulation of genes during the diauxic shift depends both on derepressing mechanisms and on gene transcription activation. The HAP4 gene product (Hap4p) is a subunit of the transcriptional activator Hap2p/3p/4p/5p CCAAT-binding complex. This complex is glucose-repressed and its derepression during the shift activates the transcription of several genes within our network [32]. The experimental data show that the overexpression of HAP4 also partially mimics the diauxic shift by triggering the overexpression of a subset of genes in the network (although some of the Hap4p direct target genes are not up-regulated). Although Hap4p is also involved in other physiological mechanisms, such as respiratory gene expression [32], we assumed this would not have a major impact on glucose repression/derepression. Using the same procedure as for the TUP1 simulations, we tested whether we could identify HAP4 as being the gene that, when activated, would best reproduce the HAP4 overexpression experimental data.

The overexpression simulation was carried out in the same way as for the TUP1 deletion simulation, but by overexpressing every gene in the network by the factor corresponding to the overexpression of HAP4 observed experimentally. We then calculated the distance between the simulated and experimental data and classified the molecules on this basis. We found that HAP4 was clearly identified as being the gene overexpressed (Figure 12.4).

12.3.2.2 Classification of Genes According to the Impact of Their Invalidation on the Fitness of the Yeast to Ethanol Medium. We trained our model to simulate the transition from growing in glucose-rich medium to the growing in ethanol-rich medium. Thus, we expected to be able to classify the knockout

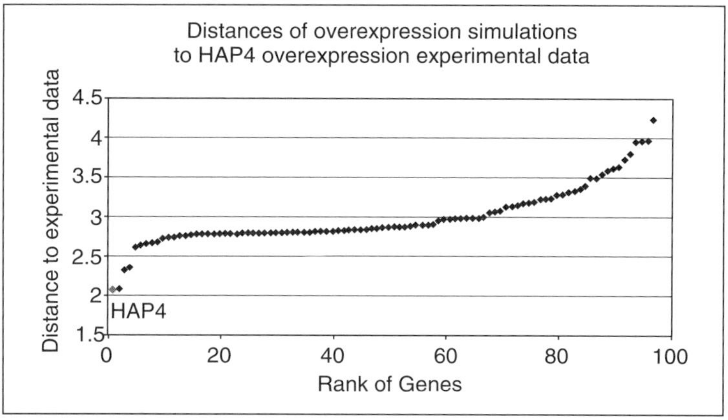

Figure 12.4 Simulation results of HAP4 overexpression. Each dot represents a gene overexpression simulation with its distance to the FTAP4 overexpression experimental data on *y*-axis and its rank in increasing order with respect to this distance on *x*-axis.

mutants in two classes according to their fitness to ethanol (good fitness/deficiency in growth). Our hypothesis was that the simulated knockout mutants that were unable to carry out the fermentation/respiration shift correctly would have growth deficiencies on ethanol. Classifying the genes according to the ability of their inhibition to modify a phenotype is particularly interesting as it is the same type of classification that would be carried out in drug targeting: identifying actions that would inhibit a given physiopathological behavior.

We tested our classification with data from studies of the ability of knockout mutants to grow on ethanol and glucose. Steinmetz et al. [30] have classed most molecules of the network according to how their being knocked out affects the ability of yeasts to grow on fermentable sugar medium (such as glucose) or non-fermentable sugar medium (such as ethanol). In our network, 30 genes have more or less serious fitness defects on ethanol, 11 mutants are unviable, 22 have fitness defects on fermentable substrates, and 43 show no real differences in fitness on different nutrients. This classification is not exclusive, as deficiencies on both glucose and ethanol are possible for the same gene.

We carried out gene deletion simulations in diauxic shift conditions (i.e. with a glucose exhaustion input) and calculated the distances between the simulated expression levels and the experimental expression levels of the shift after 690 minutes. We selected the 36 genes with the greatest distance between the deletion simulation and the shift experimental data and found that 24 of them were classified as having fitness defects on ethanol (i.e. selectivity equals 80%), whereas 12 belonged to some other class (i.e. specificity is 82%). For two of six genes, we were unable to detect; the misclassification was due to an unsatisfactory propagation of the shift signal in this region of the network (and

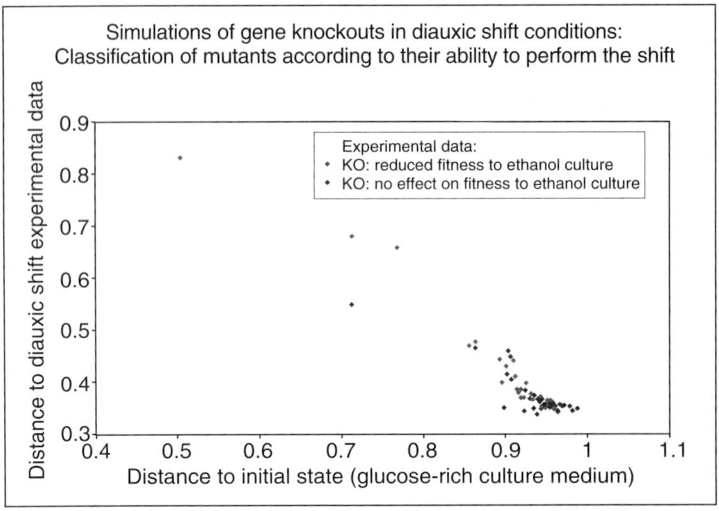

Figure 12.5 Classification of simulated KO mutants according to their fitness on ethanol-rich medium. Each dot represents a simulated gene KO. Gene deletion simulations were carried out under diauxic shift conditions. Most simulated NO mutants moving away from respiration to fermentation (initial state) are those with an experimentally confirmed deficiency in growth on ethanol. The distances have been expressed as ratios with respect to the distance between initial and diauxic shift state at time $T = 690$ minutes. See color insert.

thus insufficient fitness during learning). For one artifact (out of 12), the information on the fitness was not coherent, that is, it may not be an artifact. Thus, the selectivity-specificity percentage given above is the most pessimistic. The results are represented graphically in Figure 12.5.

12.4 DISCUSSION

An important question for the deterministic approach to the mathematical modeling of biological systems is whether the classical dynamic system paradigm, and more specifically the underlying differential formalism, is an appropriate tool, and to what extent. Classic dynamic systems give a very precise description of how the system evolves from the initial state, from a detailed understanding of every single regulatory action. However, if the number of components of the system is large, the complexity of the possible responses could greatly exceed the number of real dynamic patterns, which remains rather low. For example, even a simple mathematical model applied to a realistically complex transcriptional network (scale-free topology) might produce many more dynamic patterns (e.g. steady states, limit cycles, or even chaotic behavior) than actually observed in a living cell. In our model, we were able to control the qualitative or topological aspect of the long-term dynamic behavior by bifurcation analysis, so we avoided producing artifactual dynamics, what is crucial for the success of classification strategy we have chosen.

We achieved a two-state classification with a high selectivity-specificity score. TUP1 deletion and HAP overexpression were also successfully detected by our model.

These results were obtained despite using a biological description at the level of signal propagation rather than that of chemical reactions, the relatively poor quantitative accuracy of the experimental data used to calculate the parameters, and the linear superposition of regulatory inputs in the model.

After our systematic literature analysis, the glucose repressor/derepressor system appeared to be organized in a globally feed-forward structure: the number of feed-forward interactions from the signaling component toward the metabolic component of the network largely exceeds the number of feedbacks in the opposite direction. Even though the number of feedback loops is high, they are mostly local or non-active in the diauxic shift context, at least in our model. Physiologically, the system functions mainly in a fermentation configuration when the extracellular glucose level is high and shifts to a mainly respiration configuration only when the extracellular glucose becomes exhausted. If glucose is then added again, the system regains its initial fermentation configuration. When limited only to aspects of glucose/ethanol metabolism (i.e. taking no account of the global energetic metabolism), this may be seen as a "two-state behavior" in which both states are stable and are driven by the intensity of the same external stimulus, in this case glucose, with the response being adaptive rather than homeostatic. Such behavior can occur without strong involvement of feedbacks loops (although these may help fine-tune the mechanism). Thus, stopping the signal at one place would affect the global downstream propagation of the signal, which may have helped with the good behavior of our model. It would be interesting to measure the activity of the glucose sensor system in culture conditions with both high glucose and high ethanol levels. If in this case the system were configured like that with a high glucose level alone, this would strongly confirm its globally feed-forward functioning.

Diauxic shift also occurs with relatively simple dynamics (DeRisi et al. [2] described five different gene clusters) with no significant time delays and different timescales (stiffness). This makes the experimental behavior of the system compatible with the "up–down" dynamics of glucose, with a globally feed-forward signal propagation and a linear superposition of regulatory inputs in the model.

Our results are also consistent with the well-known capacity of simple signal transduction models to detect highly sensitive network elements. These are generally found near the input, and are highly connected or highly modulated in the training data.

Analysis of the TUP1 knockout/HAP4 overexpression experimental data shows that the major effects on gene expression are located close to and downstream from the deleted/overexpressed genes. The changes in expression level further from the deleted/overexpressed genes are smaller (quantitatively) and more difficult to detect, taking into account the low accuracy of the data. This suggests, as we have assumed, that the other biological functions of Tup1p and Hap4p would not greatly affect the glucose repression/derepression, and that our network represents a relatively independent intracellular signaling module. This is also compatible with a global feed-forward functioning of the glucose repressor/derepressor system. Our model reproduced these local changes well, and diminished in accuracy away from the deleted gene. The gap in the distance function between TUP1 (Figure 12.3) and the remaining genes is mostly due to this local detection. The knockout/overexpression experiments were from microarray techniques that were different to those yielding the data used for calculating the model parameters, which may also have adversely affected the quantitative coherence of their values. This is why we were unable to assess the performance of our modeling strategy for more global changes. Additional work on different signal transduction systems is needed to address this.

For the classification of the fitness of the KO mutants, we observed weak transitions from the shift trajectory to the initial state, being at most about ≈50% of the distance between the shift and initial state. This may be explained by network redundancy and weight homogenization due to the penalizing term in the objective function (see Material and Methods). Indeed, the average connectivity of the network and the simplicity of the training trajectory enable us to estimate the training data as insufficient for training. Penalizing the error function by the sum of squares of weights produced the unique, but highly homogenized parameter set. However, as the redundant mechanisms often act in the same direction, this artificial homogenization caused a partial repartition of the signal propagating globally in the same direction. Thus, it did not particularly affect the efficiency of the model.

In the classification of the knockout mutants into two classes depending on their fitness to culture in ethanol, we expected that knockout mutants unable to undergo the fermentation/respiration shift correctly would have growth deficiencies on ethanol. Thus, we assessed the impairment in the shift from the "global" distance to the shift trajectory during glucose exhaustion. However, the opposite may not be true: all mutants that make the system move away from the shift trajectory may not have growth deficiencies on ethanol. It would be expected that different mutants could increase this "numerical" distance by triggering transitions in different directions (steady states), some of which would not correspond to impairments in growth fitness on ethanol. For example, reinforcing the shift by increasing the up-regulation of enzymes required for growth in ethanol would not necessarily lead to growth deficiencies on ethanol but would still make the system move away from the shift trajectory. However, we still expected the mutants associated with a reduced fitness to ethanol growth to be over-represented in the corresponding class, which was indeed the case.

Overall, we show that, starting with relatively simple hypotheses that are imposed by the availability of data, the creation of models of mid-scale biological networks (100 genes/proteins) is possible by adaptive methods, provided that we accept relatively significant theoretical imperfections associated with such models. Of course, these imperfections may lead to errors in prediction. In this specific example, the classifications were imperfect, due largely to the generic formalism used and which has no formal theoretical validity. Such large-scale models can be improved by the progressive introduction of more faithful representations of biological reality, and this is possible as more accurate and extensive knowledge becomes available due to more accurate measurement techniques, completion of networks, and more precise representations of the phenomena studied including, for example, permissive or cooperative effects. However, if we consider such models as decision support tools, they can already identify effects that may lead to the biological network adopting a state of interest, with selectivity and specificity here both near 80%.

Moreover, we assessed the effect of gene inhibitions on a phenotypic behavior using as sole knowledge the description of the dynamics of the onset of the phenotype, without introducing additional biological understanding. Such situations are frequent in physiopathology research, in which it is technologically feasible to describe the onset of the pathological state at the molecular level, while the precise causal and etiological understanding of the disease is often very limited. In such cases, it is crucial to generate relevant therapeutic hypotheses of how to lead the biological system (or, technically, the network) adopting a non-pathological state without needing additional knowledge. We thus believe this type of approach has potential applications in the exploration of all diseases in which experimentation is difficult and limited, for example immunological, neurological and psychiatric diseases and, to a certain extent, cancers.

12.5 MATERIAL AND METHODS

12.5.1 Construction of the Glucose Repressor/Derepressor System

We aimed to design a model that would be as exhaustive as possible with respect to current knowledge of the glucose repression/derepression system, while remaining focused on this mechanism. We used the following general principles: (1) The molecular interaction network is represented as a directed graph. Thus, data specifying the orientation of the interaction had to exist for each interaction (for example, this kinase phosphorylates this protein, this transcription factor activates the expression of this gene, etc.). (2) We included only those interactions that were functionally validated in the literature to exclude any potential false information (noise), despite the possibility that some relevant biological mechanisms may be missing in the model due to being currently insufficiently described. These two principles mean that data from high-throughput screenings are excluded if they have not been confirmed by other methods. (3) All signal transductions represented had to occur downstream from glucose as we wanted to model the response of the system to the glucose concentration in the culture medium.

We defined the limits of the modeled network by defining an input layer (limited to glucose in this study) and a layer that is the physiological effector function of interest: the glucose metabolism in this study. This metabolic component of the network describes the use of glucose with a focus on energy metabolism. This includes the metabolism of glucose to ethanol (fermentation), the use of glucose and ethanol in respiration (TCA and glyoxylate cycles), and the storage of glucose in the form of trehalose. Glycogen synthesis through Gsy1p was represented as an output (to provide a reference during the simulations). The resulting metabolic network was very similar to that presented by DeRisi et al. [2]. We did not include the other glucose metabolism pathways (pentose phosphate pathway, etc.) because they are not yet described as playing a major role in the glucose repression/derepression phenomenon, at least in the modeled culture conditions. Once we had defined this enzymatic metabolic pathway, we systematically searched for all regulatory mechanisms within it—enzyme regulations through intermediary metabolites—and linking it to glucose–glucose sensor system and signal transduction. The choice of glucose as the only input in the network and of the metabolic pathway as the physiological effector layer unambiguously defined the limits of the modeled network. This provided a biologically sound network that had a coherent structure (continuous signaling forward and feedback loops between ligand-receptor/protein signaling/transcriptions/effector functions).

The glucose repression/derepression and metabolism system can be summarized as follows: the extracellular glucose is sensed by the membrane receptors Snf3p, Rgt2p, and Gpr1p [3]. Intracellular glucose is sensed by Hxk2p and some proteins upstream from Snf1p [3]. The osmotic sensing system involving MAP kinase cascades (Ssk1/Ssk2/Pbs2, Sho1/Stet11) converging to Hog1p [5] is an additional mechanism that is described in the context of osmotic sensing, but is putatively involved in glucose sensing due to the glucose osmotic power. These various sensing systems then coordinately transduce intracellular signals (glucose repression/derepression system): Rgt2p and Snf3p indirectly regulate the Rgt1p transcription factor [12–14], and Gpr1p activates the synthesis of cAMP, which regulates the activity of PKA [3]. PKA then regulates the transcription factors Msn2p [15], Msn4p [16], and Sko1p [17]. Snf1p regulates various transcription factors such as the Mig1p repressor [3,18], Msn2p [15], Msn4p [15], Adr1p [19,20], and Cat8p [21]. Hxk2p regulates Snf1p activity interactively with type 1 serine/threonine protein phosphatase (Reg1p-Glc7p) [22] and indirectly regulates the PKA system [3,23]. Hog1p also regulates various transcription factors [24], such as Msn2p, Msn4p, and Sko1p among others. Depending on the glucose signal, the transcription factors then regulate the expression of many genes, including genes involved in this signal transduction system, and finally the enzymes involved in glucose metabolism and more generally in carbon use [2]. There are also additional mechanisms involved, such as the regulation of enzyme activities by glycolysis metabolites [3].

Though a "classical" biochemical representation of the metabolic component of the network would interconnect the substrates and products of the enzymes and not the enzymes themselves, we chose to directly interconnect the enzymes when they are engaged in consecutive chemical reactions. Each connection was validated based on the corresponding chemical reaction from *Saccharomyces* Genome Database [11]. The arguments for this representation are detailed in the next section.

12.5.2 Experimental Data Used

12.5.2.1 Data Used for Model Parameter Calculation. We assumed that, for the diauxic shift, most measurements of gene expression were correlated to protein activity. This assumption is supported by the global coherence and temporal homogeneity of the gene expression behavior of the network during the shift (80% of the genes). Moreover, the expression of most genes was either up- or down-regulated during the shift consistent with what would be expected from their corresponding protein activity based on their function. Measurements of the expression of the network's genes during the diauxic shift were taken from DeRisi et al. [2]. For three network molecules which were not regulated at the mRNA level (TPK1, SNF1, and MIG1), we inferred the corresponding protein activity during the shift from scientific articles that have specifically studied these molecules (TPK1: Portela and Moreno [25]; SNF1: Wilson et al. [26]; MIG1: Kaniak et al. [18]; De Vit et al. [27]; Treitel et al. [28]). For ADH2, the data from DeRisi et al. [2] were inconsistent with the literature: the ADH2 gene expression should be strongly up-regulated during the shift [29], whereas it was down-regulated in the DeRisi et al. [2] dataset. This is likely due to a lack of probe specificity, as the authors used a cDNA microarray. The possible hybridization of other ADH mRNAs to the ADH2 probe (in particular ADH1 and/or ADH4) would explain this result because ADH1 and ADH4 are down-regulated during the shift. Thus, we corrected the data for ADH2 using the results of Walther and Schuller [29] and assumed that the ADH2 gene would be derepressed in parallel with the diminution of the glucose concentration in the culture medium. The final dataset used for calculating the parameters is given in the additional data available at www.heliosbiosciences.com.

12.5.2.2 Data Used for Simulation Testing. Classification of molecules according to their qualitative physiological effect. The study of Steinmetz et al. [30] has used a high-throughput method and we identified a few discrepancies between their results and current knowledge about some genes of the network, in particular from the *Saccharomyces* Genome Database [11]. Therefore, we systematically reanalyzed the results of Steinmetz et al. [30] and corrected certain results. This final classification of gene knockout mutants according to their ability to grow on fermentable/non-fermentable sugars is given in the additional data (www.heliosbiosciences.com).

Classification of molecules according to quantitative knockout/overexpression data. We took the gene expression data from TUP1 gene knockout from Hughes et al. [31] and the gene expression data from HAP4 gene overexpression from Lascaris et al. [32]. Both experiments were performed in glucose-rich medium.

12.5.3 Mathematical Model

We modeled the dynamics of the network by a system of differential equations. The underlying structure is a network of transcriptional regulation that is formalized as a directed graph. There are 97 molecules in our network (i.e. 97 state variables) for 362 interactions. Thus, there are on average 3.73 connections per gene. The number of inputs/outputs for every single gene ranges from 0 to 10, but more than 70% of genes have four or less inputs/outputs. The network is not large enough to verify some widely believed hypotheses on the topological structure of gene regulatory networks (scale-free structure), but it is clearly heterogeneous. Almost all the elements of the network are involved in feedback loops, and the length of the shortest circuit is 16 or less, with 60% of genes involved in a loop having a length of four or less. The sign of the effect (activation/inhibition) was specified for 83% of interactions.

Let $x(t) = (x_1(t), x_2(t), \ldots, x_N(t))$ be the state vector of the expression level of all genes in the network (i.e. $N = 97$) and let $x_0(t)$ be the only input in the network. The dynamics of the network are driven by differential equations

$$\tau_i \frac{dx_i}{dt} = -x_i + \sigma\left(\sum_{j=0}^{n} \omega_{j,i} x_j + b_i\right), \tag{12.1}$$

where $\sigma(x) = (1 + \exp(-x))^{-1}$ is a standard sigmoid-shape function, and T_i, $\omega_{j,i}$, and b_i are time constants, weights, and systematic bias, respectively. This simple signal transduction model is based on a linear degradation model (with the half-life proportional to time constant), a linear activation function and a production rate that can be saturated. Although the sigmoid function σ means the model is not linear, the linear superposition of regulatory inputs in its argument is a strong assumption that excludes more complex forms of joint actions such as cooperative or permissive actions. These can be taken into account by a nonlinear (e.g. quadratic) activation function.

We determined the parameters of the model using a trajectory learning procedure known as backpropagation through time (BPTT). This is a gradient descent-based method, which extends the standard backpropagation rule to the learning of time-dependent data. The error function is the standard mean square error

$$E = \sqrt{\sum_j \sum_{i=1}^{N} (x_i(l_j) - d_i(l_j))^2}, \tag{12.2}$$

where the first sum runs over all experimental times and $d_i(t)$ are training trajectories (experimental time-course data). We use also the relative error:

$$E_{rel} = \sqrt{\frac{\sum_j \sum_{i=1}^{N} (x_i(t_j) - d_i(t_j))^2}{\sum_j \sum_{i=1}^{N} (d_i(t_j))^2}}. \tag{12.3}$$

We avoided a large distribution of weights by adding a sum of squared weights to this function (penalizing term), to give the objective function:

$$E_{obj} = \sum_j \sum_{i=1}^{N} (x_i(t_j) - d_i(t_j))^2 + \frac{1}{2} \sum_{i,j} \omega_{i,j}^2. \tag{12.4}$$

We forced the gradient descent to respect the sign (activation/repression) of every interaction whenever specified. The training data are the time-course data of gene expressions during the diauxic shift. The time series consist of seven experimental time-points over 690 minutes (0, 150, 240, 360, 450, 570, and 690 minutes). We obtained the kinetics by linear interpolation, which were then normalized to give a kinetics range of 0 to 0.75. The central assumption we made about the data is that the expression of genes during the diauxic shift is mostly controlled by the glucose concentration exhaustion and no other exogenous variables (for example, ethanol concentration). This assumption was strengthened by the globally homogeneous and synchronous structure of the shift response data with the glucose concentration profile. Thus, we have only one input variable, the glucose concentration evolution as measured by DeRisi et al. [2] (normalized to (0, 0.75)), which stimulates or represses the expression of nine genes according to our physiological model of glucose signaling and metabolism (i.e. in equation (12.1) $\omega_{0,i} = 0$ for all but nine i's). This assumption also means that the initial state is stationary (steady state), corresponding to the stability of expression data observed in glucose-rich medium by DeRisi et al. [2].

All the mathematical methods used in this study were implemented in the software developed by our team.

ACKNOWLEDGMENTS

Helios Biosciences would like to thank the French Ministry of Research and the ANVAR. This work was made possible by the CETI 2002. The authors would like to thank Sylvie Dumas and Joan Brien for their contribution to this work.

REFERENCES

1. Uemura H, Koshio M, Inoue Y, Lopez MC, Baker HV. The role of Gcr1 p in the transcriptional activation of glycolytic genes in yeast *Saccharomyces cerevisiae*. Genetics 1997;147(2):521–532.
2. De Risi JL, Iyer VR, Brown PO. Exploring the metabolic and genetic control of gene expression on a genomic scale. Science 1997;278(5338):680–686.
3. Rolland F, Winderickx J, Thevelein JM. Glucose-sensing and -signalling mechanisms in yeast. FEMS Yeast Res 2002;2(2):183–201.
4. Tyson JJ, Novak B. Regulation of the eukaryotic cell cycle: molecular antagonism, hysteresis, and irreversible transitions. J Theor Biol. 2001;210(2):249–263.
5. Klipp E, Nordlander B, Kruger R, Gennemark P, Hohmann S. Integrative model of the response of yeast to osmotic shock. Nat Biotechnol 2005;23(8):975–982.
6. Sasagawa S, Ozaki Y, Fujita K, Kuroda S. Prediction and validation of the distinct dynamics of transient and sustained ERK activation. Nat Cell Biol 2005;7(4): 365–373.
7. Reijenga KA, Westerhoff HV, Kholodenko BN, Snoep JL. Control analysis for autonomously oscillating biochemical networks. Biophys J 2002;82(1 Pt 1):99–108.
8. Ruoff P, Christensen MK, Wolf J, Heinrich R. Temperature dependency and temperature compensation in a model of yeast glycolytic oscillations. Biophys Chem 2003;106(2):179–192.
9. A Database of Annotated Published Models. Available at http://www.ebi.ac.uk/biomodels/
10. The Gene Ontology Database. Available at http://www.godatabase.org/cgi-bin/amigo/go.cgi
11. *Saccharomyces* Genome Database. Available at http://www.yeastgenome.org/
12. Lafuente MJ, Gancedo C, Jauniaux JC, Gancedo JM. Mth1 receives the signal given by the glucose sensors Snf3 and Rgt2 in *Saccharomyces cerevisiae*. Mol Microbiol 2000;35(1):161–172.
13. Lakshmanan J, Mosley AL, Ozcan S. Repression of transcription by Rgt1 in the absence of glucose requires Std1 and Mth1. Curr Genet 2003;44(1):19–25.
14. Polish JA, Kim JH, Johnston M. How the Rgt1 transcription factor of *Saccharomyces cerevisiae* is regulated by glucose. Genetics 2005;169(2):583–594.
15. De Wever V, Reiter W, Ballarini A, Ammerer G, Brocard C. A dual role for PP1 in shaping the Msn2-dependent transcriptional response to glucose starvation. EMBO J 2005;24(23):4115–4123.
16. Garreau H, Hasan RN, Renault G, Estruch F, Boy-Marcotte E, Jacquet M. Hyperphosphorylation of Msn2p and Msn4p in response to heat shock and the diauxic shift is inhibited by cAMP in *Saccharomyces cerevisiae*. Microbiology 2000; 146(Pt 9):2113–2120.

17. Pascual-Ahuir A, Posas F, Serrano R, Proft M. Multiple levels of control regulate the yeast cAMP-response element-binding protein repressor Sko1p in response to stress. J Biol Chem 2001;276(40):37373–37378.
18. Kaniak A, Xue Z, Macool D, Kim JH, Johnston M. Regulatory network connecting two glucose signal transduction pathways in *Saccharomyces cerevisiae*. Eukaryot Cell 2004;3(1):221–231.
19. Young ET, Kacherovsky N, Van Riper K. Snf1 protein kinase regulates Adr1 binding to chromatin but not transcription activation. J Biol Chem 2002;277(41): 38095–38103.
20. Young ET, Dombek KM, Tachibana C, Ideker T. Multiple pathways are co-regulated by the protein kinase Snf1 and the transcription factors Adr1 and Cat8. J Biol Chem 2003;278(28):26146–26158.
21. Randez-Gil F, Bojunga N, Proft M, Entian KD. Glucose derepression of gluconeoyenic enzymes in *Saccharomyces cerevisiae* correlates with phosphorylation of the gene activator Cat8p. Mol Cell Biol 1997;17(5):2502–2510.
22. Sanz P, Alms GR, Haystead TA, Carlson M. Regulatory interactions between the Reg1-Glc7 protein phosphatase and the Snf1 protein kinase. Mol Cell Biol 2000;20(4):1321–1328.
23. Rolland F, De Winde JH, Lemaire K, Boles E, Thevelein JM, Winderickx J. Glucose-induced cAMP signalling in yeast requires both a G-protein coupled receptor system for extracellular glucose detection and a separable hexose kinase-dependent sensing process. Mol Microbiol 2000;38(2):348–358.
24. Edmunds JW, Mahadevan LC. MAP kinases as structural adaptors and enzymatic activators in transcription complexes. J Cell Sci 2004;117(Pt 17):3715–3723.
25. Portela P, Moreno S. Glucose-dependent activation of protein kinase A activity in *Saccharomyces cerevisiae* and phosphorylation of its TPK1 catalytic subunit. Cell Signal 2005 Oct 13; [epub ahead of print].
26. Wilson WA, Hawley SA, Hardie DG. Glucose repression/derepression in budding yeast: SNF1 protein kinase is activated by phosphorylation under derepressing conditions, and this correlates with a high AMP:ATP ratio. Curr Biol 1996;6(11): 1426–1434.
27. De Vit MJ, Waddle JA, Johnston M. Regulated nuclear translocation of the Mig1 glucose repressor. Mol Biol Cell 1997;8(8):1603–1618.
28. Treitel MA, Kuchin S, Carlson M. Snf1 protein kinase regulates phosphorylation of the Mig1 repressor in *Saccharomyces cerevisiae*. Mol Cell Biol 1998;18(11): 6273–6280.
29. Walther K, Schuller HJ. Adr1 and Cat8 synergistically activate the glucose-regulated alcohol dehydrogenase gene ADH2 of the yeast *Saccharomyces cerevisiae*. Microbiology 2001;147(Pt 8):2037–2044.
30. Steinmetz LM, Scharfe C, Deutschbauer AM, Mokranjac D, Herman ZS, Jones T, Chu AM, Giaever G, Prokisch H, Oefner PJ, Davis RW. Systematic screen for human disease genes in yeast. Nat Genet 2002;31(4):400–404.
31. Hughes TR, Marton MJ, Jones AR, Roberts CJ, Stoughton R, Armour CD, Bennett HA, Coffey E, Dai H, He YD, Kidd MJ, King AM, Meyer MR, Slade D, Lum PY, Stepaniants SB, Shoemaker DD, Gachotte D, Chakraburtty K, Simon J,

Bard M, Friend SH. Functional discovery via a compendium of expression profiles. Cell 2000;102(1):109–126.

32. Lascaris R, Bussemaker HJ, Boorsma A, Piper M, Van Der Spek H, Grivell L, Blom J. Hap4p overexpression in glucose-grown *Saccharomyces cerevisiae* induces cells to enter a novel metabolic state. Genome Biol 2003;4(1):R3.
33. de Jong H. Modeling and simulation of genetic regulatory systems: a literature review J Comput Biol 2002;9(1):69–105.
34. Smolen P, Baxter DA, Byrne JH. Modeling transcriptional control in gene networks: methods, recent results, and future directions. Bull Math Biol 2000;62;247–292.
35. Zhang Z, Varanasi U, Trumbly RJ. Functional dissection of the global repressor Tup1 in yeast: dominant role of the C-terminal repression domain. Genetics 2002;161(3):957–969.
36. Treitel MA, Carlson M. Repression by SSN6-TUP1 is directed by MIG1, a repressor/activator protein. Proc Natl Acad Sci U S A 1995;92(8):3132–3136.
37. Saeed AI, Sharov V, White J, Li J, Liang W, Bhagabati N, Braisted J, Klapa M, Currier T, Thiagarajan M, Sturn A, Snuffin M, Rezantsev A, Popov D, Ryltsov A, Kostukovich E, Borisovsky I, Liu Z, Vinsavich A, Trush V, Quack-Enbush J. TM4: a .free, open-source system for microarray data management and analysis. Biotechniques 2003;34(2):374–378.

13

PRESENT AND FUTURE OF PATHWAY ANALYSIS IN DRUG DISCOVERY

ANTON YURYEV

Table of Contents

13.1 RATIONAL DRUG DESIGN AND RATIONAL DRUG THERAPY

In this chapter, I define as "rational drug design" the drug development process guided by choices with clearly understood consequences. This idea contradicts, in principle, with the stochastic nature of biological processes making the drug response to be poorly predictable process. There are two major components contributing to the stochasticity of the drug response: (1) the intrinsic randomness of molecular processes that is eventually caused by the quantum nature of molecular interactions and (2) the apparent historical

Pathway Analysis for Drug Discovery, Edited by Anton Yuryev

randomness that arises from the lack of knowledge about the initial state of the organism and about the complete set of environmental conditions that can influence clinical outcome during the drug treatment. The initial organism state depends on such components as genetic background, pre- and postnatal development, and disease anamnesis. These types of information are never completely known with the accuracy necessary to predict the drug response in a given patient.

The interpretation of the drug response on the molecular level provides additional difficulty for rational drug design. Human organism is a complex system which physiological and psychological behavior cannot be completely traced to the properties of its individual microscopic components. The clinical outcome is a global property of the organism separated by at least two levels of complexity from any drug intervention rationalized on the molecular level. First, the molecular drug response is manifested on the cellular level by changing the behavior of individual cells and second, it changes the behavior of cell communities, manifesting the drug response on the physiological level of tissue function and cell interaction. Stochastic nature of the drug response, insufficient knowledge about individual patient organism, and difficulty in the interpretation of drug action make it almost impossible to calculate the clinical outcome from the basic principals as a simple sum of individual molecular responses. Hence, the drug efficacy and side effects cannot be 100% predictable.

Another consequence of complexity is the abundance of different molecular mechanisms that manifest as one apparent disease on the physiological level [1]. It is almost meaningless to discuss finding the drugs to cure cancer or diabetes these days. There are at least 230 types of cancer for different tissues and at least two types of diabetes. The exact molecular mechanisms of a disease are likely to be unique in every patient. The disease classification is usually done at physiological level and therefore separated by at least two levels of complexity from the drug intervention rationalized at molecular level. Thus, not only the drug response can't be entirely predicted at the physiological level, but the complexity of the system can disguise and confuse several different diseases under one disease phenotype.

The solution to the apparent contradiction between our desire to rationalize drug design and the stochastic nature of a biological system is in having the rational drug therapy—combining the drug treatment with constant monitoring of a patient state. The regular patient examination and diagnostics can provide a feedback about success of current treatment and suggest changes to the therapy if necessary. Such regular monitoring of the healing process or disease progression, coupled with the understanding of the molecular mechanism of a disease for each patient, can provide the foundation for personalized medicine. The "personalized medicine" is a fashionable term that assumes that every patient requires a unique treatment that fits best the molecular profile of his/her disease and his/her lifestyle habits. The unique treatment for every patient can be developed only if there is a pool of drugs that potentially fit many different patients. To enable the rational design of personalized therapy, each drug from the pool must have a clearly defined and well-understood mechanism of action. The choice of a drug or a combination of drugs from the pool should be rationalized based on the predicted drug action. The prediction can be calculated using known drug targets and the currently observed patient state on the molecular level. Due to the incomplete knowledge and the randomness of the drug response, the drug action prediction cannot be 100% accurate. It will probably have the properties of the weather forecast—the further in time the prediction is attempted, the less reliable it becomes.

The process of applying drugs based on the current state of a patient already exists for treatments of cancer and HIV infection [2,3]. The current and future efforts of drug discovery community will be aimed on further refinement of the rational therapy. The current refinement efforts include expanding a pool of available drugs, better characterization of the mechanism of action for every drug, and improving the diagnostics techniques through the development of molecular biomarkers. In this chapter, I will attempt to describe the role of pathway analysis in this process and review current examples of its application in drug discovery.

13.2 PATHWAY ANALYSIS IN MODERN DRUG DISCOVERY PIPELINE

Current drug discovery pipeline can be described in five major steps:

1. Understanding disease mechanism (diagnosis)
2. Selecting drug targets (therapy design)
3. Structural drug design: finding active molecules and lead compounds (drug selection)
4. Evaluation of drug efficacy and toxicity *in vivo* (drug action monitoring)
5. Design of combinatorial therapy (personalized therapy)

It is easy to align these steps with the stages in personalized rational therapy described in the previous section. To illustrate this point, I put the corresponding phases of rational therapy in brackets next to the steps in drug development pipeline.

Currently, the phrase “rational drug design” is used to describe the process of designing the chemical structures with biological activities based on the knowledge of target protein structure. I have called this part “structural drug design” to differentiate it from other steps of drug development-based rational choice of drug target(s). Structural drug design should eventually yield the inventory of individual drugs that can be used to design personalized therapy. Due to the druggability constrain in protein structure, not all proteins in human genome can have small molecule inhibitors or agonist. With the exception of secreted and extracellular proteins, all intracellular protein targets must have a convenient pocket or cavity in their structure suitable for binding a small molecule that can penetrate a cell membrane. Secreted proteins and some receptors can be targeted with humanized antibodies [4–8] and therefore do not need having the druggable structure. If the analysis of disease mechanism points to the optimal target that happened to be but non-druggable, the pathway analysis can help to overcome the druggability constrain by finding the alternative targets that are druggable and upstream or downstream of the optimal target. Similarly, pathway analysis can help when the desired biological effect must be the activation of a protein target, yet only inhibitors are available for this protein. In this case, target activation can be achieved by down-regulating of the negative regulators upstream of a target (Figure 13.1).

Apart from helping to overcome the druggability constrains in structural drug design, the real-life examples of pathway analysis application are avail-

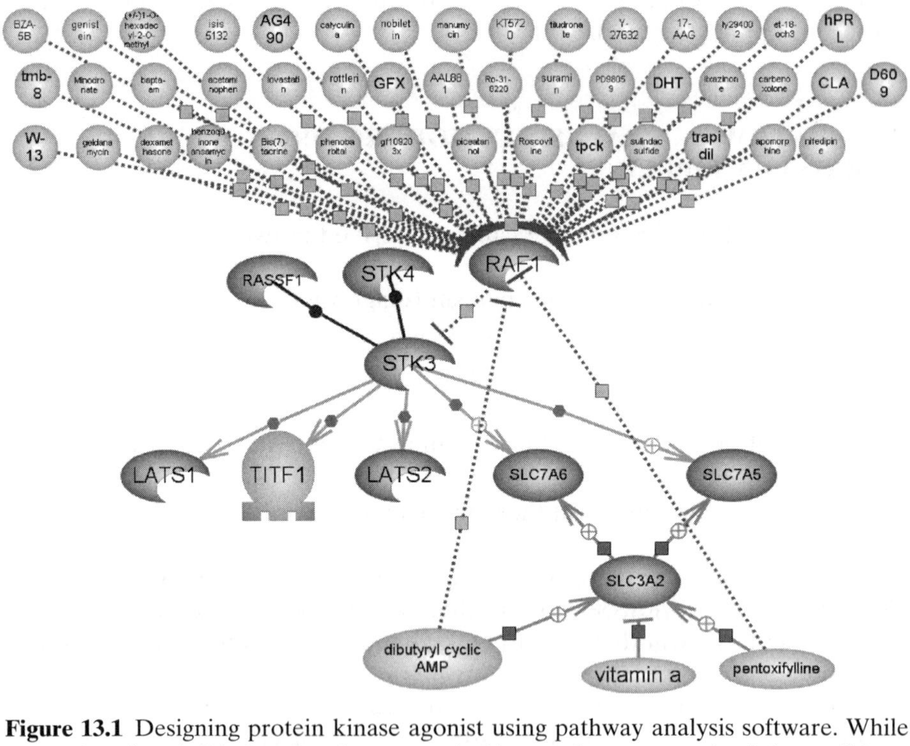

Figure 13.1 Designing protein kinase agonist using pathway analysis software. While drugs that directly bind and activate protein kinases do not yet exist, it is possible to design the protein kinase agonists using pathway analysis tools. Picture shows the activation of STK3 kinase using RAF1 kinases inhibitors. STK3 should be activated because RAF1 negatively regulates its activity and therefore RAF1 inhibitors will reduce this inhibition. While there are plenty of RAF1 inhibitors, two of them—dibutiryl-cAMP and pentoxifylline—can also activate STK3 targets via activation of the SLC3A2 expression. Thus, they not only can reduce STK3 inhibition, but also mimic the effects of STK3 activation by activating some of its targets.

able at every step of modern drug discovery pipeline. I review some of them below.

13.2.1 Pathway Analysis for Understanding Disease Mechanism (Diagnosis)

According to current view the disease exists due to the development of the robust molecular network that enables malignant information flow causing the deregulation of normal cell signaling. Such disease network must be robust structurally; therefore, it must have underlying robust physical interaction network. It also must be self-sustained and resist interventions. Therefore, it must be robust as a regulatory network. While the robust molecular networks enabling the normal function of the organism are probably unique for every individual human, a subset of these networks mediates an abnormal function causing a disease. Many molecular mechanisms enabling network robustness are yet to be determined [9]. However, it seems universally accepted that the robust network must have scale-free topology. Every known biological network has this topology, which has been shown to be the most robust among all other network topologies [10]. The significant number of recent publications describes the development of disease networks and approaches how to build them [1,11–13]. They provide conceptual foundation for new methods of disease diagnosis based on patient molecular biomarker data, transforming the medicine for the 21st century. For example, Loscalzo et al. [1] have proposed dividing disease networks into subnetworks based on the role proteins play in disease: primary genetic subnetwork of genes causing the disease, secondary genetic subnetwork of genes that can modify function of genes in primary network, stress management subnetwork containing genes responsive to stress and influencing the genes in genetic network, and environmental subnetwork of genes mediating the interaction of genes in genetic networks with the environmental factors. The combination of genes and mutations in the disease network is also unique for every patient. This causes different pathological states of a disease and manifests in different disease phenotypes [1]. The network view of a disease can only explain why disease persistently exists in the body but cannot explain what makes this network malignant. There is yet to be an effort to transform the disease networks into a set aberrant information flows or pathways that actually explain how the information flow in disease networks differs from the information flow in the normal state.

13.2.2 Selecting Drug Targets (Therapy Design)

Network biology has demonstrated that the most effective way to disrupt disease network is by targeting hubs—highly connected proteins in the network. Unfortunately, hubs in disease network tend to be the same proteins that are hubs in the normal networks even though a set of connections can be different for one hub in normal and disease networks. Therefore, targeting hubs carries the increased risk of side effects by affecting other normal pathways containing the same hub. The target-related side effects can be divided into two major classes reflecting two levels of complexity, separating molecular drug intervention from the disease phenotype: (1) side effects due to the effect on normal pathways present in the targeted tissue and 2) side effects caused by effecting pathways in other off-target tissues. To facilitate drug target selection and prioritization, any protein target can be described by the efficacy and toxicity. Efficacy indicates how much the inhibition (or activation) of the target can disrupt the disease network and redirect malignant information flow. It is proportional to the connectivity of a target in the disease network and how upstream the protein is in disease regulatory network, i.e. how many proteins in disease network will be affected by targeting this protein. The toxicity of a target is proportional to the number of off-target pathways containing the same protein, as well as its efficacies in every off-target pathway.

Another reason for side effects and toxicity of a drug is the nonselective binding to other proteins besides the intended target. This is often called the "off-target effect" of the drug. Human genome has the biggest number of large paralog families among all sequenced organisms. Because paralogous proteins have similar structure, a drug that targets one paralog is likely to bind to other paralogs as well. Even though the affinity toward these paralogs can be lower than to the intended protein target, the number of off-target paralogs can be high enough to mediate side effects [14–16]. Even though the goal of the structural drug design is making every drug as highly selective as possible, there is likely a limit to the selectivity of any drug. Therefore, side effects due to paralogous binding must be well understood rather than ignored and used for rational optimization of the drug therapy.

In summary, the goal for rational drug target selection can be formulated as finding a druggable protein target that has maximum disruptive impact on the disease network coupled with maximum correction of the malignant information flow, while having the minimum number of side effects caused by off-target activity of a drug. Here, the off-target activity includes the regulation of off-target normal pathways containing intended protein target and any of the off-target proteins.

Calculating efficacy-versus-toxicity ratio for every drug should be the major purpose of pathway analysis in drug discovery. Besides rationally selecting a drug target using pathway analysis, the drug toxicity can be minimized by selective delivery of a drug to the target tissue and/or by choosing the timing of drug treatment relative to the timing of target and off-target pathway activation. The time management can be achieved, for example, by limiting or controlling the environmental factors such as diet, lifestyle, and stress to suppress normal off-target pathways during drug intervention. Another way to control side effects is by compensating them with other drugs selected using pathway analysis. Thus, the clear understanding and rational management of all potential side effects can be done only using pathway analysis tools, together with the comprehensive collection of pathways for human organism.

13.2.3 Evaluation of Drug Efficacy and Toxicity *In Vivo* (Drug Action Monitoring)

Every drug approval regulatory agency, such as Federal Drug Administration (FDA) or European Medicines Agency (EMEA), requires the experimental evidence of drug efficacy and toxicity tests during drug submission. Since the drug response cannot be completely predicted *in silico*, these tests will remain the obligatory step in drug development and approval. They eventually should give rise to the diagnostic tests of drug efficacy for personalized drug treatment. In drug discovery, the experimental confirmation of drug efficacy and absence of toxicity is obtained using *in vivo* experiments in cell lines, animal models, and in clinical trials. A typical drug validation experiment includes treatment of a cell line or animal with a drug or lead compound, measuring the molecular profile of drug response, and showing that drug-responsive genes are associated with the biological processes relevant to the disease. The same tests should show no association of responsive genes with potential side effects. The biological association network used for pathway analysis is also used to obtain drug validation confirmation. There are plenty of examples for this application of pathway analysis [17–20]. Some of these studies also aim to further characterize the drug response, including the prediction of potential side effects in addition to the validation of the drug efficacy.

13.2.4 Design of Combinatorial Therapy (Personalized Therapy)

The idea of mixing up drugs is as old as the medicine itself. Pathway analysis and network biology can provide the 21st century foundation for this concept and enable rational design of drug mixtures using the knowledge stored in pathway analysis database. There are two major reasons why drug mixtures should work better than individual selective drugs. First, the disruption of disease network and modification of malignant information flow should be more effective if several network components are targeted at once. Each disease network must have several hubs to be robust; therefore, several hubs must be targeted to disrupt. All known pathways show the redundancy of the information flow—they have several alternative channels to transmit a signal; therefore, to inhibit the malignant information flow, it may necessary to inhibit several channels at once. Additionally, the proper combination of drugs may cancel out or reduce the side effects caused by individual drug components.

The necessity to rationally manage side effects using pathway collection and network database becomes even more obvious with the realization that essentially, every protein in human genome must participate in several functions in the organism. I showed in Chapter 1 that the comprehensive pathway collection for human organism must contain about 500,000 pathways. Since there are about 35,000 proteins in human genome, every protein must participate on average in 14 pathways. The number of signaling proteins is significantly less than 35,000; therefore, the actual functional multiplicity of any pathway component is much higher. For example, the limited collection of 712 signaling pathways available in ResNet 5.0 database from Ariadne Genomics reveals that SRC kinase can participate in 354 different receptor-transcription factor signaling transmission events. This number must be further multiplied by the number of tissues containing these pathways to get an estimate of potential toxicity associated with SRC inhibition. It is reasonable to anticipate that every protein target has multiple functions either in the same tissue or in other tissues. Therefore, targeting of any protein by a drug will have side effects. The further support for this conclusion comes from the duplication-divergence theory of evolution. The new pathways appeared in evolution by duplication of the ancestor pathway either in its entirety, after whole-genome duplication or partially using single-gene duplication events [21]. Thus, in essence, every pathway is a relative of another pathway sharing a lot of similarities with other pathways.

The previous paragraphs paint a rather grim picture: if we want to effectively disrupt the disease network and malignant information flow, we should target several network hubs. Because hubs participate in many other normal pathways with frequencies higher than average proteins, their potential for having side effects is much higher. This potential becomes even higher when several hubs must be targeted at once. Coupled with general impossibility to design 100% selective drug molecules, this leads us to an inevitable conclusion: side effects are unavoidable for any individual drug no matter what target is chosen and how selective the drug is. This conclusion is bad news for pharmaceutical industry since any regulatory agency requires demonstrating the absence of toxicity and side effects for drug approval. This is also bad news for FDA because it implies that the side effects exist for drugs that were already approved by FDA. The reason why these drugs passed FDA approval was because their toxicity was not detected during the drug testing. The recent disasters with blockbuster drugs Vioxx, Bextra, and Fen-Phen only reinforce this grim picture. This conclusion also necessitates the development of expensive tests for all kinds of toxicities that are not available at present. Because this conclusion affects negatively the public perception of essentially the entire industry, it most likely will be ignored for some time to maintain the status quo.

The first step in psychologically overcoming a fear is to consciously rationalize it rather than suppressing the fear into the unconscious. Pathway analysis offers the way to rationally approach the problem of drug safety and develop drug combinations that are highly specific for target disease and personalized for every patient, while having minimal potential side effects. The idea to use pathway analysis for design of combinatorial therapies is also not new and has been proposed in at least three publications [22–24]. All these publications designed combinatorial therapies with the goal to improve drug efficacy. It has been shown experimentally that the combination of certain drugs could be more potent than the simple sum expected from the action of two individual drugs [25]. This finding is the foundation for the technology developed by CombinatoRx, Inc. [26]. What is missing from the current discussion is the

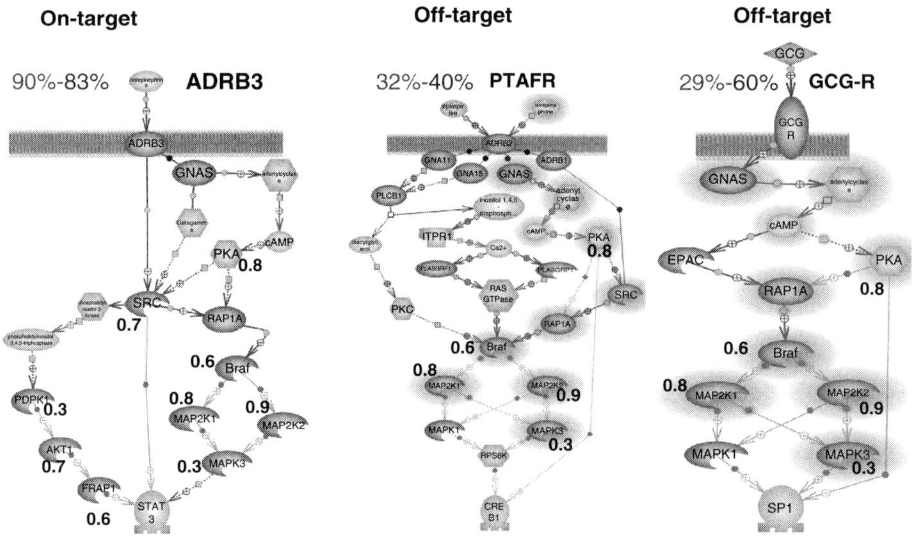

Figure 13.2 Illustration of how the rationally designed drug combination can improve efficacy-versus-toxicity ratio. The numbers next to protein targets symbolize the reduction of the information flux due to the inhibition by individual drug. In the picture, ADRB3 pathway is a target pathway that must be inhibited in disease, PTAFR and GCG-R signaling pathway share 40% and 60% of the protein components with the target ADRB3 pathway and therefore may cause side effects. The protein components common between ADRB3 and off-target pathways have a halo around them. PTFAR or GCG-R pathways are used here only as example of off-target pathways. The protein components from ADRB3 pathway that are not part of PTFAR or GCG-R are components of other pathways not shown here.

It is easy to see that inhibiting multiple targets in ADRB3 pathway allows almost complete inhibition of ADRB3 pathway (90%) while only mildly inhibiting PTAFR and GCG-R pathways. To increase ADRB3 pathway inhibition while keeping off-target pathway inhibition constant, one can add one more inhibitor into the drug mixture that inhibits one more protein component in ADRB3 pathway that does not belong to off-target pathways. To decrease the off-target effects while keeping the on-target efficacy high, one can decrease concentration of drugs interfering with off-target pathways while either adding one more inhibitor into the drug mixture or increasing the concentration of existing inhibitors that bind protein targets not belonging to off-target pathways.

The first number next to pathway name indicates the number of protein components inhibited by the hypothetical drug mixture; the second number shows the sum inhibition of the information flux in each pathway by the same drug mixture.

ability of pathway analysis to assist in designing the effective personalized drug mixtures, while rationally minimizing the side effects by calculating the efficacy-versus-toxicity ratio for each drug combination.

The strategy for side-effect minimization using combinatorial therapy is illustrated in Figure 13.2. The uniqueness of any pathway in the human organism is determined by unique combination of its components and not by the uniqueness of individual components in each pathway. Therefore, the selectivity and efficacy of a drug mixture can be improved by targeting more and more components in a target pathway. The side effects can be minimized by using individual drug components in suboptimal concentrations. This will dilute the side-efficacies of the drug mixture toward off-target pathways, while keeping the on-target efficacy high. In order to improve efficacy-versus-toxicity ratio of the entire mixture, one more drug targeting additional component of the on-target pathway must be added, while the concentration of other components can be reduced to minimize side effects.

To finish this chapter on a positive tone, I remind the readers that many drugs do work without major side effects, while side effects of other drugs can be managed by compensatory treatments [27,28]. These successful drugs so far appear as lucky charms found by the global drug discovery effort of humankind. To give you an idea of what "lucky" means in the last sentence, I can mention that the current FDA approval rate is one out of 5,000 drugs [29]. The challenge that can be addressed by pathway analysis is to increase the efficiency of the drug discovery pipeline and FDA approval rate. This can be done by terminating the discovery effort of lead compounds earlier, by early prediction of side effects, by optimizing lead compound selection using efficacy-versus-toxicity ratio, and from the realization that the real way to minimize side effects is by having the abundance of selective drugs allowing the design of combinatorial therapies to suppress and dilute the side effects of individual drugs.

REFERENCES

1. Loscalzo J, Kohane I, Barabasi AL. Human disease classification in the postgenomic era: a complex systems approach to human pathobiology. Mol Syst Biol 2007;3:124. Epub 2007 Jul 10.
2. Daskivich TJ, Regan MM, Oh WK. Distinct prognostic role of prostate-specific antigen doubling time and velocity at emergence of androgen independence in patients treated with chemotherapy. Urology 2007;70(3):527–531.
3. Antiretroviral treatment. HIV infection in adults: better-defined first-line treatment. Prescrire Int 2004;13(72):144–150.
4. Weisman LE. Antibody for the prevention of neonatal noscocomial staphylococcal infection: a review of the literature. Arch Pediatr 2007;14(Suppl 1):S31–34.
5. Ohsugi Y. Recent advances in immunopathophysiology of interleukin-6: an innovative therapeutic drug, tocilizumab (recombinant humanized anti-human interleukin-6 receptor antibody), unveils the mysterious etiology of immune-mediated inflammatory diseases. Biol Pharm Bull 2007 Nov; 30(11):2001–2006.
6. Rao AV, Schmader K. Monoclonal antibodies as targeted therapy in hematologic malignancies in older adults. Am J Geriatr Pharmacother 2007;5(3):247–262.
7. Lien S, Lowman HB. Therapeutic anti-VEGF antibodies. Handbook Exp Pharmacol 2008;181:131–150.

8. Engelhardt B, Kappos L. Natalizumab: targeting alpha(4)-integrins in multiple sclerosis. Neurodegener Dis 2008;5(1):16–22.
9. Kitano H. Towards a theory of biological robustness. Mol Syst Biol 2007;3:137. Epub 2007 Sep 18.
10. Barabási A-L, Oltvai Z. Network biology: understanding the cell's functional organization. Nat Rev 2004;5:101.
11. Cui Q, Ma Y, Jaramillo M, Bari H, Awan A, Yang S, Zhang S, Liu L, Lu M, O'Connor-McCourt M, Purisima EO, Wang E. A map of human cancer signaling. Mol Syst Biol 2007;3:152. Epub 2007 Dec 18.
12. Abdollahi A, Schwager C, Kleeff J, Esposito I, Domhan S, Peschke P, Hauser K, Hahnfeldt P, Hlatky L, Debus J, Peters JM, Friess H, Folkman J, Huber PE. Transcriptional network governing the angiogenic switch in human pancreatic cancer. Proc Natl Acad Sci U S A 2007;104(31):12890–12895.
13. Hettne KM, de Mos M, De Bruijn AG, Weeber M, Boyer S, van Mulligen EM, Cases M, Mestres J, Van der Lei J. Applied information retrieval and multidisciplinary research: new mechanistic hypotheses in complex regional pain syndrome. J Biomed Discov Collab 2007;2:2.
14. Fernández A. Incomplete protein packing as a selectivity filter in drug design. Structure 2005;13(12):1829–1836.
15. Fernandez A, Crespo A. Wrapping technology and the enhancement of specificity in cancer drug treatment. Front Biosci 2007;12:3617–3627.
16. Fernandez A, Maddipati S. A priori inference of cross reactivity for drug-targeted kinases. J Med Chem 2006;49(11):3092–3100.
17. Hanifi-Moghaddam P, Gielen SC, Kloosterboer HJ, De Gooyer ME, Sijbers AM, van Gool AJ, Smid M, Moorhouse M, Van Wijk FH, Burger CW, Blok LJ. Molecular portrait of the progestagenic and estrogenic actions of tibolone: behavior of cellular networks in response to tibolone. J Clin Endocrinol Metab 2005;90(2):973–983.
18. Mulvey L, Chandrasekaran A, Liu K, Lombardi S, Wang XP, Auborn KJ, Goodwin L. Interplay of genes regulated by estrogen and diindolylmethane in breast cancer cell lines. Mol Med 2007;13(1–2):69–78.
19. Lahousse SA, Wallace DG, Liu D, Gaido KW, Johnson KJ. Testicular gene expression profiling following prepubertal rat mono-(2-ethylhexyl) phthalate exposure suggests a common initial genetic response at fetal and prepubertal ages. Toxicol Sci 2006;93(2):369–381. Epub 2006 Jun 29.
20. Gielen SC, Kühne LC, Ewing PC, Blok LJ, Burger CW. Tamoxifen treatment for breast cancer enforces a distinct gene-expression profile on the human endometrium: an exploratory study. Endocr Relat Cancer 2005;12(4):1037–1049.
21. Ispolatov I, Krapivsky PL, Yuryev A. Duplication-divergence model of protein interaction network. Phys Rev E Stat Nonlin Soft Matter Phys 2005 Jun; 71(6 Pt 1):061911.
22. Araujo RP, Petricoin EF, Liotta LA. A mathematical model of combination therapy using the EGFR signaling network. Biosystems 2005;80(1):57–69.
23. Adjei AA. Novel combinations based on epidermal growth factor receptor inhibition. Clin Cancer Res 2006;12(14 Pt 2):4446s–4450s.
24. Huang PH, Mukasa A, Bonavia R, Flynn RA, Brewer ZE, Cavenee WK, Furnari FB, White FM. Quantitative analysis of EGFRvIII cellular signaling networks

reveals a combinatorial therapeutic strategy for glioblastoma. Proc Natl Acad Sci U S A 2007;104(31):12867–12872.

25. Lehár J, Zimmermann GR, Krueger AS, Molnar RA, Ledell JT, Heilbut AM, Short GF, III, Giusti LC, Nolan GP, Magid OA, Lee MS, Borisy AA, Stockwell BR, Keith CT. Chemical combination effects predict connectivity in biological systems. Mol Syst Biol 2007;3:80.
26. Borisy AA, Elliott PJ, Hurst NW, Lee MS, Lehar J, Price ER, Serbedzija G, Zimmermann GR, Foley MA, Stockwell BR, Keith CT. Systematic discovery of multicomponent therapeutics. Proc Natl Acad Sci U S A 2003;100(13):7977–7982.
27. Mahesh R, Perumal RV, Pandi PV. Cancer chemotherapy-induced nausea and vomiting: role of mediators, development of drugs and treatment methods. Pharmazie 2005;60(2):83–96.
28. Sagae S, Ishioka S, Fukunaka N, Terasawa K, Kobayashi K, Sugimura M, Nishioka Y, Kudo R, Minami M. Combination therapy with granisetron, methylprednisolone and droperidol as an antiemetic prophylaxis in CDDP-induced delayed emesis for gynecologic cancer. Oncology 2003;64(1):46–53.
29. Davies K, Counting the Cost of Drug Discovery. 2006. BIO-IT World. Available at http://www.bio-itworld.com/archive/071102/firstbase.html

INDEX